THE
HORSE

ALSO BY TIMOTHY C. WINEGARD

The Mosquito: A Human History of Our Deadliest Predator
The First World Oil War
For King and Kanata: Canadian Indians and the First World War
Indigenous Peoples of the British Dominions and the First World War
Oka: A Convergence of Cultures and the Canadian Forces

THE
HORSE

*A Galloping History
of Humanity*

TIMOTHY C. WINEGARD

DUTTON
An imprint of Penguin Random House LLC
penguinrandomhouse.com

Copyright © 2024 by Timothy C. Winegard

Penguin Random House supports copyright. Copyright fuels creativity, encourages diverse voices, promotes free speech, and creates a vibrant culture. Thank you for buying an authorized edition of this book and for complying with copyright laws by not reproducing, scanning, or distributing any part of it in any form without permission. You are supporting writers and allowing Penguin Random House to continue to publish books for every reader.

DUTTON and the D colophon are registered trademarks of Penguin Random House LLC.

Maps © 2024 by Dylan Browne

LIBRARY OF CONGRESS CATALOGING-IN-PUBLICATION DATA
has been applied for.

ISBN 9780593186084 (hardcover)
ISBN 9780593186091 (ebook)

Printed in the United States of America
3rd Printing

Book design by Nancy Resnick

While the author has made every effort to provide accurate telephone numbers, internet addresses, and other contact information at the time of publication, neither the publisher nor the author assumes any responsibility for errors or for changes that occur after publication. Further, the publisher does not have any control over and does not assume any responsibility for author or third-party websites or their content.

To My Longhouse:
I Love You All

Contents

Everything *Equus*: Taxonomy and Terminology   xvii

Introduction   1

PART I
EARLY INTERACTIONS

CHAPTER 1
The Dawn of the Horse: Equine Evolution and Bone Wars   11

CHAPTER 2
Straight from the Horse's Mouth: I Am the Grass, Let Me Work   32

CHAPTER 3
Eat Like a Horse: Human Hunting and Vanishing Habitats   49

CHAPTER 4
Hold Your Horses: Steppes of Domestication and the Agricultural Revolution   61

CHAPTER 5
A Horse by Any Other Name: The Indo-European Domination of Eurasia   89

PART II
FORGE OF EMPIRES

CHAPTER 6
Behold a Pale Horse: Apocalyptic Chariots and Imperial
    Ambitions   123

CHAPTER 7
Riders on the Storm: Cavalry, Assyrians, Libraries, and Scythians   143

CHAPTER 8
The Education of Alexander: Academia and Empires   174

CHAPTER 9
My Kingdom for a Horse: The Hitched Fates of the Chinese
    and Roman Empires   207

CHAPTER 10
Dark Horses: Feudal Knights and Contending Faiths   238

CHAPTER 11
Road Apples: The Medieval Agricultural Revolution and the
    Making of Modern Europe   261

PART III
GLOBAL TRAILS

CHAPTER 12
Shuttling the Silk Roads: Mongol Hordes and Eurasian Markets   279

CHAPTER 13
The Return of the Native: The Horse and the Columbian
    Exchange   302

CHAPTER 14
Big Dogs of the Great Plains: Horses, Bison, and the Downfall
    of Indigenous Peoples   324

CHAPTER 15
Spiritual Machines: The Supremacy of the Horse   368

CHAPTER 16
The Final Draft: War, Mechanization, and Medicine   400

CHAPTER 17
Equus Rising: Wild Horses, Therapeutic Healing, and
    Worldwide Sports   438

Conclusion   459

Acknowledgments   463

Selected Bibliography   467

Notes   493

Index   501

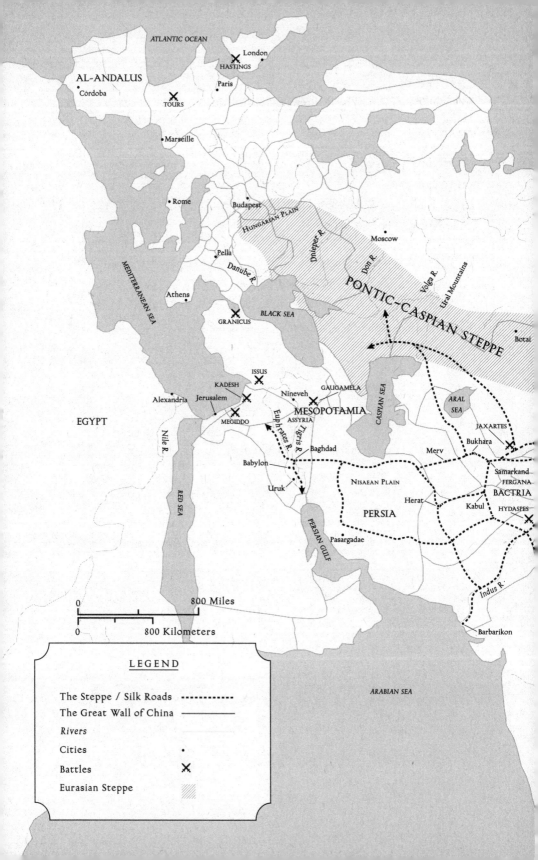

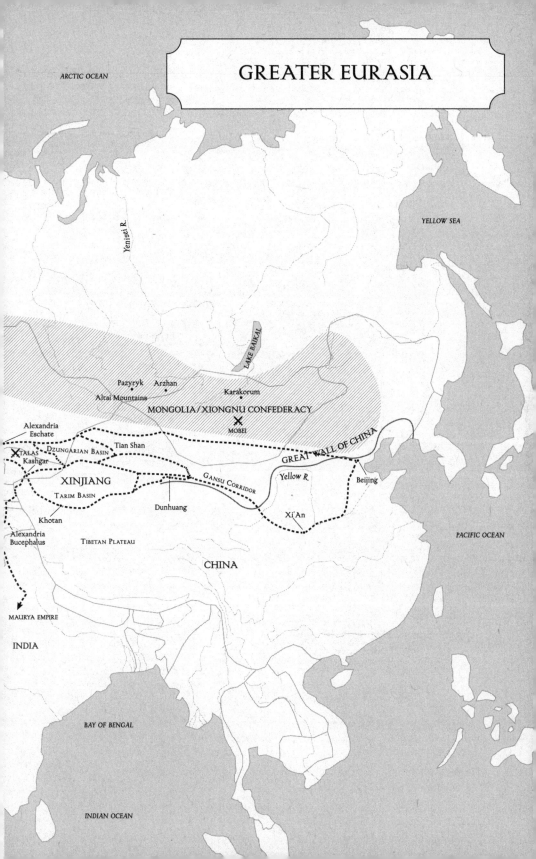

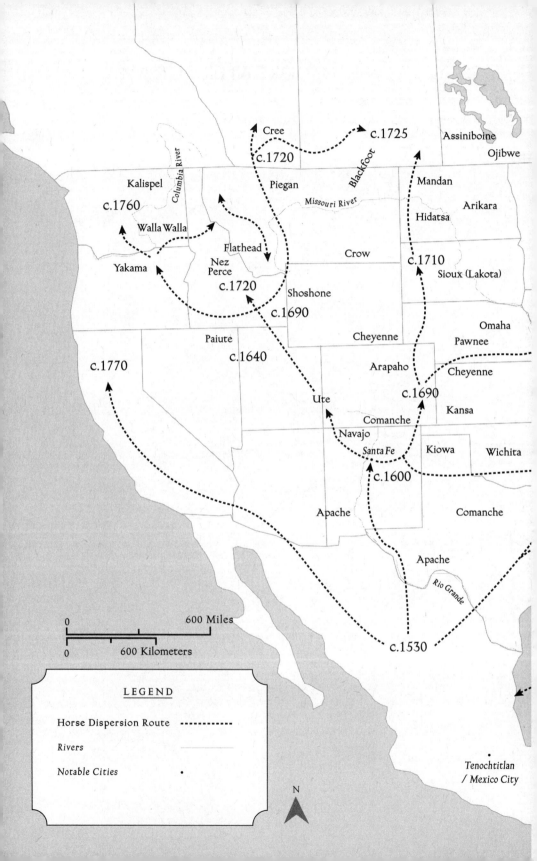

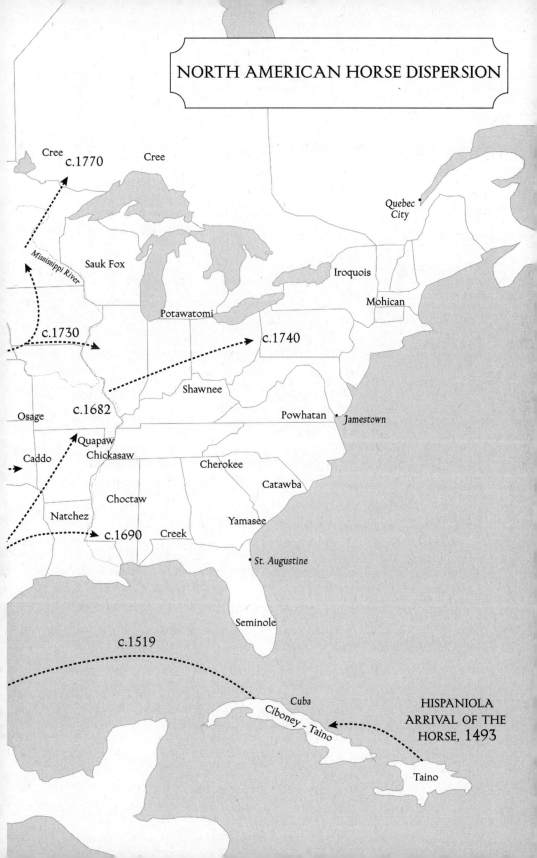

Everything *Equus*
Taxonomy and Terminology

Although this is first and foremost a history book, for ease of understanding and reading, it is important to navigate the nomenclature and categorization of horses and their relatives at the outset.

Biologists rank all organisms into an inverse pyramid of classifications. If we start at the pointed bottom, we find *species*. Animals that mate to produce fertile offspring are generally the same species. While horses and donkeys (asses) shared a common ancestor roughly four and a half million to four million years ago, and largely resemble one another, if they are forced to mate or are artificially induced, they produce sterile offspring called mules (male donkey and female horse) or hinnies (male horse and female donkey). Horses and donkeys are distinct species belonging to the next taxonomic tier of *genus*. This grouping contains different species that evolved from a common ancestor or evolutionary line.

I assume that all of you currently reading this book are genus *Homo* and species *sapiens*. This is the two-part Latin label assigned by biologists: genus followed by species. Currently, there are seven living species of the genus *Equus*: one horse, three asses, and three zebras.

Equus caballus is the binomial Latin name that Swedish zoologist and father of modern taxonomy Carolus Linnaeus gave to all domestic horses in the eighteenth century. When true wild horses were discovered, the taxonomy was revamped. All modern horses now belong to the single *Equus ferus* species. The common horse *Equus ferus caballus*, or simply *Equus caballus* (sixty-four chromosomes), represents all breeds of living horses save roughly two thousand Przewalski's horses, *Equus ferus przewalskii* (sixty-six chromosomes).*

Many view the Przewalski's horse as a subspecies of *E. caballus* and not a distinct species, since the two can propagate to produce fertile offspring. The same can be said for the now-extinct *Equus ferus ferus*, or tarpan of the western Eurasian Steppe.† A subspecies refers to natural populations of the same species generally living in isolation or distinct geographic regions with variances in morphological characteristics (predominantly external appearance and visible traits such as color, striping patterns, hair, and size). Unlike subspecies, artificially selected domestic breeds do not have distinct taxonomic entries.

In addition to the single surviving species of horse, there are three ass species in the genus *Equus*: *E. africanus*, or the African wild ass; *E. kiang*, or simply kiang; and *E. hemionus*, with its numerous aliases—Asiatic wild ass, onager, and hemione, or half-ass. Lastly, the three zebra species include the most abundant *Equus quagga* (plains zebra), the *E. zebra* (mountain zebra), and the *E. grevyi* (Grevy's zebra). Like *Equus ferus*, zebra and ass species also have subspecies.

As we move up the widening reverse pyramid, the entire genus *Equus* is part of the Equidae family, called simply the horse family. This classification includes extant *Equus* and all other related animals and extinct species. Our human family, Hominidae, for instance, includes all eight extant species in four genera (plural of *genus*) of the great apes (three orangutan, two gorilla, two chimpanzee, and one human). These all belong to the larger order of Primates and class of Mammalia.

The Equidae family belongs to the order Perissodactyla within the Mammalian class. This taxonomic order, commonly referred to as odd-toed ungulates (as opposed to the more crowded Artiodactyla order of even-toed ungulates), contains seventeen species divided into three related families: our seven species of equids, five species of rhinoceroses,

---

* These squat, stocky horses are named after the Russian colonel, naturalist, and explorer of Polish descent, Nikolay Przhevalsky (Przewalski in Polish), who obtained a specimen in 1878 from the Mongolian-Chinese border.

† *Steppe*: from the Russian for "flat grassy plain." The Eurasian Steppe is the largest temperate grassland in the world, stretching uninterrupted for more than 5,600 miles from Hungary through Ukraine and Central Asia to China.

and five species of tapirs. So, just as we are related to a wide variety of other primates, including monkeys, lemurs, baboons, and gibbons (that fall outside of our Great Ape Hominidae family), so are horses related to rhinos and tapirs.

Classification	Horse	Examples	Human	Examples
Class	Mammalia	Mammary glands produce milk for feeding; neocortex region of brain; fur or hair; three middle ear bones.	Mammalia	Mammary glands produce milk for feeding; neocortex region of brain; fur or hair; three middle ear bones.
Order	Perissodactyla	Horses, asses, zebras, rhinoceroses, tapirs	Primates	Humans, chimpanzees, orangutans, gorillas, monkeys, gibbons, lemurs, tarsiers, and lorises
Family	Equidae	Horses, asses, zebras, and extinct species	Hominidae	Humans, chimpanzees, orangutans, gorillas, and extinct species
Genus	*Equus*	Horses, asses, zebras, and extinct species	*Homo*	*sapiens, erectus, floresiensis, heidelbergensis, neanderthalensis,* Denisovans, and other extinct species
Species	*Equus ferus (caballus)*	"wild horse"	*Homo sapiens*	"wise man"

Who's Who in the Zoo

burro The Spanish word for ass/donkey. Many "wild" (see "wild horses") burros are protected and/or managed by government agencies across the Americas and Australia.

equine Relating to or describing the horse.

foal An adolescent horse, under one year old. A female is a filly, and a male is a colt.

gelding A castrated male horse.

hands The traditional measurement of a horse from the ground to the middle of the withers (the highest part of the horse's back at the base of the neck above the shoulders). Prior to standardized measurement and accurate tools, this handy system was simply the number of widths of a hand turned sideways. A hand was standardized to 4 inches in 1540 by royal decree issued by England's King Henry VIII. For example, a horse measuring 15.2 hands equals 15 hands and an additional 2 inches—or slightly over 5 feet tall. A horse standing 15 hands would be exactly 5 feet tall. *Hand* can be abbreviated to either *h* or *hh* (as in "hands high").

mare An adult female horse.

mule General cross between a horse and an ass. Male mules are infertile, while only one in two hundred thousand female mules can breed with a horse or an ass.

mustang "Wild" horse of the American West from the Spanish *mesteño* (wild or stray).

pony A horse under 14.2 hands.

stallion A non-castrated male horse.

wild horses Technically, including the captive-bred and reintroduced Przewalski's horse, there are no wild horses left in the world. All horses are the descendants of domesticated horses. Some of these domestic horses escaped or were released into the wild, where they continued to reproduce and live. These are by technical definition feral domestic horses and not true "wild" horses. The feral horses in the United States, Canada, and Australia ("brumbies") are often referred to as wild horses by the media and those government and private agencies that manage these herds. To avoid confusion, I will use both *wild* and *feral* to classify these free-roaming horses.

THE
HORSE

Introduction

Roughly 5,500 years ago on the windswept grasslands of the Pontic-Caspian Steppe, the world was irrevocably changed.* There are few instances in history where one isolated event transforms our planet forever and single-handedly leaves a deep and indelible imprint on humankind. With the first comforting hands-on encounter and reassuring whisper between one daring man and one docile mare, an unbreakable bond was forged, and the future of humanity was instantly rewritten.

This initial opportune courtship was likely the result of a swaggering teenager being dared by his snickering friends to approach a submissive or wounded mare and jump on its back. Imagine the brave stupidity and pubescent thought process of this spirited youth as he determined impulsively that attempting to mount a large, wild animal was a *good idea*! It is amazing that any of us survived to reach adulthood.

Now try to envision the strutting spectacle that unfolded as this peacocking teen led or rode this horse past the dumbfounded and slack-jawed stares of his friends and family. "Horse domestication almost certainly should be understood this way. It is doubtful that any prehistoric genius foresaw the potential capabilities of the wild steppe horse," explains equine anthropologist David Anthony. "I think the first person to climb on a horse was an adolescent or child. Some kid probably jumped

---

* An area north of the Black and Caspian Seas (within the larger Eurasian Steppe) stretching 385,000 square miles from Ukraine through southern Russia to Kazakhstan.

on the back of a mare as a prank, and everyone looked on in astonishment." Little did they know that at that very moment, this intrepid young horse whisperer had recalibrated the trajectory of human history.

The domestication of horses was nothing short of an Equine Revolution in transportation, traction, trade, and war. It was *the* complete transformative package of civilization. When the multifaceted power of the horse was finally harnessed, it permanently altered the fabric of humanity and laid the foundations of history. "The horse," declared famed eighteenth-century naturalist Georges-Louis Leclerc, "is the most noble conquest man has ever made." Surprisingly, it was an improbable candidate to claim this superpower status: the deus ex machina.

Save for small, secluded herds grazing remote pockets of the sprawling Eurasian Steppe, by 4000 BCE the horse had disappeared across global landscapes.* Domestication reined in the horse from the precipice of extinction. Without human intervention, it is likely that the horse would have followed most large mammals and other equine species into oblivion, only to be resurrected as dusty museum displays.

Imagine for a moment our human odyssey and epic sagas without the historical power of horses. I can safely say that our modern world order and sociocultural configurations would be completely unrecognizable. We might as well live on another planet in a galaxy far, far away.

With domestication, the destinies of horses and humans were eternally intertwined. Our triumphs and struggles, our accomplishments and failures, and our unfolding story have been forever fused into a single symbiotic narrative. The human-horse dyad is the most dominant animal coalition ever witnessed.

This "Centaurian Pact" combined the physical and intellectual power of two creatures into a single cohesive unit. Human beings were both literally and figuratively elevated and empowered by horses. "The cooperative union," states horseman and author Bjarke Rink in *The Centaur Legacy*, "represents a qualitative leap in human psychology and physiology that permitted man to act beyond his biological means." Horses

---

* Eurasia: the complete unbroken continental landmass of Europe *and* Asia framed by the Atlantic, Pacific, Arctic, and Indian Oceans.

allowed humanity to circumvent our physical constraints and evolutionary limitations by redefining and upgrading the capabilities and potential of our species.

Possessing a rare combination of size, speed, strength, and stamina, the horse became the pinnacle weapon of war, a political leviathan, a prime economic mover, an agricultural powerhouse, and a universal, multipurpose machine. History marched forward to the cadence of drumming hoofbeats.

For most of us alive today, however, horses were never part of our immediate daily framework or wider worldview. They are now generally relegated to high-stakes racing, equestrian events, rodeos, personal recreation, and television shows and movies depicting bygone eras or fictional fantasies. Like other fabled creatures and fantastic beasts, horses are also mascots for our schools and sports teams: mustangs, colts, mavericks, chargers, stallions, and broncos. Numerous automobile companies have also invoked the horse in branding their vehicles. Although there are still an estimated fifty-eight million horses sharing our planet, they are pragmatically redundant in our modern, mechanized society. Within our historical time line, however, this is a recent phenomenon.

It is easy to forget that cars unseated the horse only about a hundred years ago. "The automobile," laments historian Walter Liedtke, "has inevitably clouded our view of how important the horse once was in everyday life." The easiest way to highlight the status of the horse is to take a second to imagine your life today without any form of mechanical transport. Eliminate all cars, buses, subways, trains, planes, and even bicycles. The world would not extend too far beyond your doorstep and then only as far as your feet could take you. This would have been the same uncomfortable reality for our ancestors only a century ago if the horse were removed from their descending generations of humanity.

Innovations in transport are among the most dynamic catalysts of historical change. Think about what the internal combustion engine did to our societies. It drove the shift to urbanization and strip-mall suburbia, instigated a complete overhaul of infrastructure, and struck an unquenchable thirst for oil and steel. It spawned heavy industry, forged assembly-line mass production, retooled the workforce, convened a middle class, and polluted our climate-changing atmosphere. Cars became a visible display of wealth and power recognizable by the Porsche horse coat of arms or Ferrari's Prancing Horse crest. The internal combustion engine propelled modern mechanized warfare to unimaginable heights of meat-grinding slaughter. It manufactured industrial farming and the convenience of grocery stores, allowing for further technological advancements and vocational specialization. The engine connected and coupled people, places, products, pathogens, economies, armies, and ideas.

It also brazenly rode the tails of the horse, which had already blazed these familiar trails and spurred these transformations forward during its preceding 5,500-year reign. The automobile simply traversed well-trodden landscapes pioneered previously by the horse. The sociocultural and historical impact of the horse, however, lasted fifty-five times longer and was far more profound.

As much as we think of horses as organic animals, we must also view them as sophisticated machines—one of the oldest and most important inventions in human history. When the heavy burden of transport was lifted from humans and placed on the shoulders of horses and wagons, the modern age of the machine was born. The horse was so paramount and pivotal to human society that we base our units of mechanical energy or engine output on horsepower. As an evolving technology, horses, and their historic muscle, were manipulated in a living laboratory.

Under the survival pressures of natural selection, horses evolved over the course of sixty million years from a scampering fox-sized creature to assemble the unique and exploitable attributes that made them the embodiment of power and the premier human sidekick. Meticulous human-induced selective breeding subsequently promoted and enhanced desirable traits in anatomy, temperament, size, speed, and strength. This genetic engineering was enriched by cumulative advancements in training, maintenance, diet, and medicine. The utilitarian potential of this living machine was elevated further with bits, saddles, harnesses, horseshoes, stirrups, whippletrees, armor, specialized vehicles and plows, and sophisticated equine infrastructure. As a result of these natural and artificial amendments, the horse reached the apex of biotechnology.

Sometimes things are so omnipresent, so obvious, that we blithely overlook their importance. In our twittering, computerized age of artificial intelligence, we have lost sight of how essential and impactful horses have been to humanity. The horse was the pinnacle instrument of profit and power. For more than five millennia, with their unrivaled operative force, they steered and dominated every part of our existence. The horse was a source of protein, milk, and a variety of secondary products. It was a war-winning weapon, a groundbreaking agricultural engine, and a high-speed vehicle for transportation, trade, and travel. The horse was the prime mover of civilizations.

With the domestication of horses, human greed and curiosity could now be fully realized. The advent of the farming package (agriculture and the domestication of barnyard animals) sowed vocational specialization, the first urban city-states, and a capitalist surplus economy. High-speed communication allowed distant, previously isolated peoples to become neighbors, allies, or enemies. Remote, exotic lands, only whispered in legend and lore, now entered expanding—and increasingly lucrative and multifarious—trading networks within a flourishing but competitive commercial environment. The chariot and mounted cavalry enhanced significantly the ability of avaricious conquerors to establish, expand, and hold vast empires. The horse was *the* defining factor that hauled, assembled, and secured these foundational building blocks. It single-handedly created an infinitely smaller and integrated global village.

One of the reasons horses were so historically dynamic and culturally influential is that they eventually became a relatively democratic resource. For the most part, their acquisition and reproduction were outside of government control, commercial manufacturing, business monopolies, social status, and economic condition. Horses were self-reproducing, reasonably self-sufficient, and could be acquired through purchase, trade, theft, and even capture. In this sense, they leveled the playing field through their ability to gain or subvert power. Horses proffered a sense of liberty and endowed the individual with a spirit of freedom. The democracy of the horse allowed women to assume dominant political and military positions in ancient nomadic societies such as the Scythians and Massagetae, and in early dynastic China. The fabled Amazons, as it turns out, were fierce mounted female warriors of the Eurasian Steppe.

Horses transcended demographics, geography, ethnicity, gender, spirituality, class, and station. Horses pulled royal carriages and peasant carts. They conveyed humble merchants to peddling markets and gilded chariots and chivalrous knights to battlefield glory. Horses hauled plows across feudal farms and pranced in regal parades. They shuttled private coaches, hired cabs, and public transportation. Horses belonged, and belong, to all human beings.

They changed the way we hunted, traded, traveled, farmed, fought, worshipped, and interacted with one another. Horses reconfigured the global human genome and the languages we speak. They gave rise to nation-states and pulled into place modern international borders. They were a potent instrument of feudalism, land usage patterns, and European colonization during the Columbian Exchange.* Horses made vital contributions to modern medicine, lifesaving vaccines, and developments in sanitation. They even inspired recreational cannabis use, sports, invention, entertainment, architecture, furniture, and fashion. Horses are an integral part of what it means to be human.

Of the ties that bind, the oldest is that of the hunter and the hunted. Our toolmaking hominid ancestors dined on equines, and horsemeat is still on the menu for over one billion people. Like other barnyard animals, domesticated horses were initially reared for meat, milk, and a variety of secondary products. Riding fundamentally changed the rules to the game of life, and the enduring bond was expanded to traction and transport for migration, trade, farming, and, of course, war.

It is undeniable that war, in which the horse played a paramount role, is one of the most explosive catalysts of change, whether we like it or not. In a peculiar twist of evolutionary fate, as an animal built solely for flight and not fight, horse and rider became humanity's longest-serving weapon system. "I think that the most important development in history with respect to animals," asserts archaeologist and prolific author Brian Fagan, "was the adoption of the horse as a weapon of war."

---

* Coining this term with the title of his seminal 1972 work, *The Columbian Exchange: Biological and Cultural Consequences of 1492*, historian Alfred W. Crosby proposed that during the Age of Imperialism global ecosystems were forever rearranged in the largest interchange in natural and human history.

The lengthy epoch of equine-fueled warfare escorted the basic ingredients of civilization and all elements of our modern world. King Richard III's immortal Shakespearean cry shortly before his death at the Battle of Bosworth in 1485 during the English civil wars (Wars of the Roses), "A horse, a horse! My kingdom for a horse!" echoes through the bloodstained corridors of human history and five millennia of unremitting combat.

The thundering spectacle of a cavalry charge is one of the most awe-inspiring and fearsome dramas in our long history of violence. Most of us are familiar with the hoofbeat meter of Lord Tennyson's poem "The Charge of the Light Brigade." Its galloping stanzas have been quoted in classrooms, speeches, and movies ever since it was published in 1854 during the Crimean War: "Half a league, half a league, / Half a league onward, / All in the valley of Death / Rode the six hundred. . . . Theirs not to make reply, / Theirs not to reason why, / Theirs but to do and die. / Into the valley of Death / Rode the six hundred."

For more than five thousand years, horses were drafted as soldiers to fight in our military campaigns. The causes, both noble and nefarious, were nonnegotiable clauses of their conscription. Warhorses, and other animal combatants, had no say in their fate. That unalienable right escaped them.

The horse decided the destinies of immortal conquerors and momentous empires. Without its presence, the decisive battles recited in textbooks and bedtime stories would have been small skirmishes or family feuds, and the legendary exploits of Alexander the Great, William the Conqueror, and Chinggis Khan would be expunged from the historical record. Transformative equestrian cultures grazing the global grasslands of antiquity, including Indo-Europeans, Assyrians, Scythians, Xiongnu, Huns, Mongols, Comanche, and Lakota, would have remained anonymous. European colonization and the Columbian Exchange would have been strangled in their cradles.

For millennia, the horse was the invisible hand driving human history. The Indian anti-colonial advocate and ethicist Mahatma Gandhi once espoused that "the greatness of a nation and its moral progress can be judged by the way its animals are treated." Historically, horses have been treated as subjugated spectators instead of active participants. Both humans and horses were dynamic beings and wielded overriding agency within our reigning partnership and jointly forged history. Horses deserve to be treated accordingly. "The importance of the horse in human history is matched only by the difficulties inherent to its study," distinguished archaeologist Grahame Clark states aptly. "There is hardly an incident in the story which is not the subject of controversy, often of a violent nature." Given the mountain of literature dedicated to the horse and the litany of hot-button topics, this is humbly understated.

This is first and foremost a *history* book. It is not, however, a general survey of the horse, a singly focused study on any one specific element of equine research or specialist field, nor is it a treatise on cavalry or mounted warfare. While the horse was dying on battlefields from Megiddo (Armageddon) to the Second World War, it did not decide the outcome of every clash or conflict. During the American Civil War, for example, more than three million horses, donkeys, and mules (roughly equal to the number of human combatants) saw service, with half of them giving their last full measure of devotion.

Likewise, although the horse became a universal animal, its influence, like its populations, was not evenly distributed across our planet. While this is a global history, and not "centric" to one region, not all geographical spheres receive equal treatment. This is an expansive human history dictated by the pull of the horse and not by sentimental preference or egalitarian predilections. Given the ubiquitous presence of the horse, it is unrealistic and impossible to include everything in a single narrative volume. This is also not the purpose of this book. This horse-powered journey depicts the game-changing events where the impact of the horse was the definitive factor in forever altering the trails of history. We still live in a world built by horses.

Humans love horses. They hold a distinguished place in our collective heart precisely because they dominated every facet of our formative history. Across an unrivaled 5,500-year span, the horse carried the fate of human civilizations on its back. "There are countless histories waiting to be told in which horses play a major role," emphasizes Ulrich Raulff in his cultural history, *Farewell to the Horse*. "The story of technology, of transport, agriculture, energy, war and urbanization." This is the history you hold in your hand. Human history is also the history of the horse.

PART I
EARLY INTERACTIONS

CHAPTER 1

The Dawn of the Horse
Equine Evolution and Bone Wars

The enticing and seducing whisper of the Wild West was summoning Othniel Charles Marsh. The Civil War had been over for three years, and the vast, windswept prairies and majestic cloud-piercing mountains motioned for American Manifest Destiny. The war-weary nation was licking its wounds and trying to forget four years of unforgiving slaughter that left 750,000 Americans dead but ultimately unshackled 4.2 million human beings from the bondage of chattel slavery. For many like Othniel Marsh, the untamed West was the epitome of freedom and the essence of the rugged frontier spirit.

Born into modest means in Lockport along the Erie Canal in western New York, Marsh, a dour, scraggly bearded thirty-seven-year-old bachelor, had nothing tying him down. He purchased a train ticket to the newly established Wyoming Territory and methodically packed and bundled the bare necessities of a paleontologist: notebooks, pencils, shovels, picks, a well-worn straw boater hat, and, of course, a six-shooter pistol. In short order, restless men and women like Othniel Marsh transformed the West.

The providential opportunities were as endless as the horizon stretching seamlessly beyond the eternal grassland prairies and cascading foothills of the Rocky Mountains, to the sparkling, wave-crested waters of the Pacific. Fortunes awaited in the boomtown gold and silver mines. Fertile and vacant land hungered for the cultivating cleaves of the plow and the ranching hoofbeats of horses and cattle. Ferocious fur-bearing beasts howled from the rogue silhouettes of the snowy peaks. Adventure and amusement beckoned from whiskey-soaked saloons, sweaty bordellos, and dodgy gambling dens west of the Mississippi River.

Within this Gilded Age of upheaval, war, and shifting cultural and economic landscapes, railroads opened the door to westward expansion and, in the process, unlocked a window to our petrified past. The so-called Bone Wars between 1868 and 1892 witnessed a frenzy of trailblazing—and, at times, ruthlessly cutthroat—fossil expeditions and momentous discoveries. Rival fossil hunters and bone collectors scoured and combed the rich beds of (what are now) Wyoming, Colorado, Utah, and Montana.

To climb the professional ladder and attain the accompanying financial windfalls of celebrity status, callous and vindictive American and European paleontologists resorted to bribery, theft, sabotage, violence, and slander. Allegiances were fickle, and alliances fleeting. In this contentious age of embryonic Darwinian evolution, unearthing fossils meant fame, fortune, and academic immortality.

Given the increasing popularity of paleontology, and with successive and seemingly unearthly finds dominating the press and captivating popular imagination, in 1866 the millionaire financier, banker, and philanthropist George Peabody donated $150,000 ($3 million in today's money) for the construction of the Peabody Museum of Natural History at Yale University.* In a brazen act of nepotism, the institution promptly appointed its benefactor's nephew, Othniel Charles Marsh, professor of paleontology (the first such academic position in North America) and a trustee of the museum. Marsh was an unlikely candidate to seize the crown of early American paleontology.

Subsidized by his uncle, Marsh had previously studied geology, anatomy, and paleontology at Yale, followed by three years at various institutions in Germany. Although not yet established, he was talented, keen, energetic, and—most importantly, given the sizable expense of research

---

* Having no legitimate heir to his vast business enterprise, Peabody partnered with Junius Spencer Morgan in 1854. Their joint venture would eventually become J. P. Morgan & Co., the predecessor to three of the largest banking institutions in the world: JPMorgan Chase, Morgan Stanley, and Deutsche Bank. Peabody is also considered the first modern philanthropist.

excursions—fully funded. In 1868 Marsh packed his duffle, holstered his pistol, and bought passage (quite ironically as it would turn out) on one of the first "iron horse" trains to chug west on the overland route of the newly constructed Union Pacific Railroad.

During a quick pit stop on the Nebraska-Wyoming border, Marsh took the opportunity to stretch his legs and chat with the locals, explaining in the process the purpose of his trip: he was a fossil hunter. To his surprise, he was presented with fragments of bones that residents had unearthed while digging a well. He quickly assured them that they were not human remains nor, much to their disappointment, those of prehistoric saber-toothed tigers.

Marsh realized, as he later wrote, that they were "many fragments and a number of entire bones, not of man, but of horses, diminutive indeed, but true equine ancestors." Upon further examination, Marsh had acquired the bones of four separate horse species, including a small, odd-looking horse "scarcely a yard in height, and each of his slender legs was terminated in three toes." Financed by a $100,000 inheritance from Peabody, who passed away in 1869, Marsh returned to the Wild West on numerous archaeological research trips to dig fossils, hunt bison, and dodge danger.

During one of these adventures in 1870, Marsh ran into Brigham Young, president of the Church of Jesus Christ of Latter-day Saints, at the theater in Salt Lake City. The Mormon leader immediately set about interrogating him about his equine fossils. Bewildered by the intense probing, Marsh politely answered the questions while his team of Yale paleontology students, in the words of participant C. W. Betts, "flirted with twenty-two daughters of Brigham Young in a box at the theatre."

Mormon theology posits that the horse was present around 589 BCE, when the prophets arrived in the Americas. According to 1 Nephi 18:25, from the Book of Mormon, "And it came to pass that we did find upon the land of promise, as we journeyed in the wilderness, that there were beasts in the forests of every kind, both the cow and the ox, and the ass, and the horse." Young's persistent line of questioning now made sense to Marsh. He was looking for hard scientific evidence to corroborate the spiritual passage.

The seemingly lawless, no-holds-barred competition among bone collectors unfolding on the front lines of American settlement was set against the bloody backdrop of the American Indian Wars. Caught up in this cyclical frontier violence, not to mention the perils posed by rival paleontologists, Marsh and his team carried rock hammers in one hand and, in true Indiana Jones fashion, Colt revolvers or carbine rifles in the other.

During one expedition in 1872 in southwest Wyoming Territory, for example, Edward Drinker Cope, *the* rival paleontologist, had been spying on Marsh's team from a craggy outcropping. At dusk, when the team packed up, Cope crept down to the site. Emerging from the skulking shadows, he was elated to find a skull fragment, several teeth, and other assorted bones. The unique combination of anatomical features represented in his looted specimen all pointed to a new, undiscovered dinosaur species. Instant fame and the satisfaction of publicly humiliating his archnemesis Marsh was as good as carved in stone.

What Cope did not know until after he had touted the discovery and published an academic paper on his groundbreaking new fossil was that he had been set up, duped, and played for a fool. Marsh had been aware of Cope's clumsy espionage and deceptively instructed his diggers to "salt" the ground with the skull, teeth, and bones of miscellaneous fossil species. The credulous Cope arrogantly took the bait. When his blunder was publicly exposed, he had to recant his findings and, in the process, tarnished his reputation. Marsh got the better of his chief adversary in this round of the Bone Wars.

In addition to clashing with Cope, Marsh was often a spectator to the fierce skirmishes between the US Cavalry and mounted Indigenous warriors during the last colonizing campaigns of American progress. On one harrowing expedition near the Bighorn Basin of northwest Wyoming, Marsh hired a young army scout named William Cody as a tracker. Before arriving at their destination, the future Buffalo Bill galloped away to investigate a scuffle with Pawnee warriors. Later in the day, these same Pawnee horsemen became boisterously captivated when Marsh, an influential advocate for Indigenous rights, showed them some ancient fossils of their most prized possession: horses.

On another occasion, Marsh met and chatted cordially with the famed Oglala Lakota (Sioux) leader Red Cloud. Marsh listened with solemn interest as Red Cloud methodically articulated the plight and starvation of his people. The combative atmosphere on the Northern Plains was aggravated by the encroachment of settlers onto unceded Sioux land and the corruption of federal politicians and agents known as the Indian Ring.

Furious with the treatment of Red Cloud and the Sioux, Marsh traveled to Washington and met personally with high-ranking members of the Ring, who brushed him off like an old fossil and stonewalled him at every opportunity. They even tried to discredit Marsh by circulating falsehoods that he was a drunkard who committed scandalous and depraved sexual acts with his "Yale Boys" while excavating fossils out west.

Eventually Marsh secured an audience with President Ulysses S. Grant, alerting him to the frontier tinderbox fueled partially by the corruption of his own cabinet. The Sioux, stressed Marsh, received nothing more than "frayed blankets, rotten beef, and concrete-hard flour." His humanitarian efforts on behalf of Indigenous peoples did not go unnoticed. "I thought he would do like all of the other white men, and forget me when he went away," Red Cloud remembered. "But he did not. He told the great father [President Grant] everything just as he promised he would, and I think he is the best white man I ever saw." By this time, however, Grant had lost control of his own administration, and his subordinates, including those in the Indian Ring, were running roughshod over his policies and his presidency.*

Othniel Marsh, an eccentric and awkward paleontologist, had headed out west to find and catalog equine fossils. In the process, he helped to bring down one of the most infamous and corrupt political rings in American history. In 1875, after Marsh's tireless petitioning on behalf of Red Cloud and all Indigenous peoples, and a series of scathing newspaper reports, the Indian Ring, or "Trader Post Scandal," was finally

---

* The seventeenth-century term "running roughshod" described a horse that wore shoes with projecting nailheads. This gave the horse better traction while also creating a more lethal trampling weapon. Over time the term evolved to mean "attaining one's goals or desires by completely ignoring the opinions, rights, or feelings of others."

exposed. Numerous members, including prominent politicians, were forced to resign from their positions or faced impeachment trials. Appalled by the brazen embezzlement of the Ring, in the spring of 1876 Lieutenant Colonel George Armstrong Custer even testified on behalf of the Sioux. Three months later, he would die at their hands.

While this encounter in June 1876 along the greasy grass of the Little Bighorn River in Montana Territory may have been Custer's Last Stand, it was also, in a sense, the last stand of Indigenous autonomy and self-determination. The Sioux won the battle, but with their Ghost Dance massacre at Wounded Knee, South Dakota, in 1890 at the hands of the US Cavalry, they lost the war, sealing the fate of Indigenous peoples across the United States.

By 1900, their total population hit its historic nadir at 237,000 survivors, and their land holdings had been reduced to 52 million acres from 155 million only fifteen years earlier. With Indigenous peoples removed, relocated, shattered, and subjugated, the West was open for business for miners, farmers, ranchers, trappers, and fossilists.

As the ongoing Bone Wars intensified, with the aid of Peabody's money, Marsh bribed the diggers of other paleontologists, including Cope, to reroute the most exotic, complete, and valuable fossils to him at Yale. Ironically, or perhaps hypocritically, like the corrupt Indian Ring he had helped to unmask, Marsh was not above using unscrupulous tactics to further his own financial and professional agendas. "Within a few years, Marsh had so many crates of fossilized bones stored in the attic of Yale's Peabody Museum," writes Pulitzer Prize–winning author David Philipps in *Wild Horse Country: The History, Myth, and Future of the Mustang*, "that he had to prop up the ceiling with extra beams." Following his death in 1899, it took a team of paleontologists over sixty years to unpack and inventory his enormous collection of fossils.

Having assembled this vast trove, which included the partial remains of more than thirty ancestral species of the horse, Marsh began the arduous task of trying to piece together the complex puzzle of equine evolution. In doing so, he catapulted British naturalist Charles Darwin's audacious new theory of evolution forward by cataloging the fossilized

Lakota leader Red Cloud visits paleontologist Othniel Marsh at Yale University, 1883. *(Smithsonian National Portrait Gallery)*

ancestral sequence to the modern horse. He started the hereditary branch at one hoof, the most obvious recent relative, and worked backward to those species with multiple toes. "The line of descent appears to be direct," Marsh announced, "and the remains now known supply every important form." He labeled his early, small, three-toed Wyoming specimen *Eohippus,* or "dawn horse."

Inadvertently, and initially unknowingly, Marsh became entangled in the rancorous and heretical debate surrounding evolution among the British scientific community following Darwin's 1859 publication of *On the Origin of Species by Means of Natural Selection, or the Preservation of Favoured Races in the Struggle for Life.* We must remember that in Darwin's day, evolution was a perverse and profane assault upon the doctrines of Christianity. To many, including academics, it was blasphemous pseudoscientific rubbish.

Equine evolution was originally thought to be an orderly and direct chronological branch, staunchly obedient to the simple, uncomplicated organic laws of natural selection. Distinguishable anatomical modifications and adaptations were easily identified, specifically those amendments to feet and teeth, with successive specimens stacked like Russian nesting dolls. This galloping journey from ancient ancestor to existing *Equus* was frequently touted as the literal textbook example of unambiguous "straight line" evolutionary progression and the ever-popular exhibit A for museums.

This evolutionary canard is known as orthogenesis: an outmoded theory that variations follow a particular direction and are not merely sporadic or fortuitous. In short, orthogenesis would imply that evolution has a clear objective, a single-minded commitment to a desired end state or biological ambition, and that each extinct relic somehow gave rise to a superior replacement. "The orthogenetic template has . . . influenced millions of lay people," concedes distinguished paleontologist at the University of Florida Bruce MacFadden, "many of whom visit natural history museums with turn-of-the-century exhibits that convey 100-year-old ideas." Naturally, evolution is an interlaced forest of tangled trees, full of branches, twigs, blind turns, and, more often than not, dead ends.

Darwinian natural selection, or "survival of the fittest," is driven by various external pressures, including changing climate, environmental catastrophes, sustenance variation and accessibility, and reproductive cycles, within ever-shifting and fluctuating local ecosystems.* We tend to forget that evolution is not the engineer of inevitability. According to an orthodox quotation attributed erroneously to Darwin, "It is not the strongest of the species that survives, nor the most intelligent that survives. It is the one that is most adaptable to change."†

---

* The expression "survival of the fittest" is commonly, but mistakenly, attributed to Darwin. The English biologist and anthropologist Herbert Spencer coined the catchphrase in his 1864 book *Principles of Biology* after reading Darwin's *On the Origin of Species*. Darwin then borrowed the term from Spencer for the fifth edition of his book, published in 1869.

† This often-referenced quotation does not appear in any of Darwin's published writings, journals, or letters.

Evolution is driven by immediate survival needs, not the preordained perfection of species over millions of years. Many species existed for millennia before survival pressures condemned them to extinction and an ignominious eternal night at the museum. According to recent estimates, more than fifty billion, or 99.99 percent of all species that once occupied the terrestrial stage, have gone extinct.

Natural selection is a one-way process of trial and error. Once traits, both detrimental killers and advantageous saviors, have been bestowed and propagated, it is very difficult to reverse course. "In each generation, natural selection can only use the raw materials that previous generations of selection have left it. Thus, over long periods of time, certain branches will prove more successful than others; certain trends will emerge," recaps historian and author Stephen Budiansky in *The Nature of Horses*. "But, again, it is always worth remembering that these trends reflect what are in effect the sum total of lucky guesses or accidents. They are the long-term consequences of short-term 'decisions,' consequences that could not have been anticipated at the time those choices were made. Most extinct species were victims of their own success—they had the misfortune of being supremely adapted to a niche that did not last." As Darwin surmised, the adaptable survivors "breed out" those who do not possess favorable traits—simple and uncomplicated survival of the fittest.

As a young naturalist sailing aboard the HMS *Beagle* on a survey expedition, in 1833 Darwin was "filled with astonishment" to find a horse's tooth at Santa Fe, Argentina, in the same soil stratum as the fossilized remains of giant armadillos. Upon his return to England three years later, his presumptions were confirmed by Richard Owen, the renowned biologist, anatomist, and paleontologist.

Despite producing a remarkable life's work on a zoological Noah's Ark of animals, both extinct and living, and founding the British Museum of Natural History in 1881, Owen is now best remembered for coining the immortal word *Dinosauria*, from the Greek *deinos sauros*, or "terrible lizard." As a rabid disciple of anatomy, Owen had procured right of first refusal to any dead animal or carrion from the London Zoo—an arrangement that vexed and infuriated his wife, who once had the pleasure of entering her front parlor to the cordial reception of a decaying rhinoceros.

Owen recognized that Darwin's horse tooth belonged to an extinct equine species, noting: "Every point of comparison that could be established proved it to differ from the tooth of the common *Equus caballus* only in a slight inferiority of size." Adhering to the undisputed universal doctrine that horses evolved in Eurasia, he went on to state: "This evidence of the former existence of a genus, which, as regards South America, had become extinct, and has a second time been introduced into that Continent, is not one of the least interesting fruits of Mr. Darwin's palaeontological discoveries." This amiable collegiality between the two men was a mirage, however.

Owen would eventually become a harsh critic of Darwin's theory of evolution, championing instead a philosophy that all living matter was arranged in an "organizing energy," or life force, that encoded and oversaw the origin, growth, and decay of living tissue. This arrangement determined the anatomical blueprint and life span of individual animals within the larger life cycle and framework of their species. Owen's concept is eerily similar in construct to *Star Wars* guru George Lucas's spiritual canon and metaphysical creed of "The Force" and its energy-binding "midi-chlorians" cellular interface.

Upon the publication of his seminal treatise *On the Origin of Species*, Darwin courteously sent Owen a complimentary copy, acknowledging that to Owen "it will seem an abomination." Within the dog-eat-dog world of academia and personal ego, Owen classified Darwin's work as an "abuse of science" in a scathing review. "The Londoners say he is mad with envy because my book is so talked about," Darwin responded later. "It is painful to be hated in the intense degree with which Owen hates me."

Owen's envious attacks were not limited to Darwin. He also slandered "Darwin's disciples," including Joseph Dalton Hooker, Charles Lyell, and the eccentric biologist Thomas Henry Huxley, whom he admonished as an "advocate of man's origins from a transmuted ape." Huxley was an evangelical crusader for evolution, which earned him the understated nickname "Darwin's Bulldog." His initial response to Darwin's idea of natural selection was simply "How extremely stupid not to have thought of that!" Huxley successfully recruited converts to Darwinian evolution and was relatively successful in publicly spurning Owen's persistent attacks.

As an unforeseen appendage to the controversial yet proliferating natural sciences, in the priggish horse-powered society of Victorian England, "fossil hunting" had become a trendy pastime. For upper-crust day-trippers, it offered an adventurous respite from the elysian day in the life of British aristocracy. Fashionable for both noble gentlemen and ladies, these nature walks were purported to be morally elevating and spiritually inspiring by exulting in the breadth of God's divine handiwork.

In 1839 William Richardson was engaged in just such an outing, foraging for fossils in the famous London Clay beds of Kent County representing the earliest stratigraphic stage of the Eocene period (fifty-six million to thirty-four million years ago). Richardson recalled that he was scavenging with "strong expectation for the evidence of some form of animal life." His excursion proved fruitful, yielding the front half of a tiny skull, which eventually found its way into the hands of none other than Richard Owen.

Based on the teeth and the forward-facing position of the eye sockets, Owen, who had studied and cataloged thousands of extinct species, was perplexed, noting, "The teeth, instead of great, ridged, grinding prisms of our present horse, were small, low, and cusped, really more like monkey teeth than horse teeth. The little skull, with its relatively large eyes set about midway from snout to ears . . . like that of a Hare or other timid Rodentia." He decided that the contours of the teeth bore striking similarities to existent hyraxes.* Fittingly, Owen called his intriguing new specimen *Hyracotherium* ("hyrax beast").

Although Owen categorized the odd-toe ungulate taxonomic order Perissodactyla, he never entered what turned out to be the earliest known ancestor of the horse into this branch. As a reminder, there are seventeen living species of Perissodactyla from three related families:

---

* These rather unassuming, chubby, furry critters—weighing five to ten pounds and resembling a marmot or a groundhog—are found only in Africa and a small swath of the Middle East. Quite unexpectedly, given their rodent-like appearance, hyraxes are closely related to elephants, aardvarks, and manatees. Evolution works in mysterious ways.

seven Equidae (one horse, three asses, and three zebras), five Rhinocerotidae, and five Tapiridae.

In the meantime, while Owen was pondering hyraxes, Huxley had begun the painstaking process of assembling a vague linear evolution of equine fossils. Key to his findings was that horses came from a small tapir-like animal, with multiple toes and low-crowned teeth. Despite academic scorn and ridicule, Huxley and supportive British and Russian scholars sequenced a partial array of fossils they believed represented the lineage of the horse, acknowledging that it was spotty and missing links between specimens.

What Huxley and his colleagues could not have known at the time is that their mainline evolutionary limb was incomplete, and would always be incomplete, because this anatomical journey took place in the remote continent of North America—not in Eurasia, as was universally and unwaveringly believed. After all, when Europeans first arrived in the Americas, there were no horses to be seen. None. Across the entire Western Hemisphere, the horse was completely absent from the landscape.

A parallel study of equine evolution was taking place in both Europe and the United States independent of each other. The left hoof was not in sync with the right. Getting in step would eventually happen with a meeting of the minds (and fossils) between our resident Englishman, Thomas Huxley, and our American archaeological adventurer, Othniel Marsh, whose forays into the Wild West had produced a brimming collection of ancient horse fossils.

Huxley sailed to the United States during its centennial year of 1876 to embark on a lecture tour promoting Charles Darwin's controversial new theory by highlighting his evidential fossil record of equine evolution. The first appointment on Huxley's busy agenda, however, was a visit to Marsh and his museum at Yale. As you might expect, the two scruffy oddballs hit it off immediately.

They analyzed fossils and compared notes by day, continuing their jovial academic banter by horse-drawn carriage at night. After spending a week chatting with Marsh and immersed in his fossils, a giddy Huxley animatedly reported to his wife, "His fossil collection is the most wonderful thing I ever saw." Huxley's wife seems a mite better off than that of his archnemesis Owen, who received a decomposing rhino instead of a riveting letter.

Astounded by the breadth of American specimens, Huxley solicited Marsh repeatedly to produce equine fossils exhibiting specific anatomical features. With each request, Marsh turned to his assistant and recited a box number to retrieve from the dusty, and recently reinforced, shelves. This game seemed to last for days until Huxley turned to Marsh and declared, "I believe you are a magician; whatever I want, you just conjure up." Utilizing Marsh's technique of working backward from specimens with one toe to three toes, although lacking a fossilized example, they postulated that the original member of the horse family had four toes, which Marsh theoretically named *Eohippus*. Huxley drew his new friend a satirical cartoon he labeled "Eohippus + Eohomo," depicting a petite early hominid joyfully riding this ancient multi-toed horse.

Marsh recalled: "He then informed me that this was new to him, and that my facts demonstrated the evolution of the horse beyond question, and for the first time indicated the direct line of descent of an existing animal. With the generosity of true greatness, he gave up his own opinions in the face of new truth and took my conclusions." Huxley confided in Marsh, "The more I think of it, the more clear it is that your great work is the settlement of the pedigree of the horse." Huxley rewrote his upcoming lectures to include this new chain of evidence as irrefutable proof of evolution.

These discoveries were also immediately embraced by Darwin, who wrote to Marsh with applauding praise. "Your work on these old birds & on the many fossil animals of N. America," he extolled, "has afforded the best support to the theory of evolution, which has appeared within the last 20 years." While buttressing the theory of evolution, the findings of Marsh and Huxley simultaneously refuted the established theory that horses evolved in Eurasia. Overwhelming evidence now pointed to a North American evolutionary origin of the species. European and Asian specimens were episodic immigrants from the main North American epicenter of equine evolution.

This was *the* issue that had perplexed Darwin ever since 1833, when he had stumbled upon the Argentinian horse tooth during the *Beagle* voyage. "Horses appeared and disappeared in Europe with an absolutely rude abruptness, presenting a kind of magical 'now-you-see-'em, now-you-don't' aspect of which Darwin thoroughly disapproved. It wasn't just that the horses behaved oddly across the ages. It was that when they reappeared after long absences of millions of years, they were *different*," explains Wendy Williams in *The Horse*. "Darwin was thinking about all this long before we knew about genetics or DNA, and from the European point of view, the evolution of horses flew in the face of logic."

This intermittent occurrence of horses in the Eurasian fossil strata, with extended interruptions of millions of years—and the glaring specimen gaps in Huxley's original fossil sequencing—was now easily explained. Only certain North American horse species made the trip to Eurasia, where they eventually died off. Now it all made sense. For Darwin and his disciples, including Huxley and Marsh, evolution prevailed once again, and logic and reason were restored.

Shortly after his return to England, Huxley received a letter from Marsh informing him that he had found their four-toed *Eohippus*. It had been in the museum all along. Marsh apologized to his colleague, lamenting that he had so many crates of bones pouring in from the western fossil beds that he hadn't had time to even open, let alone catalog, them properly before Huxley's visit.

Neither Huxley nor Marsh recognized that the *Eohippus* was the same creature that Richard Owen had dubbed *Hyracotherium* some thirty-five years earlier. Separated by the middle passages of the Atlantic, the left hoof was still not quite in sync with the right. The study of equine evolution stumbled on until 1932, when Sir Clive Forster-Cooper of the British Museum finally identified that the European *Hyracotherium* and the American *Eohippus* were the same species.

By convention of zoological nomenclature, Owen's *Hyracotherium* (1840) is the official designation, since it was the initial taxonomic entry. This tongue-twisting classification, reasoned pioneering equine paleontologist George Gaylord Simpson (1902–1984), was "not likely to win so many friends." By this time, however, *Eohippus* (1876), with its endearing translation of "dawn horse," had already become entrenched in books and

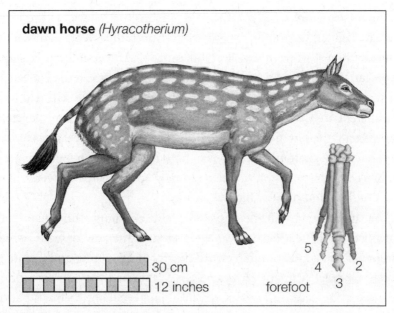

The dawn of the horse: Anatomical reconstruction and dimensions of *Hyracotherium*. *(Universal Images Group North America LLC / Alamy Stock Photo)*

museum displays. Although it is a rescinded junior synonym, *Eohippus* is still used widely today, especially in the United States and Canada. After all, the horse was a gift from North America to the rest of the world.

Inhabiting a vast quadrant of the planet roughly fifty-seven million years ago, the *Hyracotherium* was a small, fox-sized mammal between twenty-five and thirty-five inches in length and between twelve and twenty inches in height (or three to five hands, in horsespeak). Its small, flexible frame, weighing ten to twenty pounds, was supported by four toes on its front feet (the large fifth "thumb-toe" was off the ground) and three on its hind feet (with the fifth and vestigial first toes off the ground), all tipped with padded paws, or soft proto-hooves. Its back legs were longer than its front, allowing it to scamper and bound (like a deer), and browse on fruits, leaves, and other soft foliage with its short snout containing forty-four low-crowned teeth. The fact that its five toes were morphing into three and four toes is an indication, as postulated by Marsh and Huxley, of an earlier evolutionary transfiguration from a five-toed ancestral forerunner.

This progenitor of *Hyracotherium* remains a mystery, as the post-dinosaur epoch beginning roughly sixty-five million years ago was the transcending shift from reptiles to mammals.* At this embryonic stage, however, the small ratlike mammals that paleontologist at the University of Bristol Christine Janis dubbed "artful dodgers" were still relatively undifferentiated, with a sparse evolutionary tree. Although our skeletons have been pulled in different directions by the survival pressures of natural selection, anatomical vestiges in both horses and humans point to a shared common ancestor, or "stem animal," somewhere between a hundred million and sixty million years ago.

The global environments these adaptable and burgeoning mammals patrolled would be unrecognizable to us today. The now-desolate, windswept plains and dusty tumbleweed deserts of Wyoming and Colorado where Othniel Marsh's *Hyracotherium* scampered fifty-seven million years ago, for example, were a wet, lush, jungle canopy. It was a period of global warming that paleontologists refer to as "Greenhouse Earth," when crocodiles basked on sunny, breeze-kissed beaches and took leisurely naps under the shade of swaying palm trees—in the Arctic. Landscapes were dominated by forest and jungle, and not by the savannas, prairies, grasslands, steppes, and tundra of today.

As the temperature and ecosystems of the Earth fluctuated, *Hyracotherium* was forced to adjust, and through an assortment of evolutionary offspring—and many more dead ends—gradually transitioned into modern *Equus*. These adaptations allowed the modern horse to not only survive but also, eventually, with human intervention, thrive. It is important to consider, as paleontologist Richard Hulbert at the Florida Museum of Natural History stresses, that the evolutionary journey of the horse is a "web of hypotheses and theories, with many points of contention and controversy. No two paleontologists interpret the fossil record exactly alike, and this is certainly true in the case of horses. . . . Nevertheless, because of our long-term relationship with horses, their fossil record will always have special significance."

---

* Dinosaurs ruled the Earth for an astounding 165 million years—six hundred times longer than our own hubris-driven reign as *Homo sapiens*.

For our purposes, however, we must condense the fascinating and complex bloodlines of equine evolution and corresponding anatomical adaptations. Only those vital and exploitable anatomical, biological, and behavioral attributes (and events) on the evolutionary ladder that allowed humans to harness the unrivaled potential and historical influence of the horse will be detailed. Without these critical steps and traits, our dominant, earth-shattering relationship would have been untenable. To initiate this evolutionary sequence, we must travel back to the dawn of the horse.

Although the supercontinent Pangea, encompassing almost all the Earth's land mass, slowly began to break apart 175 million years ago, most cross-continental land routes were still accessible. A high-altitude land bridge connecting North America to Europe via the Arctic islands of Canada, Greenland, and Scandinavia allowed for the extensive migration of animals across current continental divides. By fifty-five million years ago, *Hyracotherium* was established throughout Europe, Asia, India, and North America, particularly in Wyoming, Colorado, and New Mexico.

Around fifty million years ago, the Earth began to cool, and humidity levels dropped. As jungles and forests retreated slowly, grasses filled the ecological vacuum. *Hyracotherium* became extinct between thirty-four million and twenty-three million years ago, save for North America, where they continued to evolve. According to archaeologists David Webb and Andrew Hemmings, during this middle Eocene period, "North America became isolated from other northern landmasses. In Europe the family . . . diverged from the true Equidae, which persisted only in North America. Thereafter North America had hegemony over equid evolution." By thirty-two million years ago, North America was producing equines that, although much smaller in stature, had a similar silhouette to modern-day horses. The 120-pound *Miohippus*, for example, which lived roughly thirty-two million to twenty-five million years ago, would be easily recognized today as a miniature horse.

Throughout these evolutionary trails, equine species would emigrate from North America across Beringia to Eurasia and begin a separate evolutionary path, only to become extinct, while the main American line continued its own progression. "This pattern of global radiation and Old World extinction would occur repeatedly," explains anthropologist Pita Kelekna, "before modern *Equus* finally evolved in North America." For example, eleven million years ago, a remarkably successful globetrotting species known as *Hipparion* fanned out from North America across the Arctic, Asia, Europe, the Middle East, and the Indian subcontinent. These animals persevered in Africa until four hundred thousand years ago, long after the extinction of their kin across Eurasia and North America.

At Laetoli, Tanzania, roughly thirty miles south of the infamous Olduvai Gorge, a surface known as Footprint Tuff records the earliest known interaction between the ancient ancestors of horses and humans. Among the numerous footprints excavated between 1976 and 1978 by renowned archaeologist Mary Leakey are the side-by-side, juxtaposed steps of *Australopithecus* and *Hipparion*. As they walked across, and imprinted their mark upon, this muddy, ashen surface some 3.6 million years ago, these progenitors could not have imagined that the eventual dynamic union of their offspring would irrevocably change the world. "Wherever man has left his footprint in the long ascent from barbarism to civilization," acknowledged late-nineteenth-century historian and author John Trotwood Moore, "we will find the hoofprint of the horse alongside."*

The most significant global equine dispersion involved an early species of the modern genus *Equus*, which DNA research suggests evolved in North America some 4.5 million to 4 million years ago.† Traversing the newly formed Panamanian land bridge, or Isthmus of Darien, about 2.7 million years ago, these horses were among the first mammals to enter South America during a wave of accelerated migration known as the Great American Interchange, which included the ancestors of

---

* While Moore was a prolific writer and journalist, and the state librarian and archivist for Tennessee, he was also an unabashed racist and apologist for the Old South and Confederacy.

† Some argue *Equus* is as old as seven million to eight million years.

the heavyweight *Arctotherium,* a hulking four-thousand-pound short-faced bear!

Successive waves of *Equus* also wandered across the Bering Bridge to Siberia and Asia and eventually established themselves across Europe and the Indian subcontinent by 2.6 million years ago. These multiple migratory populations created a deep genetic rift and deviation among modern horses, asses, and zebras. *Equus* found its way down to Africa, displaced the resident *Hipparion,* and, in time, gave rise to various groups of zebras. At roughly the same time, the *Equus* subgenus *Asinus* (asses) evolved and scattered across Eurasia and the Middle East.

According to a recent archaeological study by Peter Mitchell from the University of Oxford, the common ancestor of both zebras and asses "likely coexisted with the earliest horses in North America before dispersing into the Old World a little before 2 million years ago. A recent whole-genome study (as opposed to ones targeting only some parts of an individual's DNA) places the separation of zebras and asses at 1.99–1.69 million years ago." Both zebras and asses continued to diverge into subspecies until roughly 150,000 years ago.

During the Pleistocene epoch (roughly 2.6 million to 12,000 years ago) no fewer than fifty-eight species of *Equus* are present in the North American fossil record, with dozens living simultaneously. This is certainly not "straight-line" evolution at play. The direct descendant of modern caballine horses showed up in North America loaded with genetic adaptations around 1.2 million years ago and its offspring in Europe as early as 1 million years ago. With the advent of DNA testing, genomic analyses of the seven living species of the genus *Equus* have revealed a more composite schematic blueprint of equine evolution than thought previously.

Shadowing and interwoven within the fabric of these migrations and extinctions was a series of successive adaptations fashioned by the gradual climatic and ecological transition from jungle and forest to grasslands and tundra, or what influential nineteenth-century poet Walt Whitman called the "vast, trackless spaces" within North America. Natural selection is not all that complicated. It has one superseding prerequisite: the need to eat and not be eaten. This simple equation dictated most equine evolutionary modifications from *Hyracotherium* to horse spanning a period of roughly thirty-five million years.

An era of global warming beginning roughly twenty million years ago kicked off an evolutionary explosion. This "Great Transformation" produced numerous competing equine species in North America. While most were evolutionary dead ends, those inheriting specialized genetic mutations to teeth, the digestive tract, feet, legs, and brain survived and thrived in a capricious ecosystem being overrun by grass. "The evolutionary saga of the horse is a kind of cast-out-of-Eden story," writes Philipps. "The horse started in a lush paradise that it lost as the global climate changed. . . . That evolution helps explain why the animal made such a tight bond with humans, and why it was able to spread all over the world in the modern era when so many other animals disappeared."

It is precisely these biological features, developed within an adapt-or-die evolutionary arms race with grass, that set the horse apart from all other animals and engineered its matchless potential as a living machine that would eventually come to dominate our planet and our shared history.

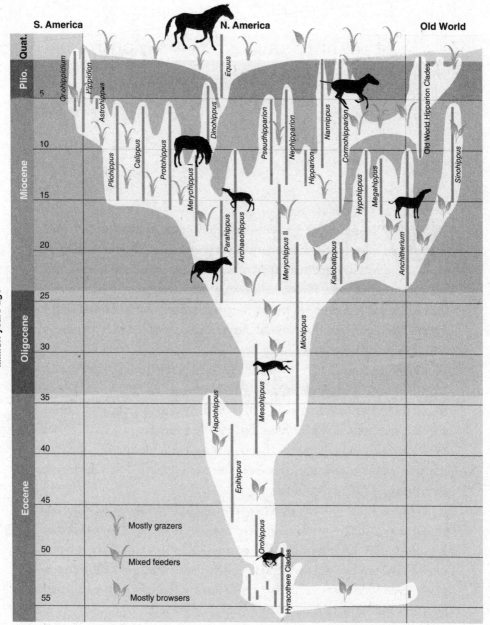

From *Hyracotherium* to horse: The global evolution and dispersion of *Equus*. Note the survival pressure shift from browsers to grazers. *(American Association for the Advancement of Science)*

CHAPTER 2

Straight from the Horse's Mouth
I Am the Grass, Let Me Work

While I didn't grow up around horses, they were not completely absent from my hockey-dominated childhood in the small Canadian town of Sarnia, Ontario, nestled idyllically on the shores of Lake Huron. They were, however, enough of a novelty that when my parents took me to the local petting zoo with the simple Orwellian nickname the Animal Farm, I would stare, mouth agape in childlike wonderment, at the horses.

I loved to feed them our sour backyard apples from my shaky outstretched hand and giggle at their greedy tongues and big teeth. I loved to pet them and feel their coarse hair on my youthful hands while they snorted, nickered, and whinnied. But mostly, I loved to just stare at their regal elegance and raw power. I was, and still am, mesmerized by their massive, rippling muscles. The sheer force of strength contained under that shiny coat is awe inspiring.

When I was a kid, my mind would race with a single question that would skip through my thoughts like one of my dad's worn-out Gordon Lightfoot, Bob Dylan, or Beatles records: "How can they have such big muscles, when they only eat grass?" Maybe you asked yourself this very question as a kid too—or, perhaps, still do. I will finally answer my own question and satisfy the curiosity of my younger self.

The explanation is not quite as simple as the eat-but-not-eaten equation and revolves around numerous interconnected and intertwined evolutionary attributes. Like Othniel Marsh's overflowing crates of equine fossils in the bowels of the Yale Peabody Museum, history also does not warehouse well in neatly labeled boxes, for events do not exist in quarantined isolation. Historical episodes are rarely built on the grounds of a single foundation. Most are the product of a tangled web of influences and cascading cause-and-effect relationships within a broader historical narrative. The anatomical development of the horse, and its association with humans, adhere to this paradigm.

In the process of answering my childhood question, we will navigate the complex evolution of the horse and the arrival of associated adaptations that elevated it to the historic heights of animal super status within human society. We will start, and end, eventually, with grass. "I am the grass; I cover all," wrote poet and three-time Pulitzer Prize–winner Carl Sandburg. "I am the grass. Let me work." The evolution, and eventual human utility, of the horse is all about grass.

We take grass for granted. Sure, we play sports or wrestle with our children and dogs on it, spread out a blanket or some lawn chairs on its surface to picnic or watch a concert, enjoy looking at its verdant hue—or maybe, if you are like me, even find it relaxing to mow and take in its sweet, aromatic post-sheering perfume. All of this occurs on only 20 percent of the grass. The remaining 80 percent observes us from below. If it could speak, I imagine that grass would have a remarkable and bestselling tale to tell.

As grass began to cover increasing swaths of the planet, including North America, it worked directly on equine evolution. Currently, an emerald sea of roughly ten thousand types of grass, some rising ten feet or more, covers more than 30 percent of our planet. I suppose Carl Sandburg was 30 percent right. Although, prior to human intrusion and the advent of the Agricultural Revolution twelve thousand years ago, grasslands would have been far more abundant and ecologically authoritarian. Like humans and horses, grass hedged obscurity to achieve unprecedented global hegemony and dominion.

This climate-changing landscape triggered an evolutionary arms race between horses and grass, lighting the fuse for a life-or-death struggle of natural selection and survival of the fittest. Rather than browsing for an increasingly limited and dwindling supply of fruits, leaves, and pulpy shrubbery proffered by thinning and retreating forests, some equines became grazers. Grazers of grass. But this presented a problem. In the world of eat or be eaten, grass developed an ingenious defense.

Most of us have chewed on a piece of grass or blown across it as a woodwind reed tucked in between our nimble fingers, and inevitably cut our lips or hands. A *blade* of grass has sharply honed, knifelike edges. It is also a tough chew. Grasses are relatively devoid of nutrients, and the beneficial complex sugar, cellulose, is locked inside the cell walls. These walls also contain the woody, ridged structural material lignin and phytoliths of razor-sharp silica sand (also used to make glass), which are drawn from the ground as it grows, making it indigestible to most mammals. As a side serving to this main course, gritty soil and sand are inevitably taken up in the process of cropping grass close to the ground. All of this is tough on teeth. Grass also requires a lot of chewing. This is why we humans are not grass eaters.

Some mammals, however, evolved to inherit a few counterpunches to these cunning evolutionary fortifications. While most browsers, including those equines, died off roughly twelve million years ago, grazing horses circumvented these obstacles cultivated by grass, positioning themselves with a greater ability to exploit the new steppe habitat and environmental prairie niche. "The story of the horse," states paleontologist Kenneth Rose, "is really the story of its teeth. They have changed so much you wouldn't be able to recognize the species if you couldn't connect them over time." In short, as the masticating molars, or cheek teeth, expanded and widened, they were accompanied by forceful jaws and reinforced eye sockets to absorb the shock generated by more vigorous and violent grass grinding powered by robust mandibular musculature—notably the masseter complex of the rear cheek or jaw hinge (the big, round, can't-miss muscle on the side of the face).

As a result, the skull and muzzle needed to stretch and deepen to accommodate these modifications. The eye of a horse is eight times the diameter of that of a human and the largest of any land mammal. Adhering to the principles of survival of the fittest, over successive generations these eyes shifted upward and farther back on an elongated muzzle, head, and neck, producing a wide sweep of almost 350-degree vision and allowing horses to spend most of their time facedown in the grass.

This evolutionary adaptation enabled horses to graze or drink with their eyes above the grass line, while dramatically increasing their peripheral vision without having to move their eyes or their head to spot predators such as saber-toothed tigers and dire wolves. The common turn of phrase "putting the blinders on" refers to coverings that keep a horse's eyes and focus trained forward, preventing it from getting "spooked." These outcomes were advantageous characteristics for both grass grazing and vision. It was an evolutionary win-win. With these adaptations, the horse could eat without being eaten.

In a corresponding and connected stroke of natural selection, the horse developed sophisticated dental alterations to adapt to the abrasive qualities of its new diet. Eating siliceous grass is the literal equivalent to chewing sandpaper. Eventually, if the horse lives long enough, it will wear down teeth to worthless stubs, leaving the animal unable to eat, and inevitably starving to death. Inuit women of Arctic Canada, who use their teeth to soften leather, for example, often wear their teeth down to the gums if they reach an advanced age. To combat this erosion, horses came up with a brilliant retaliatory strike against grass.

All mammals have two sets of teeth: milk, or deciduous, (baby) teeth and permanent (adult) teeth. Adult male horses generally have forty teeth, and mares, thirty-six to forty. The twelve incisors at the front (the big, flat teeth you see when a horse "smiles") are used for clutching and tearing grass. These are followed by four canines, sometimes called bridle teeth or tushes (thickset in males and comparatively small or more commonly absent in females), used as a weapon and for grooming and communicating.

At the back are twenty-four molars, which are the all-stars of the lineup. Over time, the twelve premolars acquired the same shape and size as the twelve standard molars, doubling the surface area for prolonged grass grinding.* Although the four cheek teeth arcades are composed of six individual molars, they erupt in a tightly packed unit, acting

---

* Wolf teeth (originally the forward premolar that did not become fully "molarized" like the other twelve) are extremely small teeth in front of the first upper molar. As an evolutionary dead end, not all horses have wolf teeth, and those that are present are not fully erupted.

essentially as a single lateral grinding tool to break down tough grasses. The molars are the workhorses of the mouth.

During adolescence, roughly four to five years old, a horse's permanent teeth are fully formed at five inches long. Apart from the visible crown, most of the tooth remains hidden below the gums, reserved for future use. As the exposed area wears down, teeth continue to erupt from the gums at a rate of two to three millimeters per year (much like the lead in a mechanical pencil), giving rise to the expression "long in the tooth" to signify old age. From horse teeth also come the sayings "Don't look a gift horse in the mouth" and "Straight from the horse's mouth." Both refer to the ability of skilled and seasoned traders or handlers to judge the age, and the corresponding value, of a horse by the wear of its teeth.

Compared with mammals with low-crowned teeth—such as humans and *Hyracotherium*—who generally do not feast on grass, high-crowned horse teeth also have an extra outer bony layer of cementum, in addition to the enamel coating and dentin core, to compensate for the increased wear and tear of a pre-domesticated diet consisting of 90 percent to 95 percent grass. Cementum is the glue that binds the entire tooth system together and fills in the ridges, peaks, and valleys of the outer shell.

The final characteristic of equine dental schematics is also the most indispensable for human manipulation. Separating the front grouping of incisors and canine teeth from the back molars is a vacant space or gap known as the diastema, where the bit, or mouthpiece, of the bridle is placed.* I hazard to guess that without this anatomical convenience, the mounting and riding of horses by humans, or, more importantly, *controlling a horse while riding*, would have been far more difficult, if not impossible. Correspondingly, the widening of the diastema also elongated the muzzle, beneficially expanding the distance between the mouth and the eyes mentioned earlier.

While the horse's combined first-strike capabilities of mouth, jaw, teeth, and mandible muscles have won the first battle against grass, the war is far from over. Eating grass is one thing; absorbing its relatively

---

* The bridle is the headgear used to direct a horse, consisting of the buckled straps and pieces to which the bit and reins are attached. Assorted tools of horsemanship, including bits, crownpieces, and cheekpieces, were stored together, hence the term "bits and pieces."

low nutrient content is quite another. Natural selection awarded the horse another offensive weapon against grass in the form of a unique digestive tract, or cecum.

Like the diastema bit gap, this distinctive "hindgut" digestion was also an expedient and exploitable trait for the utilitarian adoption of the horse. More importantly, it finally answers my nagging boyhood question, one that Master Yoda also asked Luke Skywalker upon their first meeting in the swamps of Dagobah in *The Empire Strikes Back*: "How you get so big eating food of this kind?"

The dietary requirements of a horse are bewildering. Large carnivores such as polar bears and tigers weighing roughly a thousand pounds consume upwards of eighty-five pounds of high-protein meat in a day, whereas a horse of equivalent size ingests only fifteen to thirty pounds of low-grade grass. Horses have adapted a few evolutionary tricks to sustain themselves on this inferior vegetarian menu.

A cellulose-heavy diet is the favored fare and standard cuisine for all hoofed animals except pigs. They all use the same basic symbiotic strategy by employing bacteria and other microorganisms to break down and unlock the tough cellulose into digestible enzymes. "Grass is to provide a food source for its real meal—the bacteria," explains biogeneticist Sam Westreich. "It's the bacteria that break down the hard-to-digest cellulose in grass and convert it into a plethora of different amino acids, which in turn become the building blocks for creating a 1,200-pound animal." The crucial difference, and the key to the increased functionality of horses compared with other hoofed animals, is where and how they ferment and absorb the nutrients of their roughage.

Foregut fermenters such as cattle, sheep, goats, deer, camels, hippos, giraffes, and most other Artiodactyls (even-toed ungulates) are patient food-processing ruminants.* To maximize the nutrition in grass, they chew, digest, regurgitate, "chew the cud," and redigest their more selective tender grasses through four stomach chambers over a prolonged period. They are limited by the carrying capacity of their digestive systems

---

* *Hippopotamus* comes from the ancient Greek "river horse."

and need to rest or ruminate after packing it in, when an internal valve literally shuts off the intake flow.

Think of the synonyms in the English language for *ruminate*. By nature, their meanings all revolve around resting, sitting, or maintaining a stationary posture allied to some sublime form of trancelike introspection, meditation, or namaste nirvana. Hindgut herbivores, such as horses, are far less transcendental and do not have the same ruminating constraint.

For horses, after the lateral grinding of grass by those miraculously evolved teeth, the stomach is the first quick pit stop in the digestive system, followed by the intestines. The real show happens in the cecum, where the fermentation or nutrient breakdown occurs thanks to a cocktail of bacteria, protozoa, and fungi. We, like many mammals, also have a cecum—a vestige of our shared ancestral heritage. Tucked in between the small and large intestine, ours is a tiny 2.5-inch pouch of little consequence, whereas the cecum of a horse is more than four feet long, with a tank capacity of eight gallons (approximately thirty liters).

Finicky ruminants like cows spend roughly half their time nibbling meticulously and the other half ruminating, whereas horses spend roughly 70 percent to 80 percent of their time grazing assiduously over vast ranges. "The horse's digestive system is like a conveyor belt," writes Wendy Williams, "where food moves in at one end, passes through at comparatively high speed, and comes out the other." Comparatively high speed to ruminants, that is. From mouth to anus, grass will enjoy a streamlined, express-lane 130-foot tour of the equine digestive tract. Proportionate to total body weight, this is about one-third the size of the digestive system of a cow.

Grass passes through horses much quicker, so they need to eat more grass. It takes twice as long for ruminants to empty their fermenting tanks (ninety hours) when compared with horses (forty-eight hours). There is a tradeoff, however. Compared with horses, ruminants are on average 40 percent more economical in extracting usable energy from a given weight of food—and for cows specifically, this increases to 70 percent. To sustain itself with hindgut digestion, a horse must eat a greater quantity of food than would a ruminant of equivalent body mass (roughly

20 percent to 65 percent more, depending on the specific animal), which explains the augmented feeding time.

In summary, horses get less energy from every mouthful, so they take more mouthfuls, giving rise to the turn of phrase "eating like a horse." Horses need a daily forage intake of 1.5 percent to 3 percent of their body weight. For the average thousand-pound American quarter horse, this equates to fifteen to thirty pounds of grass per day, in addition to six to ten gallons (twenty-five to forty liters) of water. Benefiting human pursuits, horses can survive up to a month without food and six days without water.*

The advantage of a horse's digestive system is simply that it can "eat and run." Unlike its ruminant cousins, it does not need to rest to digest. Across all human horse-driven enterprises—from agricultural pursuits to long-range mobility for trade, travel, and war—this was a vital time-saving (and moneymaking) attribute. Grass, however, has not yet finished its evolutionary manipulation of the horse.

The requirement of horses to push more food through the cecum to obtain sufficient nutrients also had a butterfly effect on grass quality, grazing range, and, ultimately, size. "Per unit of time, if not per weight of food, a horse can get more energy out of a low-quality diet than can a cow of the same weight," affirms Stephen Budiansky. "Equids have chosen an ecological niche that allows them to avoid competition from other species. Their niche is the poorest-quality vegetation. . . . So, having adopted a strategy that depends on eating the lowest-quality, most energy-poor stuff, equids continue to gain a further competitive advantage over ruminants by getting bigger. . . . [T]he horse's strategy was thus to avoid competition by choosing a diet too fibrous for ruminants to cope with at all. And that made getting big a necessity." The answer to my boyhood question continues to grow, although it seemingly always comes back to grass.

With faster metabolisms, smaller animals need to consume more

---

* The modern aphorism "You can lead a horse to water, but you can't make it drink" is a modified version of what may be the oldest English proverb still in common use today. First recorded in 1175 in Old English *Homilies*, the original phrase, "Hwa is thet mei thet hors wettrien the him self nule drinken?" reads "Who can give water to the horse that will not drink of its own accord?" Modern translation: You can present someone with a beneficial opportunity, but you cannot force them to take it.

energy per pound than large animals, based on exposed surface area and heat retention. Following a mathematical formula, an animal that is eight times the size of its companion has only four times as much surface area. In a small object, such as a golf ball, everything inside is closer to the edges, as opposed to a large object like a beach ball, where more of the interior is buried deep within, protected from the surface elements.

The petite *Hyracotherium* ate fruits, berries, leaves, and shoots, providing a higher caloric intake. Over time, as forests gave way to prairie, this diet shifted to grasses, and, in leaner times, included samplings of bark, branches, and other scrub brush. According to Bergmann's rule, as the horse grew larger, it could also exploit colder climates, expanding its natural habitat and increasing its stamina for long-range grazing in search of food resources during various seasons.* In pursuit of suitable silage, or fodder, such as hay and corn, horses have been known to graze extreme distances of up to thirty-five miles a day. Which brings us back to the answer to my childlike wonderment. In short, based on ecology, diet, teeth, and digestive tract, and to maximize energy to ensure survival, horses grew larger and acquired those rippling, swollen muscles.

Another key component of size is longevity. Larger animals usually live longer.† The average life span of *Hyracotherium* was about four years, compared with roughly twenty-five to thirty years for a modern horse.‡ Why spend years training (and large sums of money) on an animal only to have it die immediately after it is formally educated and fully skilled? Longevity was yet another equine feature in humanity's favor. Think of Alexander the Great's beloved and brave horse Bucephalus. For twenty years and through twenty battles, the black stallion faithfully carried his young horse-whispering prodigy to "the ends of the Earth" and beyond the

---

* The theory posited by the nineteenth-century German biologist Karl Bergmann that in order to conserve heat, warm-blooded animals at higher latitudes (colder climates) are larger and thicker than those closer to the equator.

† It is an amazing fact, and possibly sheer coincidence from a scientific standpoint, that virtually all mammals live an average of 1.5 billion heartbeats, whether that be a pygmy shrew's 1,300 beats per minute (bpm), a human's 60 bpm, a horse's 44 bpm, or 4 to 8 bpm for the blue whale.

‡ Horses, along with humans, chimpanzees, orangutan, gorillas, dogs, and elephants, belong to a select group of mammals that experience hair graying with age.

summits of supremacy, prestige, and legend. Which leads us to another powerful *and* energy-saving evolutionary apparatus: the feet and legs.

During the long and winding evolutionary road from *Hyracotherium* to horse, multiple toes were amended or amalgamated into one hoof. The benefits of this, and its correlation to ligaments and bone structure in the legs, are immeasurable. I will put it to you this way: there is no Kentucky Derby for cows, no Royal Ascot for rhinos, no Breeders' Stakes for goats, no Australian Derby for giraffes, and no Triple Crown winners among pigs, sheep, deer, or hippos. Please do not get me wrong, though: I would absolutely *love* to watch all these races. In accordance with the most unalienable right of the "Seven Commandments of Animalism" in George Orwell's 1945 political satire *Animal Farm*—that "All animals are equal"—I also fully support equal opportunity and inclusion for all of evolution's miraculous creatures.

Throughout our shared history with the horse, however, just like in *Animal Farm*, the Seven Commandments can be edited to just one: "All animals are equal, but some animals are more equal than others." After all, the main principle of evolution and natural selection is precisely the promotion of advantageous *differences,* not equality. There are no unalienable rights endowed by any creator in evolutionary natural selection, only a blind progression of biological chance and capricious, mutable characteristics. Within these natural genetic hereditary laws and the twisted double helix of DNA, there is certainly no liberty afforded to organisms to choose their own destiny or to pursue evolutionary happiness. To be truthful and accurate, then, perhaps the slogan should read "All animals are created equal but evolved uniquely."* Enter the horse.

On the soft, muddy ground of jungles and forests, multiple toes like those of the *Hyracotherium* were an advantage to splay and displace weight, much like a snowshoe. They also provided for a wider, more stable stance, enabling the ability to scamper through the boggy undergrowth for protection. On the open plains and hard pastures, with no cover from predators, survival necessitated drastic modifications to this design.

---

* For an optional homework assignment, read historian and author Yuval Noah Harari's take on this in chapter 6, "Building Pyramids," of his internationally acclaimed bestseller *Sapiens: A Brief History of Humankind*. It is humorously brilliant.

While still possessing multiple toes, by thirty million years ago equine weight was balanced on the middle, third toe, supported by stronger ligaments rather than *Hyracotherium's* doglike underpads. This alteration led to enhanced elastic energy and a more efficient springlike action transferred to the tip of the foot with each stride, ultimately generating a faster gait. Over time, this middle, weight-bearing toe grew, while the others became withered, vestigial "splint bones" or simply disappeared.*

Horses became monodactyl with one quick-striking hoof made from keratin, the same fibrous protein as our fingernails and hair. Given that horses do not have claws and rely on an herbivore's dental plan created for grinding grass rather than the vampiric puncture and ripping configuration of carnivores, in the fight-or-flight animal arena of eat and not be eaten, horses honed, perfected, and patented the flight response. With multiple evolutionary overhauls and selective tweaks, they became assembled for pure speed.

In most mammals, the two bones in the arm or shin are equal and can rotate around each other, in addition to swiveling ankles and wrists. Grab your lower arm just above your wrist and rotate your hand. You can feel the crossing and uncrossing of the radius and ulna. This helps us grasp and manipulate objects with our multi-fingered hands. Roughly ten million years ago, horses were evolving into specialized runners on one toe and no longer required this anatomical feature.

To increase efficiency and drive, their limbs needed only to move in a forward-backward motion. In the forelimb, bones in the foot, ankle, and leg lengthened and fused, effectively transforming into one long, slender, solid, stilt-like lower limb, supporting 60 percent to 65 percent of body weight. In the hind limb, the shin bone, or tibia, became the primary weight-bearing crutch, while the fibula became a slight splint. As a balance, the bones and musculature in the upper limb were updated to compensate, creating a more powerful stride. These lower-body transformations had yet another energy-efficient upshot unique to equids.

Fully developed in the first species of *Equus*, roughly 4.5 million to

---

* There are two small, hairless, callous outgrowths on the legs known as chestnuts and ergots, which are theorized to be the remnants of toes.

4 million years ago, this distinctive alignment of the tendons and bones allows the limbs to lock while a horse is standing or grazing, nullifying the need to expend energy to fight the genius of Sir Isaac Newton's gravity. Given that horses spend 80 percent to 90 percent of their time upright, this is a clever and adroit adaptation. In fact, horses expend 10 percent *less* energy standing up than they do lying down.*

In addition to being a nutritiously economical stance, it is also a healthier defensive posture. It is much quicker to spot and flee predators while standing as opposed to lying prone. Horses, therefore, prefer to stand and, like Newton, can see farther. By comparison, cattle and sheep use 10 percent more energy standing than lying down, which is why cows are often viewed lazily ruminating, musing, and daydreaming while lying down in emerald fields of grass.

In concert with these anatomical alterations, the rounded back of the *Hyracotherium* gradually straightened as the horse grew in stature and matured from a scamper to a gallop. According to evolutionary biologist Martin Fischer, horses have "dorsal-stable locomotion. . . . This is a very peculiar thing. Particular to horses. And it's the prerequisite to riding." Unlike other mammals, the back of a modern horse is relatively level, and is both stiff and supple, allowing for a firm *and* flexible point of balance and a sustainable, fairly straight seat. The cartoon doodle of "Eiohippus + Eohomo" that Thomas Huxley etched for Othniel Marsh would have been anatomically impossible. The jubilantly blithe Jawa-like prehominid rider would have been hurled clear over *Hyracotherium's* head.

Having acquired the complete evolutionary package detailed above, horses can sustain a gallop reaching upwards of forty miles per hour, with a top clock speed of fifty-five. This combination of size and speed is a rarity in the animal kingdom. The peregrine falcon is the fastest animal

---

* Horses sleep, idle, rest, or doze for three to five hours a day—usually at night and typically while standing. Despite the common mythology that horses do not lie down to sleep, they will take two prone postures: sternal recumbency (lying on their chest, belly, and folded limbs) and lateral recumbency (on their side) to enter REM sleep for roughly thirty minutes per twenty-four-hour cycle, usually in spurts of three to ten minutes. They take turns lying down or dozing, relying on herd mates to keep watch. As with humans, this REM sleep is extremely important for the overall physical and mental health of horses, and disruptions to this sleep pattern can adversely affect their well-being.

on the planet, clocking in at a staggering breakneck speed of 240 miles per hour, but only when engaged in a straight predatory dive or hunting stoop. The cheetah is the fastest land animal, accelerating from zero to sixty miles per hour in a mind-boggling three seconds and reaching a maximum velocity of seventy miles per hour, but only for short bursts. Other comparable animals, such as the blue wildebeest and the lion, can hit bursting speeds of fifty miles per hour but cannot maintain that pace for any substantial length of time. The same can be said for smaller animals like the pronghorn, springbok, and ostrich, all of which can crank it up to fifty-five miles per hour for brief intervals.

The horse is an anomaly. It possesses a freak-of-nature fusion of size, strength, stamina, and speed, all of which would eventually be harnessed for multiple human uses. "The bigness of the horse is an oddity among mammals," concludes Budiansky. "Size and swiftness are most likely accidental by-products of its ecological niche—at the low end of the herbivore diet." Accidental by-products of grass, that is.

Finally, the central computer controlling this big body also underwent a reboot upgrade. From complementing and regulating this coalition of equine evolutionary adjustments, the brain became rapidly larger, specifically the neocortex. Unique to mammals, this integral circuit board of our highly sophisticated brain is responsible for higher-order functions such as sensory perception, cognition, generating motor commands, spatial reasoning—or, more generally, learning and coordinating or networking multiple corresponding sensory inputs and entries.

Our human brain, which takes about twenty-five years to reach maturity, for example, accounts for roughly 2 percent to 3 percent of our total body weight. But it requires 15 percent of the body's blood supply, 20 percent of the body's oxygen supply, and even when the body is stationary and at rest, it exhausts 25 percent of our energy. It is the motherboard of our mammalian computer and provides us with good ol'-fashioned "horse sense."

Horses, however, are not actually as intelligent as we might want to believe. They have small brains roughly the size of a human child's: 1.5 to 2 pounds. The principal cerebrum is about the size of a walnut. An adult human brain weighs about 3.3 pounds. Horse behavior is linked primarily to innate instincts, social herd etiquette, and classical conditioning repetition, recall, and reflex.

Researchers have also hypothesized that, in horses, this cerebral development provided a competitive advantage over other herbivores, as it was crucial to rapid-response decision-making to predatory threats on the vulnerable and exposed grasslands. In the eat-and-not-be-eaten paradigm, horses need only be the second-to-last species to flee or elude hungry stalkers. Being quick on their feet ensured their survival. In addition, it allowed horses to navigate the elaborate spheres of herd dynamics.

At this point, it should not come as a surprise that these complex social structures were predicated upon a steppe habitat and a grass diet. Generally, animals that are grassland grazers, including horses, American bison (buffalo), and other ungulates, do not defend territory, as they are migrating, long-range, opportunistic feeders based on season, climate, and availability. Unlike selective feeders, they are not territorial creatures and do not engage in hostilities to "hold ground."

Having to rely on sparser and sporadic food sources, horses (and plains and mountain zebras) form into harem herds of one polygamous stallion with multiple mares (usually two to six) and their offspring for both protection through safety in numbers and to ensure reproduction. The gestation period of eleven to twelve months sees to it that the single foal is born relatively physically mature to promote quick integration and mobility. Newborns can stand and move within an hour of birth and graze within a few weeks, although they continue to nurse for up to a year. Adolescent mares, who are reproductive from the age of two through the late teens, can leave voluntarily if courted or be abducted from their parental bands by upstart bachelor stallions looking to assemble their own clan or by established stallions eager to augment their harem.

Young yearling (between one and two years old) males are forced from their family herds and find refuge and fraternity in bachelor groups as small as two or as large as twenty. These herds of misfits and outcasts, however, can also include older or infirm males, creating quite the motley gathering. This expulsion, estrangement, commandeering, or appropriation of younger horses from the natal group is a socially ensured genetic safeguard to reduce direct inbreeding. Upon reaching sexual maturity at roughly the age of five, males will venture out to start a harem of their own.

Although surrendering (frequently overlapping) territory to hostile invaders, stallions will fiercely defend their harems and will mate only if their band is safe and secure. Any disturbance or threat can cause them to become nonbreeding lone males. Unique among hoofed animals, horses lack antlers or horns. While a convenient attribute for eventual domestication by humans, it leaves horses lightly armed. The weapons of war available to a stallion include rearing, kicking, and biting with those canines mentioned earlier. Cornered females will also fight back. Unlike males, they utilize their hind legs for kicking while showering an attacker with urine. Remember, evolution manufactured horses for flight, not for fight.

While horses are mirthfully gregarious and enjoy "horsing around" in everyday "horseplay," within these long-term and tight-knit family bands there is an authoritative hierarchy and power structure. To keep a watchful eye on his harem, the stallion leads from the rear during seasonal migrations or the search for resources, while the clan mother, or dominant mare, tracks from the front as the leader of the band. The remaining members fall in and follow her lead in the rank-and-file order of their herd status. Foals will follow their mothers in ascending order from youngest to oldest. Horses instinctively understand submission, signals of supremacy, and the chain of command. If forage is limited, only the dominant horses partake, working down the pecking order.

This preadapted trait is paramount to their unrivaled capabilities as transport vehicles and weapons of war. "These behavioral patterns explain how a convoy of horses or a cavalry charge can be held together and kept moving by a human rider taking over the position of the dominant leader whom all the rest will blindly follow," explains pioneering zooarchaeologist at the Natural History Museum in London Juliet Clutton-Brock. "This type of social organization is an adaptation to migration and to environments with unpredictable conditions as well as to a regularly changing but constant food supply." Again, it comes back to grass. "This adherence to an order of dominance whilst moving from place to place," adds Clutton-Brock, "explains how a convoy of domestic horses or a cavalry charge can be held together with very little effort by the human rider or driver who assume the position of the stallion." This is certainly an attribute required for the multiple invaluable roles the horse would play throughout human history, including trade and travel caravans, packtrains, and its numerous game-changing military occupations including chariotry, cavalry, and service and supply.

A poignant and heartbreaking story illustrating the innate and coached behaviors of cavalry horses is that of wounded British mounts following the Battle of Waterloo in 1815 during the Napoleonic Wars.* More than twenty thousand horses died during this clash, which saw Napoléon Bonaparte decisively defeated by the British-Prussian coalition led by the Duke of Wellington, who spent seventeen hours straight in the saddle on his horse Copenhagen. Those injured horses lucky enough to survive were retired from military service through public auction.

Sir Astley Cooper, chief surgeon to King George IV and Queen Victoria, purchased a dozen of the most seriously maimed and mutilated animals, welcoming them to his home at Hemel Hempstead. Dr. Cooper and his team of dedicated physicians and servants spent weeks painstakingly removing musket balls, grapeshot, and shrapnel from the disfigured, blistered, and twisted bodies of his adopted equine patients. In time, they all regained their physical health and were set loose to spend their remaining years in peace and solace, roaming and grazing the verdant sprawling grounds of the estate.

Not long after, while gazing out the window across the morning mist–covered downs, an emotionally tearful Cooper witnessed a remarkable and tender scene. As if on command, the twelve veteran cavalry horses mustered into a perfect shoulder-to-shoulder military line and proceeded to charge forward at a paced gallop. After a few sustained strides, they quickly halted, performed an about-face in formal drill,

---

* During his ill-fated Russian campaign of 1812, Napoléon lost 450,000 men and 200,000 mounts. He largely replaced the human losses within the year but could not replace his horses (and wagons), procuring only 29,000 replacements. The dearth of horses was a main reason for his defeats of 1813 in Germany, including the Battle of Leipzig—by far the largest single battle of the Napoleonic Wars, which raged intermittently from 1799 to 1815.

before breaking formation to prance and play freely across the rolling English plains. Every morning thereafter, like a proud parent, Cooper watched his scarred and battered old warhorses perform their cathartic ritual cavalry charge on his sanctuary haven fields in Hertfordshire, far removed both physically and mentally from the traumatic horror and haunting memories of the bloody Belgian valley at Waterloo.

Not all the king's horses were this well trained, however. For the 1761 coronation of George III, the unstable villain of the American Revolution personified in the musical *Hamilton*, a horse was tutored specifically for the event. It was taught to walk backward, so that, after performing its role, it could gracefully back away, sparing the eccentric young king the sight of its ass as it left the room. Apparently, the horse was either not very well coached or wanted the new king to be the butt of a joke. The moment it entered the grand hall, it turned around and walked all the way to the throne ass first.

This unique equine package of traits and abilities constructed by natural selection, however, evolved over millions of years far removed from human agency or manipulation. The modern horse represents a splintered twig on a massive fifty-seven-million-year-old tangled family tree. Extreme and rigorous selective survival pressures, including grass, constructed a creature with all the right stuff—size, speed, strength, stamina, and exploitable anatomical and behavioral attributes—that ensured its eventual domestication and historic ascent to global domination. When the horse came under human control roughly 5,500 years ago, the rule books and programming to the game of life, world of warcraft, and civilization were completely revamped.

This Centaurian Pact of domestication, however, is a relatively recent fellowship and encompasses an extremely small fraction (roughly 2.75 percent) of our two-hundred-thousand-year modern relationship as *Equus caballus* and *Homo sapiens*. For all human history, the horse has served another primary purpose. Our initial bond was not a communion, nor one of camaraderie, utility, service, or codependency. It was a cat-and-mouse game of predator and prey. Hungry horses eat grass, and hungry humans eat horses.

CHAPTER 3

Eat Like a Horse
Human Hunting and Vanishing Habitats

Most of us in the Western world cringe at the thought of eating what we now deem to be a noble animal, a loving companion, or a cherished pet. I have traveled across our global village and sampled a wide variety of local fare that many would consider repulsive or taboo: dog, cat, pigeon, snake, spider, lizard, beetle, grub, various insects, and, yes, horse. And while you might not want to hear this, more than a billion people still consume this surprisingly delicious meat. For most of us, however, butchering a horse is deplorable and a crime against nature. But this virtuous stance is a relatively recent social and spiritual convention.

The ancient Hindu text Rig Veda condemns the killing of humans and horses. While Hinduism does not require a vegetarian diet per se, many adherents choose this lifestyle to practice nonviolence against all living beings and to purify the body, mind, and spirit. Similarly, while not all Buddhists are vegetarian, and interpretations differ as to the sanctity of eating meat, the consumption of ten specific animals, including humans and horses, is prohibited. The three Abrahamic faiths also oppose eating horsemeat.

Certainly, under the Kashruth dietary laws of Judaism, the Israelites were forbidden from eating any animal that did not both chew its cud *and* possess a cloven hoof, as codified in both Leviticus 11:3–8 and Deuteronomy 14:4–8 of the Old Testament. Although a relative rarity in the

Levant at the time of their composition in the seventh century BCE, horse was nevertheless removed from the menu.*

Debate surrounds the eating of horses and other species of *Equus* in Islamic tradition. The Koran mentions that Allah created "horses, mules, and donkeys, for you to ride and use for show," which has been understood as a prohibition against consumption, while others disagree with this interpretation. Generally, however, eating horsemeat is discouraged in both Shia and Sunni Islam.

Catholic doctrine is unambiguous. Pope Gregory III believed the consumption of horsemeat to be a pagan abomination and officially outlawed it in 732. "You say, among other things, that some eat wild horses and many eat tame horses," he wrote to Saint Boniface, the Anglo-Saxon missionary to the Frankish realm. "By no means allow this to happen in future, but suppress it in every possible way with the help of Christ and impose a suitable penance upon offenders. It is a filthy and abominable custom. . . . We pray God that . . . you may achieve complete success in turning the heathens from the errors of their ways."

There were also far more lucrative reasons to not eat horses: they were moneymakers and simply too valuable for work and war to be wasted for meat. Horses boosted agricultural yields, enriching the tithing (in Islam, zakat) harvests of the church, monarchy, and feudal lords. As our story will later reveal, the bedrocks and bludgeons of feudalism—proficient cavalry, heavy knights, and the wealth they plundered—were increasingly reliant on horses.

Military concerns surrounding the Frankish duke Charles Martel, his defeat of the Moorish army at the Battle of Tours in 732, and the ensuing rise of qualified cavalry under his grandson King Charlemagne, also played a strategic role in the Pope's edict banning horsemeat and the perpetuation of this practice. These sanctions curbed, but did not eliminate, horse consumption, and it remained an important gastronomic resource for our ancestors across many terrestrial and spiritual domains.†

---

* The Levant refers to the region of the Eastern Mediterranean consisting of Syria, Lebanon, Jordan, Israel, Palestine, the Sinai, Egypt, and portions of southern coastal Turkey.

† Not every culture of antiquity dined on equine. Lacking suitable pasture, the horse was a relatively exotic animal across the deserts of the Middle East, and the earliest records from

While I appreciate that this might be hard for some readers to digest, the unappetizing truth is that for most of our hominid and modern human history, the horse was appreciated and prized *only* for its meat and the secondary products fashioned from its hide, bones, fat, hair, sinews, and hooves. Like the 150 multidimensional uses for every diverse part of the bison by the resourceful Indigenous peoples of Canada and the United States, all components of the horse also found a practical application. Human consumption of horses has been continuous from first contact to the present day.

Large mammals, including horses, roamed great swaths of the planet for the majority of the Paleolithic period, or Stone Age, roughly shadowing our own evolution and migrations between 3.3 million and 10,000 BCE. In the Americas, horses, reindeer, saber-toothed cats, bison, dire wolves, ground sloths, short-faced bears, and mammoth patrolled ranges from the Arctic to South America. Horses were also present in varying but substantial numbers across the vast Eurasian landmass. In Europe, the fossil record is dominated by reindeer, bison, aurochs (wild cattle), ibex, mammoth, antelope, deer, and horses.

Equine meats have been a staple in our diets since *Homo erectus* was using tools to carve up zebras (whether hunted or scavenged is a topic of fierce debate) in Africa some 1.6 million years ago, or possibly as early as 2.5 million years ago in the Awash valley of Ethiopia, where a *Hipparion* femur shows the unmistakable signs of tool-made cut marks. A site in southern China dating to 1.7 million years ago also indicates a correlation between horse bones and *Homo erectus*. Horses were commonly hunted by *Homo heidelbergensis*, an extinct subspecies of archaic human, and by *Homo neanderthalensis*, as evidenced by archaeological sites in England, Germany, Spain, and France dating back 500,000 years. It is generally accepted that we began our rapid ascent as modern *Homo sapiens* ("wise man") somewhere between 300,000 and 200,000 years ago. At any rate, we are a relatively new kid on the Darwinian block.

Like their upstanding progenitor *Homo erectus,* who embarked on a global tour before them, the first inquisitive *Homo sapiens* left our

---

Mesopotamia and Egypt reveal a consistent lack of consumption. Likewise, for the Greeks and Romans, horses were prized as expensive beasts of burden or highly skilled mounts for war, not destined for the meat hook or butcher's block.

ancestral African homeland roughly a hundred thousand years ago and continued the practice of hunting and eating horses. In fact, global fossil records reveal that until the end of the last Ice Age, roughly twelve thousand years ago, horses (and reindeer) dominated the diets of most peoples across the Northern Hemisphere, while other equids were hunted in the Middle East, Asia, and Africa. There is good reason why we humans and our hominid forebears have been feasting on horsemeat for millions of years: it is an extremely nutritious superfood.

Heavily saturated with blood, horsemeat contains 50 percent more protein and 30 percent more iron than the leanest beef. It is also rich in vitamins, minerals, and the only two essential fatty acids vital to human health: linoleic (omega-6) and alpha-linolenic (omega-3). These nutrients are rare in the meat of ruminants, including cattle.

Given that horsemeat is extremely lean, to maximize the nutritional depth, it would have been best to slaughter horses in the late autumn, when they possessed the maximum fat content stored for winter. Smoking strips of meat or jerky, first evidenced 12,500 years ago in a mural-adorned cave in France, would preserve the kill for the lean winter months. Freezing large caches of meat in carefully demarcated pits, secure from scavengers, served the same purpose.

Hunting techniques included stalking individual prey with throwing spears or using natural features and terrain to corral and dispatch large herds. The great horse-kill site at the Rock of Solutré in east-central France, for instance, was used for twenty thousand years, producing a massive shed of bones. During their seasonal migrations, horses followed their unsuspecting lead mare through a natural corridor between two limestone ridges. Using convergent lines, generations of savvy hunters drove the panicked horses into a cul-de-sac—facing a cliff—where their comrades lay in ambush. This two-and-a-half-acre site was operational from 32,000 to 12,000 BCE and contains the remains of up to a hundred thousand horses.* Similar natural horse cordons and kill zones from this period have been found in Germany, Russia, and Ukraine.

---

* Despite reports to the contrary, these horses were *not* driven off the cliff like the bison-harvesting practice of Indigenous peoples in Canada and the United States immortalized by the famous Head-Smashed-In Buffalo Jump site in southern Alberta.

In addition to the archaeological record, our Paleolithic ancestors left us with intricate and finely crafted artistic representations of their association with horses in the form of paintings, engravings, plaques, and effigies splashed across sites dotting the European landscape. "At a time before domestication," writes University of Kansas zooarchaeologist Sandra Olsen, "early European hunters had a perspective on horses very different from that of modern humans, and yet these people captured the beauty, strength, and grace of horses in their cave art and horse effigies." Numerous caves and rock shelters, including the renowned Chauvet, Lascaux, and Pech Merle caves, all in southern France, and Altamira in northern Spain, house extraordinary, emotive, and inspiring paintings dating broadly from thirty-six thousand to seventeen thousand years ago. Of the roughly two thousand strikingly accurate sketches of numerous animals, more than six hundred are of horses.

These well-preserved prehistoric art galleries, primarily in France and Spain but also found in Portugal, Italy, Russia, and Mongolia, provide us with a window into the past. It is obvious that our artists had an

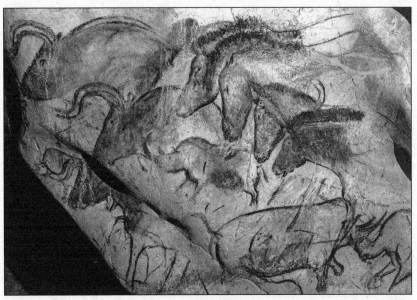

The panel of four horses: Chauvet cave rock art depicting wild horses alongside aurochs and woolly rhinoceroses, France, circa 32,000 BCE. *(Wikimedia Creative Commons)*

intimate knowledge of the anatomy and behavior of horses. "Through their art," explains zoologist Dale Guthrie, "scholars have an empirical way of looking into the minds of Paleolithic peoples, and it is clear that one of the things on their minds was the horse. . . . Artists normally drew horses in natural everyday postures such as standing, feeding, or sometimes even urinating or defecating. They also drew them in active poses—threatening, fighting, running or trotting, jumping, and even copulating. . . . It is clear from the images that the main mental preoccupation of these artists was with hunting." While they valued horses for their meat, hides, and other by-products, these hunters also seem captivated and emotionally engaged with the spirit of horses and the essence of *Equus* just like we are today. After all, horses are an integral part of what it means to be human.

These were some of the first *Homo sapiens* in Europe, and, in a way, they were also the first zoologists, ethologists, and hippologists.* Although they certainly ate horses, it appears that, like us, they also admired horses and enjoyed watching them. As they sat back and stared at the grazing wild herds, I wonder if they also asked themselves my boyhood question about how horses got so big eating grass. I like to imagine that they did.

Cave paintings, engravings, and pictographs were not the only mediums the ancients used to immortalize horses. The Paleolithic period is also represented by horse figurines, pendants, and plaques fashioned from various types of ivory, antlers, stones, and bones, including a 35,000-year-old German statuette made from mammoth tusk and an elaborate 14,000-year-old carving of a horse head from France. The oldest known work of art in the United Kingdom, coinciding with the uninterrupted human occupation of Britain, is a stylish 12,500-year-old depiction of a galloping, flared-nostril horse engraved on an equine rib bone found in Robin Hood's Cave, Derbyshire.† By this time, however, as the last Ice Age melted and the glaciers retreated between fifteen

---

* Ethology: the study of animal behavior. Hippology: the study of horses.

† The imposing pictogram known as the Uffington White Horse, carved into a hillside in Oxfordshire, England, roughly 3,400 years ago and then filled with white chalk, measures 360 feet long and 3 feet deep.

Lascaux cave canvas: A herd of wild horses, France, circa 15,000 BCE. *(Wikimedia Creative Commons)*

thousand and ten thousand years ago, horses began to vanish from the global landscape.

As the Earth warmed, staggered climate change and shifting ecologies created new survival pressures. The once-sprawling grasslands and tundra gradually gave way to woodlands of oak, ash, hazel, beech, elm, and alder. For the horse in both the Americas and across Eurasia, its wide-open spaces were quickly shrinking. Horses are not like elephants, who act as stewards of their sacred pastures. Elephants are savvy land managers, halting the encroachment of forests by purposefully browsing on saplings thereby safeguarding their precious grass. As the steppes and prairies melted away, most global horse populations followed their evolutionary environment into oblivion. To make matters worse, the number of hungry humans was correspondingly increasing.

The Upper Paleolithic period, from 35,000 to 15,000 BCE, witnessed the widespread migration and global domination of *sapiens* as the premier predatory species on the planet. Increasing population pressures saw hunter-gatherers expand their foraging farther afield to include a wider selection of plants, including gritty wild grains and grasses that would later come to dominate the cupboards of domestication. Dwindling herds of large game (or diminishing regional and seasonal availability) also initiated a hunting trend toward smaller animals, birds, and fish. These shifting patterns in hunter-gatherer dietary habits toward domesticated husbandry—dubbed the broad spectrum revolution—coincide with the Pleistocene extinctions of large mammals, including horses, particularly in the Western Hemisphere.*

When the First Peoples arrived in the Americas at least twenty-three thousand years ago from Siberia via the Bering land bridge or by riding the currents off the Pacific coast, they were unquestionably greeted by horses.† These equine populations, however, were already in precipitous decline *prior* to the appearance of humans. Some estimates place the total number of horses at only 1.2 million, which is shockingly low considering the immense size of the American continents. Given the limitations of reproduction mentioned earlier and natural predation by large carnivores, a population this small would already be teetering on the threshold of extinction.

Although horses were comparatively sparse, and bison, mammoth, and mastodon were certainly the prey of choice for hungry humans, there is still ample evidence from at least twelve thousand years ago of horse hunting from Alberta to Argentina. These interactions are substantiated by horse blood residue on tools, cut marks and evidence of butchery or burning on horse remains, and bone daggers. Contrasting Eurasian sites such as Solutré, these locations contain the remains of only an individual or a handful of horses. Changes in climate had already spun the perfect storm for eliminating numerous large mammals during the great Late Quaternary extinction event.

This Darwinian devastation peaked between fifteen thousand and twelve thousand years ago, shortly after the earliest human migrations to the Americas, where an estimated 78 percent of large mammals

---

* I refrain from using the terms Old World and New World, as these assign artificial importance and rank under false concepts and flawed constructs of European colonization.

† Like all cultures, Indigenous peoples have their own creation stories and oral histories, which I do not presume to prejudice or dishonor.

disappeared. Horses, tapirs, camels, saber-toothed cats, dire wolves, mammoths, mastodons, short-faced bears, and giant armadillos, ground sloths, cave bears, and beavers—in addition to a natural history museum of others—all met the same catastrophic fate.* "North America lost thirty-four out of its forty-seven genera of large mammals," explains Yuval Noah Harari. "South America lost fifty out of sixty."

Most of these animals, including horses, had disappeared by eleven thousand years ago. The last surviving equine species, known as *Equus lambei* (or the Yukon horse, after a well-preserved Canadian specimen), hung on in North America for another two thousand years, marking the final sound of fading hoofbeats on the horizons. By this time, horses stood on the precipice of not only American annihilation but also global extinction.

It was once assumed that as human populations fanned out across the planet in ever increasing numbers, hunters annihilated one herd of horses after another, driving them to extinction in the Americas and across most of Eurasia. This rapid blitzkrieg, or human-overkill hypothesis, suggests that our species—armed with relatively primitive projectile weapons and with no quick-strike transportation other than their own two feet—systematically eradicated not only horses but also a wide catalog of hefty, hostile megafauna.

During this alleged genocidal all-you-can-eat buffet, humans took on massive, highly aggressive animals armed with sharp claws, tearing teeth, and hundreds if not thousands of extra pounds. Based on the dearth of current archaeological evidence and various demographic models, this overkill scenario is not plausible. We forget that our predecessors were only one predator among a larger spectrum of predators. Humans were both the hunters and the hunted.

The extent to which humans are culpable cannot be precisely known and is still fiercely contested, but it is safe to say that we were certainly not the exclusive cause of these extinctions. The Indigenous hunter who

---

* These animals were massive. Short-faced bears weighed upwards of 4,500 pounds, and cave bears, 2,200 pounds. Ground sloths, measuring twenty feet long, tipped the scales at 9,000 pounds. Armadillos weighed 5,000 pounds (roughly the same weight as a midsize SUV), while beavers reached the size of modern black bears (280 pounds).

killed the sole remaining free-spirited horse, giant ground sloth, or hulking short-faced bear in the Americas did not let loose an arrow or a spear armed with the knowledge that he just slaughtered the last of its kind. These mass extinctions were also not isolated to the Americas and can be traced with greater accuracy in Australasia, Eurasia, and Africa.* The destruction of one species can upend the ecological balance, precipitating a domino effect of drastic consequences for other organisms that share that environmental niche.

This *trophic cascade* is defined as "an ecological phenomenon triggered by the addition or removal of top predators and involving reciprocal changes in the relative populations of predator and prey through a food chain, which often results in dramatic changes in ecosystem structure and nutrient cycling." If large herbivores disappeared for lack of roughage, top-tier carnivores would follow suit for want of prey. When mammoths vanished, for example, so too did the hapless varieties of specialized ticks that fed on them.

The Pleistocene, or Quaternary, extinction adhered to the simple equation of natural selection: adapt or die. For horses in the Americas, extinction was the result. After more than fifty million years of occupation, the ancestors of *Hyracotherium* vanished from their evolutionary homeland. "It is one of the greatest ironies of history—equine and human—that the continent on which the horse was born was also the continent on which it died out," laments University of Oxford archaeologist Peter Mitchell in his indefatigably researched *Horse Nations: The Worldwide Impact of the Horse on Indigenous Societies Post-1492*. "Sometime between 12,000 and 7,600 years ago, the last truly wild horse in North America was no more." Well, no more for the time being.

To the detriment of Indigenous peoples, the horse eventually found its way back to its birthplace in the Americas courtesy of one of the most celebrated *and* abhorred characters in history: Christopher Columbus. Horses served at the sharp end of European expansion as a crucial and potent conscript of the Columbian Exchange.

---

* Australasia encompasses modern Australia, New Zealand, New Guinea, and surrounding islands.

Horses also gradually disappeared across Eurasia, with equine archaeological deposits becoming increasingly sporadic around twelve thousand years ago. Save for a few small, secluded pockets in Spain and the Central Plains, between 7000 and 5500 BCE, the horse had largely vanished from Europe. By this time, there is no hint of horses in most of the midden-rubbish pits and bone pile deposits. In those of Central Europe, for instance, horses represent a paltry 0.3 percent of bones recovered. The horse faced the same dire situation in eastern Asia. There was, however, one last refuge for the world's last stalwart horse populations.

The rolling Eurasian Steppe corridor, the largest expanse of temperate grassland prairie on the planet—stretching uninterrupted for 5,600 miles, from the Great Hungarian Plain to the gates of Manchuria—provided asylum for lingering horse herds. This vast emptiness intersected by soaring mountain ranges and great snaking rivers provided horses with a haven relatively devoid of human footprints. "On the Eurasian steppe, and especially in the more arid, treeless zone of the steppe, horses were in their element," explains historian Robert Drews of Vanderbilt University in *Early Riders: The Beginnings of Mounted Warfare in Asia and Europe*. "Perhaps most important for the proliferation of horses on the steppe was the widespread absence of humans. People found the steppe, except for its river valleys, difficult terrain in which to hunt, and too dry and too intractable for agriculture. . . . Hundreds of thousands—and possibly millions—of wild horses continued to roam the steppes long after the horse had all but disappeared in Europe and in the Near East." When domesticated, these steppe herds would transform history.

The thawing period at the end of the Ice Age created formidable environmental pressures on flora and fauna across the planet. It was truly a period of survival of the fittest. Some species vanished; others, like the horse, teetered on extinction, while select creatures, such as *Homo sapiens*, not only survived but thrived. Within this vacillating warming epoch, in between chasing waning herds of large game, humanity gave birth to modern civilization with the advent of the international Agricultural Revolution. Between twelve thousand and six thousand years ago, there were at least eleven independent sites of agricultural origin covering all continents. Survival needs, rather than choice or invention, necessitated experimentation with other modes of acquiring food. Hunger makes people desperate and resourceful.

For a handful of select animals with preadapted domesticable traits, their bounty transitioned from hunted to husbandry, providing soaring human populations with meat and hide on the hoof. As a literal *live-stock* of meat, milk, eggs, and other ancillary resources during what renowned University of Oxford archaeologist Andrew Sherratt coined the Secondary Products Revolution, domesticated animals represented an itinerant form of wealth and a secure source of protein. Although a later addition, like all animals cultivated within the farming package, the primary purpose of horse domestication was dinner.

Numerous researchers, including Juliet Clutton-Brock, suggest that the "horse only escaped extinction through domestication." Human intervention both reined in and rescued the horse. "Abrupt climatic changes at the end of the Ice Age some 15,000 years ago," surmises Stephen Budiansky, "drove the modern horse to extinction in North America and within a hair's breadth of extinction in Europe and Asia as well. Were it not for domestication, *Equus caballus* would have gone the way of the *Hyracotherium* and all the other ancestral horses that are testimony to the inevitability of extinction." As inconceivable as it is to imagine the past 5,500 years of global history without horses, that is very nearly what happened.

CHAPTER 4

Hold Your Horses
Steppes of Domestication and the Agricultural Revolution

I enjoy shopping for groceries. I know, it's bizarre, but I find it relaxing. Some people meditate, hike, horseback ride, or practice yoga. I do groceries. Arriving at my local Safeway, I secure my earbuds and casually push my cart to the soothing and comforting screams of Guns N' Roses. Taking the sage counsel of "patience" espoused by the esteemed philosopher Axl Rose, I leisurely saunter up and down the aisles, taking in the astonishing assortment of products.

I read the labels and marvel at the fact that I have a choice between ten brands of canned corn (one of which has the face of Napoléon Bonaparte staring back at me), eight blends of wheat flour, eleven varieties of beans, seven grains of rice, five types of potatoes, and thirty-one allegedly delicious flavors of food for my dog, Steven. I gently guide my cart through the global village of groceries, bumping into produce and sundries from every pocket of our planet. With every new shelf of goods, I think to myself that indeed the world is now a small place and that we are the preeminent species.

The grocery store should be a mandatory elementary school field trip. Doing groceries is a history lesson that bridges a wide swath of civilizations, just like the surfeit of agricultural produce that navigates the Earth to line the shelves of my Safeway. The lists of items and ingredients on my supermarket receipts represent an astounding array of cultures and can all be boiled down to one event: the Agricultural Revolution. As the famed British author H. G. Wells put it, "Civilization was the agricultural surplus . . . [and] the power of the horse." Our modern world was built on farming foundations and edible scaffolding, and it galloped forward on the back of a horse.

Most of us do not know where our food comes from or how our corn, bread, beans, rice, potatoes, or kibble gets to our dinner plate or into Steven's silver dog bowl. Most of us hardly take the time to read, let alone understand, the numbers on the stickers applied to fruits and vegetables. Aside from the type of apple or pepper, those numbers contain a hidden world of information, including country of origin, how the item was grown—organically or with pesticides—and whether it is genetically modified. As part of the recent "shop local" consumer trend, some stores, like my Safeway, opt to advertise the source in larger print or signage as a reminder of our dietary carbon footprint.

I can safely say that most of us do not grow our own food, and if we do, we do not grow enough to be self-sustainable. If by some natural, pathogenic, or nuclear disaster we were forced to return to a hunter-gatherer lifestyle, 95 percent of our fellow human beings would perish. Personally, my family would starve on our meager backyard garden yields of tomatoes, carrots, beans, and my young son's carefully staked rows of chili peppers, including the blistering Carolina Reaper.

In the United States only 2 percent of the population are currently engaged in some form of commercial agriculture or stock raising (compared with 25 percent a century ago), yet they can produce enough food to not only feed the other 98 percent of Americans but also yield an enormous exportable surplus. In terms of output, US industrial farmers and ranchers are the most productive humans in history. By simply clattering a few keys on my computer or cell phone, I *could* have my all-time favorite food of blueberries delivered instantly to my front door before I could even salivate my Pavlovian response. (I *could*, but remember: I enjoy grocery shopping.)

In the historical time frame of humanity, this is a relatively new phenomenon. "The past 200 years, during which ever-increasing numbers of Sapiens have obtained their daily bread as urban labourers and office workers, and the preceding 10,000 years, during which most Sapiens lived as farmers and herders," explains Harari, "are a blink of an eye compared to the tens of thousands of years during which our ancestors hunted and gathered."* When the first seeds planted this new domesticated era of human history, the Earth was home to roughly five million to eight million nomadic hunter-gatherers.

Under this form of subsistence, our planet's maximum estimated human carrying capacity falls somewhere between ten million and one hundred million people. The Earth is now crawling with more than eight billion humans. If we condensed our modern existence of roughly two hundred thousand years into one hour, we have been domesticated by plants and animals for only about three minutes (or 5 percent of our stay as *Homo sapiens*), and this farming package has been the sole human provider of sustenance for around one minute (1.7 percent). Farming is a peculiar institution and intrinsically unnatural.

When we tamed plants and animals, they also tamed us. The English word *tame* traces its origins to the root Proto-Indo-European (PIE) term **dem~*, meaning "to build." This definition is reflected in all its linguistic descendants—from Albanian, Bengali, and French, through Greek, Hindi, and Persian, to Russian, Urdu, and Welsh—in numerous words, including *diminish*, *domesticate*, *dominate*, *tame*, and *timber*. The Latin stem of the word *domesticate*, for instance, is *domus*, or house. It did just that.

We farmed, settled down, and built dwellings, massive structures, and monuments. "Domesticated plants and animals form the very foundations of the modern world," reports award-winning author Tom Standage in *An Edible History of Humanity*. "Throughout our history, food has done more than simply provide sustenance. It has acted as a

---

* Hunter-gatherers enjoyed a more relaxed, more peaceful, healthier, and happier life than their agricultural heirs. The swelling Rolodex of zoonotic (animal spillover) pathogens was still nonexistent. They had significantly more leisure time and a more diverse and nutrient-rich diet by gathering plants and hunting protein across a wider cross section of environments, promoting superior physical and mental health. Hunter-gatherers were, on average, five and a half inches taller than early farmers. Only recently have we matched their average heights of five foot nine for males and five foot five for females. High infant mortality skews the overall longevity of hunter-gatherer populations. If you survived past 15 years of age, your average life expectancy was between 70 and 72 years old, just shy of the United Nations' 2023 global life expectancy of 73.16 years. They also had a greater say in the democratic decision-making process within nomadic groups as small as an extended family of fifteen to larger bands of no more than one hundred people, which foraged across ranges anywhere from seven square miles to five hundred square miles.

catalyst of social transformation, societal organization, geopolitical competition, industrial development, military conflict, and economic expansion." The Agricultural Revolution led to the creation of modern city-states with escalating population densities. By 2500 BCE, some cities in the Middle East topped twenty thousand residents. Neolithic agriculture led to a tenfold increase in the number of people inhabiting the planet.*

Between 10,000 and 5000 BCE, the global population soared from roughly 5 million to 50 million. By 1000 BCE, it had doubled again to 100 million, and, by 1000 CE, had reached 250 million. "This increase can be explained, to a certain extent, by the extension of slash-and-burn cultivation," state Marcel Mazoyer and Laurence Roudart in *A History of World Agriculture*, "but also by the development of large societies based on hydraulic agriculture in the valleys of the Indus, Mesopotamia, and the Nile . . . organized around floodwaters and irrigation, which were organized in these privileged valleys."

By 6000 BCE, we have definitive evidence of irrigation, or the "domestication of water," at Çatalhöyük in southern Turkey, as well as other sites dotting the major rivers and tributaries of the Middle East. "When they succeeded in controlling the water, however, the results were spectacular," writes historian Daniel Headrick in *Technology: A World History*. "Whereas Neolithic farmers in the Middle East might hope to reap four or five grains of barley for every grain they planted on rain-watered land, in a valley, a grain of barley receiving the right amount of water during the growing season could yield up to forty grains." Domestication was a global effort.

Across the world, in at least eleven independent regions, farmers instigated a revolution despite being completely isolated from one another. Amazingly, in the context of our larger history, they all picked up a stick and drilled seed holes at roughly the same time, transforming all facets of humanity in the process. Between 9000 and 8000 BCE, pockets of peoples in the Fertile Crescent stretching along the Tigris and

---

* The Neolithic Period, or "New Stone Age" (roughly 10,000 to 3000 BCE), is characterized by the global human shift from hunter-gathering to fixed pastoral settlements, including the domestication of plants and animals, and advancements in polished stone tools.

Euphrates Rivers from southern Turkey through the Levant and Mesopotamia—including the ancient city of Qurnah (the purported site of the Garden of Eden)—had domesticated wheat, rye, peas, lentils, barley, goats, sheep, and pigs. The fields of Asia were graced with rice and millet around 7500 BCE, while maize (corn) and potatoes were proliferating in the Americas between 7000 and 5500 BCE.

Intensive agriculture was being practiced in the Middle East, Egypt, China, India, Africa, Europe, and the Americas by 4000 BCE, giving rise to all the trappings of modern civilization. By this time, other, lesser plants had been added to the inventory, including avocados, beans, cabbage, dates, grapes, olives, peanuts, peppers, plums, squash, and tomatoes.

Nonedible crops such as cotton, flax (linen), ramie, and hemp were raised for fabrics, while llamas, sheep, alpacas, and silkworms provided valuable fiber for textiles.* Bottle gourds were grown as food and containers—an organic, renewable, and biodegradable Hydro Flask. The earliest domesticates, however, were the mainstays of our grain-based civilization. We modified these plants to suit our needs, and in the process, they irrevocably changed us.

Modern maize or corn, for example, is the direct descendant of teosinte, a simple wild grass that bears absolutely no resemblance to the strange, yellow, sweet mutant corn we now gnaw off the cob and pick out of our teeth.† Through a series of random genetic mutations and designed human modifications, what was once a small, inedible wild grass is now a plant that could not survive on its own without human intervention and management. Of the roughly forty-five thousand items currently filling the shelves of the average American supermarket, more than 25 percent contain corn. This includes both edible and nonedible products.

Many of those strange words on food labels that we simply shrug at and stumble to pronounce—in addition to the familiar sweetener high-fructose corn syrup—are forms or by-products of corn sugars or corn

---

* Cotton has multiple-site domestic origins: India and Pakistan, Peru, Mexico, and possibly Madagascar and Mesopotamia.

† The word *teosinte* ("divine maize") is from the Nahuatl language of the Uto-Aztecan family. English has adopted a handful of words from Nahuatl, including atlatl, avocado, chia, chili, chipotle, chocolate, coyote, guacamole, mezcal, mesquite, peyote, shack, and tomato.

starches: ascorbic acid, caramel color, citric acid, crystalline fructose, dextrose, glucose syrup, lactic acid, lecithin, lysine, maltodextrin, maltose, monosodium glutamate, polyols, xanthan gum—and the list goes on. From beer, pop, coffee creamer, and Twinkies, to Cheez Whiz, waffles, salad dressings, and frozen yogurt, you are eating or drinking corn. It is also in everything from batteries, garbage bags, and charcoal briquettes, to shampoos, toothpastes, and diapers, to linoleum, fiberglass, adhesives, and pesticides—even in the wax that gives shine to the covers of magazines and books and gloss to other fruits and vegetables. Corn has literally become the universal cornucopia for human society.

Once corn and these other staple foods were inside us, their roots drove deep into the foundations of not only our rapidly transforming existence but also our very survival. We had unwittingly been tamed by our own agricultural undertakings. At some point, when we realized there was no turning back, we simply shrugged our shovel and dug in for the long haul.

According to acclaimed anthropologist Jared Diamond in his groundbreaking 1997 bestseller *Guns, Germs, and Steel: The Fates of Human Societies*:

> Of the 200,000 wild plant species, only a few thousand are eaten by humans, and just a few hundred of these have been more or less domesticated. . . . Most provide minor supplements to our diet and would not by themselves have sufficed to support the rise of civilizations. A mere dozen species account for over 80 percent of the modern world's annual tonnage of all crops. Those dozen blockbusters are the cereals wheat, corn, rice, barley, and sorghum; the pulse soybean; the roots of the tubers potato, manioc, and sweet potato; the sugar sources sugarcane and sugar beet; and the fruit banana.

If we take Diamond's list one step further, roughly 90 percent of the world's total caloric intake is still supplied by the original six major domesticates from between 9500 and 3500 BCE: wheat, barley, rice, millet, maize (corn), and potatoes.

These crops birthed civilization, colonized the Earth, and continue to buttress its growth. Wheat, for example, now covers upwards of nine hundred thousand square miles of the planet. In tangible terms, by landmass, this is equivalent to the tenth largest country in the world, or approximately the same size as India, Argentina, Algeria, or Kazakhstan, and almost ten times the size of New Zealand or the United Kingdom.*

The advent of agriculture led to a surplus of crops and an accumulation of wealth. Control of food is a projection of power and a surefire way to guarantee political influence. This innate human avarice for prosperity and power led to complex social stratification; specialized skills and occupations; sophisticated and tiered spiritual, legal, and political structures; slavery, tributes, taxation, zoonotic diseases, and, most significantly, to trade and war.† Cultivation was shackled to a corpse.

Statistically, throughout our existence, societies that produce greater surplus and engage in elevated trade also have a higher propensity for waging war. As a result, conflict became less unwieldy and evolved into a habitual commercial enterprise and an ingrained human ritual designed to produce a surplus of wealth and power.‡ "Warfare and trade are not only related, but depend on each other," stress archaeologists Christian Horn and Kristian Kristiansen. "Warriors both protected and benefited from trade, which therefore increased the demand for warriors. Ultimately, it developed new incentives for warfare to control trade routes." The reality of economics is unfortunately quite simple: Why trade when you can invade?

On a more intimate level, specialized skills and occupations allowed people to increasingly rely on, and blindly trust, the expertise of others. We all became specialists at one thing or another. As a history professor, author, and hockey coach, my survival is, thankfully, not dependent on how well I know my way around the human heart or which way to swing a screwdriver or turn a hammer.

---

* Kazakhstan is the largest landlocked country in the world.

† *Zoonosis* means "animal sickness" in Greek.

‡ In 2022, American taxpayers ponied up $822 billion (or 16 percent of the federal tax budget) to defense-related expenditures. This equates to roughly 38 percent of total global military spending, which topped $2 trillion for the first time.

The earliest human writing (cuneiform) inscribed on clay tokens and tablets was a result of vocational specialization, trade, bookkeeping, and accounting for wealth. In fact, 90 percent of those recovered document some component of occupational administration or money management. Clay tablets dating to 3500 BCE from the Mesopotamian city of Uruk—then the largest on the planet, boasting a population of forty thousand to fifty thousand—reveal consistent cuneiform text called the "Standard Professions List," which was used to teach scribes. It is an early and concrete indication of the agriculture-induced stratification of society. The list contains 129 professions, always written in the same order and succession of importance.

The top entries include political, spiritual, and military positions before descending through skilled trades and agricultural pursuits. Conspicuously, there is no mention of equine-related occupations, establishing that horses were not yet present across the greater Middle Eastern expanse. Roughly 80 percent of the residents were farmers. While this is still a considerable portion of the Uruk populace, it is far less than 100 percent of hunter-gatherers. These farmers freed up 20 percent of the inhabitants for other economic, cultural, artistic, technological, and military pursuits.

The fabled city of Jericho in Palestine, dating to 10,500 BCE, might be one of the first such sites to exemplify the major themes germinating from the farming package: urbanization, commercialism, trade, and war. When the original walls of Jericho, enclosing an area of ten acres, were erected in 8000 BCE, domesticated wheat, barley, peas, lentils, and goats provided its 2,500 inhabitants with more than 80 percent of their food. The ramparts, including the famous tower of Jericho, were defended by upwards of six hundred adults to protect both people and agricultural profit. Sedentary populations commanded ownership of their resources and wealth with ditches, fences, palisades, and walls. Similar agro-urban-military sites appeared across the Middle East by 9500 BCE and in Europe by 7000 BCE, as farming tended to its siblings of market economics and mounting war.

This Neolithic agricultural ripening also led to the housebreaking of barnyard animals and beasts of burden, including the horse, which accelerated and powered our fateful decision to domesticate. Although the horse was a later addition to the modest collection of domesticated animals, it would far outpace all others in importance and influence. Once domesticated, the horse radically altered all established traits of human society ingrained by the Agricultural Revolution. "Considering the large numbers of mammals that have been exploited by humans, it is, at first glance, surprising that so few species have been fully domesticated," acknowledges Juliet Clutton-Brock in her playfully titled book *The Walking Larder: Patterns of Domestication, Pastoralism, and Predation*. "The answer lies in the behaviour of the species of the animals that undergo domestication and in the inherent diversity of the genetic constitution of wild mammals. It is only those species that have inherited behavioural patterns that correspond to those of humans that can survive the process." Most large animals, as Charles Darwin's cousin Francis Galton recognized in the 1880s, were "destined to perpetual wildness."

Of the two hundred genera of animals weighing more than one hundred pounds that existed prior to the late-Pleistocene mass extinctions, roughly one hundred remained at the dawn of the Agricultural Revolution. Of these, just fourteen were domesticated between 8000 and 2500 BCE in concert with agriculturally fueled sedentary societies. Currently, only five are of any true global significance: cows, sheep, goats, pigs, and horses, while two others, camels and donkeys, attained regional and time-specific importance. In most Western societies, the donkey has been relegated to petting zoos, caricatured by Winnie-the-Pooh's gloomy friend Eeyore, or Shrek's wisecracking sidekick.

You might be wondering why canines are absent from the hundred-pound weight class. Although modern designer dogs can top this threshold, with mastiff and Saint Bernard breeds tipping the scales at over three hundred pounds, the first domesticated Paleolithic wolf dog weighed around seventy to eighty pounds. These dog descendants of the Eurasian gray wolf were the first to be tamed, or tamed themselves, by loitering for scraps. As a criterion of domestication, both dogs and humans learned to manipulate and take from each other what they wanted and needed in a mutually beneficial relationship. This longer association has afforded humans and dogs a much deeper understanding, devotion, and friendship than any other fellowship save horses.

The subsequent domestication of wolf dogs occurred independently, and on numerous occasions, across Eurasia between thirty-two thousand and eighteen thousand years ago. Dogs accompanying the First Peoples from Siberia to the Americas at least twenty-three thousand years ago were used to pull travois sleds or were eaten.* More generally, we received a valuable hunting partner, protection, and companionship in return. Unlike the horse, however, if we dissect the dog out of our historic journey, our world would perhaps look somewhat different—and certainly lonelier—but it would not be turned upside down.

It is important to differentiate between tame and domesticated. The Carthaginian general Hannibal's shivering elephants that crossed the Alps in 218 BCE, those employed as laborers in Southeast Asia, or heartbreakingly imprisoned as shackled attractions in Brutal Bob's Exotic Elephant Bazaar are tame, but elephants as a collective species have never been domesticated. While select, more passive, animals must be tamed before domestication can proceed (as in our wolf dog scenario above), taming is by no means an automatic guarantee for eventual domestication.

Generally, any young animal removed from its parent(s) and collective can be tamed by humans. We see this with pets of all kinds, from koalas, chimpanzees, and bears, to Koko the gorilla, ex-boxer Mike Tyson's tiger, and Hannibal's alpine elephants. The capture and subduing of young wild animals for pets is an enduring practice across all cultures.†

As with dogs, this custom of pet keeping, however, led directly to the domestication of select animals at several sites, particularly sheep, goats, pigs, cattle, and fowl, between 8000 and 6000 BCE. Similar to the numerous independent root sites of agriculture, genetic patterning reveals multi-origin domestication for these animals. Cattle were domesticated

---

* A travois is a type of A-frame sledge consisting of two poles attached to a dog or horse used by Indigenous peoples, primarily on the Great Plains, to haul possessions and supplies.

† Currently, Argentina holds the top position for companion pets per capita with 80 percent of people reporting at least one animal, 66 percent of which are dogs. The United States comes in at 70 percent, with dogs and cats representing 50 percent, with an annual pet care price tag of over $50 billion. Half of British homes have a pet, while South Korea at 37 percent and Japan at 31 percent are among the lowest. Interestingly, although we treat our pets as children and part of the family, globally only 22 percent of dogs and 10 percent of cats have human first names. My dog, Steven, is in the minority, unlike my cat, Scout.

independently in Southwest Asia, India, and Africa. The pig was domesticated at numerous sites across Eurasia, while sheep have two and goats have three distinct domestication pathways.

Evolutionary biologists stress that only certain animals, or those original candidates for domestication, have the "preadapted" package of characteristics in wild populations that make them predisposed or prime candidates for domestication: temperament, disposition, social herd behavior, hierarchy structure, diet, climate, reproduction, growth rate, panic level, and fight-or-flight responses. These animals pass through the blurred stages and distinctions of wild, captive, tame, and domesticated.

In all cases, young animals were captured and kept instead of being killed and eaten right away. They were incorporated into the social structure of their abductors, while becoming quasi-independent objects of human ownership in a barnyard adaptation of Stockholm syndrome. With successive generations of breeding, they became more submissive and dependent on their caretakers for the basic physiological and safety needs at the bottom of psychologist Abraham Maslow's pyramid-shaped hierarchy, including food, water, shelter, and, in some cases, reproductive sex. The psychological tier of affection and social belonging afforded by a herd collective, which could include their human hosts, is an important, although secondary, component of domestication for certain pack animals.

As Sandra Olsen stresses, "[A]nimals that end up in close contact with humans either select themselves or are selected (consciously or unconsciously) by humans because of their ability to adapt to close proximity with our species [and] have some genetic predisposition to tolerance of human behavior." Eventually they lost any fight-or-flight response to people, and each generation became more docile, malleable, and, in some cases, trainable. As Clutton-Brock reiterates, "A tame animal differs from a wild one in that it is dependent on man and will stay close to him of its own free will." Why run away when you can eat, sleep, play, and stay?

Morphological changes occur when this "breeding group" or "founder population" of tame animals has been separated from its wild lineage.

Both artificial and natural selection alter the physical and behavioral traits of these animals over time. All domesticated plants and animals are in effect man-made creations. They are products of our manipulation either directly by targeted human selection or indirectly through organic natural selection. This controlled and handpicked evolution occurs within an artificial domestic environment as opposed to those spontaneous and unprompted Darwinian adaptations responding to ecological survival pressures. Wild populations spawn subspecies, whereas domesticates produce breeds.

The anatomy of domesticated animals changes to reflect their work routine, skeletal and muscular pressure points, and diet. When humans shifted from hunting-gathering to farming, our skeletal remains also left visible clues and telltale indicators. "One way of determining whether a woman who died nine thousand years ago was living in a sedentary, grain-growing community as compared with a foraging band was simply to examine the bones of her back, toes, and knees," writes political scientist at Yale University James C. Scott in *Against the Grain: A Deep History of the Earliest States*. "Women in grain villages had characteristic bent-under toes and deformed knees that came from long hours kneeling and rocking back and forth grinding grain. It was a small but telling way that new subsistence routines . . . shaped our bodies to new purposes, much as the work animals domesticated later—cattle, horses, and donkeys—bore skeletal signature of their work routines."

In ridden horses, for example, the wearing of a bridle bit in the mouth causes a distinctive rubbing erosion on molars. To relieve pressure from the tongue and soft tissues of the toothless diastema—that vacant space between the front incisors and back molars that houses the bit—horses (not the rider or driver) tend to elevate and push the bit back to the first molars with their tongues, leaving characteristic abrasions, chafing, or beveling on these teeth. This "bit wear" is *the* primary archaeological evidence that a horse has been ridden.

This twitchy mannerism gave rise to the axiom "champing at the bit" to describe our annoyance and impatience at being restrained from moving. In wild horses, however, champing is a submissive behavior whereby a horse will rapidly open and close its mouth while drawing back its lips to reveal those big incisors. Much like a military salute, the champing horse is communicating a signal of recognition that the receiving horse is dominant and has a higher rank.

At its core, domestication is designed to produce beneficial outcomes such as sheep with more wool, stronger or faster horses, meatier pigs, cows with higher milk yields or meliorated marbling, and chickens that lay more eggs. As part of a larger capitalist economy, these primary and secondary animal products deliver continuous profit. I defer to Clutton-Brock's definition of a domesticated animal as "one that has been bred in captivity for the purposes of economic profit to a human community that maintains complete mastery over its breeding, organization of territory, and food supply." Like fields of tended crops, domesticated animals also tied down people to well-defined territories. Land quickly became the common currency and represented a form of fixed wealth that could be passed down generations within rapidly expanding commercial markets.

The final stages of domestication involve adaptations, codependency, and a brand of shared communal and social coevolution. Domestication is a contractual partnership entered into willingly, if at first begrudgingly, by both partners. It is an arranged marriage for mutual benefit. The acceptance, or more often rejection, of these wedding vows is precisely why a handful of large animals were domesticated and the majority were not. Most animals simply do not want to be tied down. As hard as we have tried with a zoological Noah's Ark of would-be life partners, they doggedly and violently repulse our attempts at assimilating them into our domesticated human collective. For most animals, resistance was not futile.

Within this paradigm, my students always ask the same question during our discussions on the domestication of animals within the larger Agricultural Revolution: "Why were horses domesticated but not zebras?" This analogy is commonly referenced to explain the larger processes of domestication. It also moves us one step closer to the moment when humans irrevocably altered history by bringing horsepower under their control.

The simple answer is that an exceedingly small percentage of large animals possess those domesticable traits. The horse, as we have seen so far, with its evolutionary adaptations, anatomy, social structures, and proximity to humans, had all the right stuff for domestication. A zebra simply does not. It is not for lack of effort on our part. Given our curiosity, greed, adventurous spirit, and generic "just because" anti-logic, if it were solely up to humans, myriad animals would be domesticated.

Ancient Egyptians attempted to break ibex, antelope, gazelles, and hyenas. Several cultures, including the Indigenous Ainu of Japan and the First Peoples of the Americas, kept bears as pets, but they never made the jump to domesticates, for obvious reasons. There have been numerous recent scientific studies and fruitless (or meatless, in this case) efforts to domesticate moose, elk, eland, musk oxen, and zebras.

Every attempt has been made to domesticate zebras. They have been hitched to pull-carts. Daredevils have tried to saddle and ride them. "One is tempted to speculate what would have happened if the zebra had been amendable to domestication by Africans," speculates anthropologist Pita Kelekna. "Surely the history of Africa would have been very different if advancing slavers had been met by zebra cavalry." In South Africa, the Afrikaner Boers and the British endeavored vainly to use zebras as both military and draft animals during the eighteenth and nineteenth centuries. "The zebra is," concluded a Cape Town chronicle in 1819, "wholly beyond the government of man."

Attempting to prove that they indeed could be tamed, in 1895 the eccentric zoologist and politician Lord Walter Rothschild paraded through the streets of London in a carriage harnessed to four zebras. Famed American biologist, explorer, and father of taxidermy Carl Akeley argued rightly that all attempts to domesticate the zebra are "done only for the amusement it affords, because the zebra, like all wild animals, has never quite enough of . . . a domesticated horse to make him useful."

For one thing, zebras grow increasingly nasty and aggressive as they age, their teenage angst and ornery unpredictability only worsening over time. Compared with horses, they are also prolific and talented biters. Not only do they enjoy biting, but when they bite, they do not let go. Intractable zebras injure more professional zookeepers than do big cats.

Zebras also have a lasso-flinch factor that prevents even the best rodeo cowboys from roping them. Possessing supernatural peripheral vision like horses, even in direct sunlight, they have the uncanny ability to track the lasso and duck their heads at the last second. Harnessing the power of a cranky zebra, whether for cavalry or traction, is not biologically feasible and is just plain dangerous.

While initially animals were domesticated primarily for their meat, they soon offered far more capital on the hoof. When compared with hunting and immediate slaughter, domesticated animals could be harvested at an expedient time or bracketing sexual maturity, thereby exploiting reproductive seasons, ages, and maximum meat yields. Domesticates offer a relative safety net to starvation. They also provided traction and transportation, a catalog of secondary material products, and sustainable foods such as milk (and dairy derivatives), eggs, and a quick-prick blood meal in times of salt or iron deficiency. This wholesale commercial exploitation of livestock, known as the Secondary Products Revolution, bolstered their capital worth through marketable wares, while augmenting their political, military, and agricultural relevance and standing.

Although they provided a stable supply of protein, traction, and secondary products, these animals were also teeming reservoirs of zoonotic "spillover" disease.* As celebrated historian Alfred W. Crosby argues, "When humans domesticated animals and gathered them to the human bosom—sometimes literally, as human mothers wet-nursed motherless animals—they created maladies their hunter-and-gatherer ancestors had rarely or never known." Of the roughly 1,450 pathogens known to infect humans, approximately 62 percent made the zoonotic jump from other animals. Humans and horses, for example, currently share more than thirty-five diseases.† And things are only getting worse.

Over the past thirty years, 75 percent of new or developing pathogens

---

* While farming flourished in South America and Central America as early as ten thousand years ago, as we will see, unlike the rest of the world, it was not accompanied by extensive domestication of livestock. As a result, Indigenous peoples of the Americas remained sheltered from the storm of all zoonotic diseases—for the time being.

† These include influenza, tuberculosis, pneumonia, anthrax, glanders, thrush, rabies, hoof-and-mouth disease, tetanus, diphtheria, typhoid, and a variety of worms, bacteria, and mosquito-borne encephalitides.

are spillovers. According to a 2021 report from the Centers for Disease Control and Prevention (CDC), "In the last two to three decades, the world has experienced six major viral disease outbreaks caused by emerging zoonotic viruses." With COVID-19 and other infectious spillover agents consuming our planet, zoonosis-related issues have made global headlines, and we are slowly waking up to the reality that our decisions have long-term implications—whether that be the Agricultural Revolution and epidemic diseases or the Industrial Revolution and climate change.

Future generations are left holding the mop and bucket to clean up the messes their ancestors made. As William Shakespeare foreshadowed in *The Merchant of Venice*, "The sins of the father are to be laid upon the children . . . for truly I think you are damned." Perhaps the recent surge in natural disasters, which have increased fivefold over the past fifty years, will remind us that we are not quite as clever or omnipotent as we often believe. Nature always has a way of bringing us, and our hubris, back down to earth.

The intelligent Houyhnhnm horses of Jonathan Swift's 1726 satirical novel *Gulliver's Travels* would undoubtedly disparage our own haughty opinion of ourselves. The Houyhnhnm are rational equine beings who live in a peaceful, utopic society based solely on reason. Their obedient and orderly culture is free of political, religious, covetous, and immoral concepts. The Houyhnhnm contrast sharply with the filthy Yahoos, a savage humanoid species consumed by lust for "pretty stones." Yahoos personify "a brute in human form" and the follies of lust and greed wrought by the Agricultural Revolution.

Like the crowning anthropomorphism of the Houyhnhnms in Swift's fantastic worlds, the status of the horse also surpassed all animals across every facet of our own civilization. The domestication of the horse redesigned our shared history, redefined what it meant to be human, and irrevocably transformed the world. Given the significance and implications of this monumental event, as an important disclaimer, there are still no definitive answers as to where and when horse domestication took place.

Recent advancements in archaeology, dating techniques, and DNA analysis, however, have markedly focused the microscope, and we are now working off a smaller map. The immense volume of academic research dedicated to domestication only reinforces the historical impact of the horse. For our purposes, a summary of the salient points will suffice.

Like most other farmed animals, the horse was first domesticated as a sustainable, low-maintenance food source, particularly during winter. The logical assumption is that domestication was a by-product of taming through frequent hunting association. Prior to the widespread practice of castration, horse taming would have been a Herculean task, which explains partially why the horse was the last of the classic barnyard animals to come under human management.

While taming individual horses was difficult, true domestication presented a whole new set of complications, starting with capturing and holding live horses. Given the social structure of herds and the reproductive tendencies of stallions, breeding horses in captivity would also have been challenging. An advanced and refined familiarity with tamed horses by a skilled cadre of "horse whisperers" would have preceded sophisticated domestication and controlled breeding over successive generations.

The other constraint was the dearth of horses themselves. As the Agricultural Revolution unfolded, changes in climate had already taken a drastic toll on horse habitats and populations. As mentioned, horses are generally absent from the archaeological record between 12,000 and 4000 BCE, straddling the initial push of animal domestication. This scarcity, however, also helps to narrow the location of initial domestication, the heart of which lies in continuous and close association between horses and humans. No one is going to attempt to domesticate an animal of which they have no prior experience or behavioral knowledge. Where were horses prevalent enough to be viewed as more than just an inadvertent target of opportunity and exotic meat?

The Pontic-Caspian Steppe, a crescent-shaped area north of the Black and Caspian Seas stretching 385,000 square miles from Ukraine through southern Russia to Kazakhstan (within the larger Eurasian Steppe), marks the bold-font X on the map. The archaeological site at Botai in Kazakhstan has dominated the discourse over the past few decades. "The mid-fourth-millennium BC Botai culture of northern

Kazakhstan fixes (or was at least very close to) the time and place in which domestication first happened," contends Peter Mitchell. "Genetic data, too, point to the steppes of northern Kazakhstan, Ukraine, and southern Russia as the area in which horses were first domesticated." The current academic crosshairs remain trained on this region.

Around 3500 BCE, the human occupation at Botai ballooned from four shoddy dwellings to a thriving settlement of at least 158 semi-subterranean pit houses covering 22.2 acres. A permanent steppe settlement of this size is incompatible with customary nomadic hunting, herding, and migration. This urbanization was also unmistakably allied to an economy that revolved entirely around horses.

Small corrals or pens designed for milking mares have been uncovered adjacent to houses, evidenced by thick dung deposits and distinctive post molds.* On the stark and treeless steppe, horse manure moonlighted as high-performance home insulation and, like "buffalo chips" on the North American Plains, as a valuable source of fuel. Shards of pottery found within these enclosures reveal the presence of animal fats and the hydrogen isotope deuterium specific to horse milk. While the inference here is obvious, I will say it anyway: milking a wild mare is impossible.

Although naturally lean, with 1 percent to 2 percent milk fat, horse milk contains five times more vitamin C than cow's milk, with substantial yields of vitamins A, B_1, B_2, B_{12}, D, and E. While dependent upon the size of the breed, health of the mare, and needs of the foal, one mare can produce anywhere from 3.25 to 4.75 gallons (12 to 18 liters) of milk per day. Horse milk quickly became a staple on the Eurasian Steppe and is still consumed in its slightly fermented alcoholic form, called koumiss (in the Kazakh language) or *airag* (in Mongolian), made famous by Chinggis Khan and his mounted Mongol hordes.

Koumiss was reported among the "mare-milking, milk-drinking Scythians" of the steppe as early as the fifth century BCE by the Greek historian Herodotus:

---

* A post mold is an archaeological term for an infilled soil cavity interpreted to be the result of a human-made post or stake associated with dwellings, fences, palisades, pens, and other structures that was once embedded in the ground.

Now the Scythians blind all their slaves, to use them in preparing their milk. The plan they follow is to thrust tubes made of bone, not unlike our musical pipes, up the vulva of the mare, and then to blow into the tubes with their mouths, some milking while the others blow. They say that they do this because when the veins of the animal are full of air, the udder is forced down. The milk thus obtained is poured into deep wooden casks, about which the blind slaves are placed, and then the milk is stirred round. That which rises to the top is drawn off, and considered the best part; the under portion is of less account. Such is the reason why the Scythians blind all those whom they take in war; it arises from their not being tillers of the ground, but a pastoral race.

Stirring or churning like butter, mentioned by Herodotus, is necessary for the fermentation process to activate the lactobacilli bacteria to acidify the milk while yeasts create carbonated ethanol. The alcohol by volume can be increased by a process called freeze distillation—a cyclical repetition of freezing, thawing, scraping away the ice crystals and water content, and refreezing until the desired alcohol potency is reached.

To suit a nomadic steppe lifestyle, small-batch koumiss was brewed in leather sacks hung from a waist belt or a horse. Passersby or others in the traveling party would periodically punch the bag to beat the milk and activate or accelerate the fermentation process. Given its alcoholic nature, koumiss can be stored longer than fresh milk, another benefit for people on the move. This method was still in practice 1,700 years later when the Flemish missionary William of Rubruck traversed the ancient Silk Road trade routes of the Mongol Empire. "As the nomads churn and beat the milk, it begins to ferment and bubble up like new wine," he reported in 1253 CE from the steppes. "Makes the inner man most joyful!"

Due to its high lactose content, unfermented mare's milk is avoided except as a medicinal purgative and laxative for constipation. Horse milk contains 6.2 percent lactose compared with 4.9 percent in cow milk. Although humans are naturally lactose intolerant, secondary milk products became a staple commodity, not only on the Eurasian Steppe but also among all peoples who domesticated cows, goats, and horses. Genetic mutations promoting lactose tolerance occurred randomly across these populations.

In areas where milk was an important commodity and consumed in higher quantities, natural selection would certainly have backed those with this hereditary advantage. Recent genetic research has revealed that the tolerance mutation quickly became dominant across the Eurasian Steppe between 4600 and 2800 BCE. These dates align with the emergence of an explosive dairy revolution engulfing the steppe, including horse domestication and milking at Botai.

Of the three hundred thousand animal bones unearthed at Botai, 99 percent are from horses. In some cases, slaughter was done using a poleax to the forehead, suggesting a domesticated, rather than a hunted, source of meat. Research into the robusticity of recovered leg bones by equine anthropologists David Anthony and Dorcas Brown "has produced a clear metric separation of the Botai horses from the wild populations, grouping them instead with domestic horses. This is the earliest metric change in horse bone proportions indicating domestication." A lengthy catalog of items and tools made from horse bone has also been unearthed at Botai—all representative of a horse-based culture.

The only human burial is that of two men, a woman, and a child surrounded by the arched remains of fourteen horses. This mass sacrifice certainly implies a domestic herd. It seems implausible that those conducting the ritual went to the extraordinary measures of tracking, killing, and transporting (*without* the aid of horses) more than fourteen thousand pounds of wild horse carcass over great distances to one committal. While Botai boarded only horses and dogs, other sites from this period contain additional domesticated animals.

By 3500 BCE, for example, a nomadic Proto-Indo-European herding culture known as the Yamnaya had become entrenched on the Pontic-Caspian Steppe. While these itinerant bands functioned as commercial intermediaries and engaged in intermittent, small-scale agriculture along the snaking river systems, they primarily hunted while tending to their cattle, goats, sheep, *and* horses.

On the blustery, ice-covered steppes, adding horses to the mix significantly benefited the seasonal survival of other livestock and, in turn, their human masters. During the frozen winter months, horses can fend for themselves in a way that other animals, including cattle, sheep, and goats, cannot. While attempting to browse on frigid terrain, both cows and sheep push their snouts through the abrasive snow, leaving them frostbitten and incapacitated. Their lower anatomical posture also prevents them from moving through high snowpack. Without fodder and intervention by humans—or horses—they will starve.

With their higher gait and forward power train, horses can maneuver through snow with greater ease, plowing paths for other animals. Horses also belong to a different taxonomic order and evolved to possess that highly sophisticated single toe. To find or uncover fodder and water in winter months, horses use their tough hooves to scrape away snow and punch through ice to eat and drink, paving the way for other animals, such as sheep and cattle, to do the same.

In the Canadian and American West, open-range cattle have survived blizzards and sudden whiteouts by joining wild mustang herds. The feral horses and burros that currently roam the deserts of the western United States and Mexico are ecosystem engineers and use their hooves to dig borehole wells, some more than six feet deep. Thanks to the evolution and ingenuity of horses and donkeys, these equine-quarried wells (and ass-holes) provide lifesaving water for an astounding fifty-six additional animal species ranging from deer, bighorn sheep, and bears, to badgers, toads, and an assortment of birds.

Domestication was the first step to unlocking the power of the horse and igniting the Equine Revolution. "There can be little doubt that horse domestication originated on the Pontic-Caspian Steppe, in the areas where communities habitually hunted horses," writes University of Oxford archaeologist Sir Barry Cunliffe in his study *By Steppe, Desert, and Ocean: The Birth of Eurasia*. "The next crucial step for the herding community was to isolate the more docile beasts and to learn to ride them.... The enterprising youngsters who first dared each other to jump on the backs of wild horses started a revolution." With the domestication of horses, our Centaurian Pact was forged and the fates of two distinct species forever intertwined. Riding reinforced and elevated this unrivaled dyad and eternally revised the course and curriculum of history.

Riding coincided with, or was an immediate corollary to, initial domestication. Based on the practices of horse cultures both past and present, most researchers agree that domesticating, herding, and managing horses in any substantial numbers without their being mounted is an impossible feat—even with the use of dogs. Of the horses at Botai, 70 percent stood between 13 and 14 hands, easily large enough to ride and slightly bigger than the average Roman cavalry horse. The mounts utilized by the consummate Indigenous horse nations of the Great Plains, including the Comanche and the Lakota, were slightly under 14 hands. For comparison, modern recreational horses average around 15 hands, and Thoroughbreds, 16 hands.*

Riding had numerous other pastoral benefits. One man on foot with a trained dog can herd roughly two hundred sheep or goats. But on horseback, with his trusted canine still corralling animals, this number increases to five hundred. The horse more than doubled the efficiency of livestock droving. Mounted herdsmen, hunters, and foragers could now regularly cover forty to fifty miles a day; their pedestrian counterparts could not keep pace or compete. Applying a later but topical example, Nathaniel Wyeth, an American fur trader practicing his craft across the Northern Plains in 1851, wrote in his journal: "Men on foot cannot live, even in the best game countries, in the same camp with those who have horses. The latter reach the game, secure what they want, and drive it beyond the reach of the former."

As a result, a mounted lifestyle spread rapidly across the Pontic-Caspian Steppe before radiating to Europe, Asia, and the Middle East. "His name is not Wild Horse any more, but the First Servant," wrote Rudyard Kipling in his beloved children's story "The Cat That Walked by Himself" (1902), "because he will carry us from place to place for always and always and always." On horseback, previously unattainable,

---

* The tallest (and heaviest) horse on record is an English shire horse named Sampson (later renamed Mammoth), who stood 21.25 hands and weighed 3,360 pounds. The smallest horse is an American miniature sorrel mare named Thumbelina, measuring 4.375 hands, or 17.5 inches.

distant resources and markets were now within reach and available for exploitation. Boundaries of the "beyond" were stretched and broadened in all directions. As both Europe and China would soon find out, new neighbors—either potential friends or prospective foes—appeared unexpectedly at frontier approaches and undefended doorsteps.

Horse-powered expansion of traditional territorial limits was extended to secure new pastures for larger herds, cultivating political and commercial competition. "It is clear that there was an important change in steppe ecodynamics at this time (from around 5000 to 3000 B.C.)," acknowledges archaeologist Marsha Levine. "Horses were becoming much more common in archaeological deposits. Important cultural, social, and economic changes were taking place." In time, commercial and territorial rivalries led to mounted raiding and more refined methods of war, which included horse-drawn chariots and cavalry.

While no definitive pictorial or written evidence exists for horseback riding prior to seals from Mesopotamia and Afghanistan dating from 2300 to 2100 BCE, Anthony and Brown point to bit wear or beveling on the molars of horse remains at Botai. "We are convinced that people were riding in the Eurasian steppes certainly by 3500 BC," says Anthony, "and probably earlier." Similar evidence of riding has been verified at neighboring sites dating from 2800 to 2600 BCE in Ukraine, Kazakhstan, and Russia.

More telling is a 2023 study of Proto-Indo-European (Yamnaya) remains from Romania, Bulgaria, and Hungary dating to 3000 BCE. According to the international team of researchers, these skeletons clearly display "changes in bone morphology and distinct pathologies associated with horseback riding. These are the oldest humans identified as riders so far." Seeing as Yamnaya Indo-Europeans were westerly conquering migrants from the Pontic-Caspian Steppe, early riding predates these archaeological finds in eastern Europe.

All available evidence points to a diffusion of horse domestication, riding, and emergent selective breeding from the Pontic-Caspian Steppe. Based on the comparison of sequenced genomes from ancient Eurasian and modern horses, an impressive international team of researchers led by paleo-geneticist Ludovic Orlando proposed recently that "all modern horses can be traced to ancestors from 4,200 years ago." The DNA profiles also suggest that this population of domesticated horses from the Pontic-Caspian Steppe was selectively bred to promote two desirable variant genes.

Amazingly, almost immediately, humans were purposefully genetically engineering this living machine to produce advantageous utilitarian traits. These characteristics made horses even more indispensable and serviceable to humans for high-capacity traction, travel, trade, and war. The first is linked to more docile behavior, an increased ability to cope with stress and anxiety, and a decrease in fear, or "spook," memory recall. The second produced a stronger backbone.

These advantageous traits facilitated not only relatively painless taming (making subsequent selective breeding easier) but also engineered horses with increased stability, and superior weight-bearing, endurance, and high-speed riding capabilities. "These qualities may explain why the new horse type had such a global success. . . . The genetic data also point to an explosive demography at the time, with no equivalent in the last 100,000 years. This is when we took control over the reproduction of the animal and produced them in astronomic numbers . . . to supply increasing demands for horse-based mobility," says Orlando. "Everyone wanted a horse."

Given their superior genetic attributes, these new domesticates, in combination with riding, wagons, and spoke-wheeled chariots, replaced all other horse populations, both wild and domestic, across Eurasia within five hundred years of initial dispersal from the Pontic-Caspian Steppe. The migration "was almost overnight," states Orlando. "This was not something that built up over thousands of years." Steppe horses led this galloping charge of global change.

The DNA analysis of modern horses also reveals a high diversity of matrilines—meaning a large pool of female genetic diversity with limited male infusion. "Horse domestication might have depended on lucky coincidence," states Anthony. "The appearance of a relatively manageable and docile male in a place where humans could use him as the breeder of a domesticated bloodline. From the horse's perspective, humans were the only way he could get a girl. From the human perspective, he was the only sire they wanted."

At a specific site such as Botai, for example, herders continued to introduce wild females into established harems to mate with domestic stallions in a process known as female-mediated introgression. From a pragmatic standpoint, this makes sense. Stallions are far more aggressive and irascible, making them more difficult to capture and tame. "Given the high degree of human migration, trade, and diffusion that took place all along the Eurasian steppes," clarifies Sandra Olsen, "domestic horses could have been traded from the culture that developed the first herd.... These domestic herds could then be infiltrated with genes from wild individuals."

As horse husbandry radiated from this single "founder" population, local wild mares were incorporated into an ever-increasing number of domestic clusters, mimicking and reconstructing the natural order, female relocations, and social structures of wild horse herds and stallion harems. According to geneticist Vera Warmuth at the University of Munich, "Our best-fitting scenario further suggests that horse domestication originated in the western part of the Eurasian Steppe and that domestic herds were repeatedly restocked with local wild horses as they spread out of this area.... Considering the exceptionally high levels of matrilineal diversity in horses, we suggest that introgression from the wild was mainly female mediated." In time, we may find the ancestral *Equus* Eve or *Equus* Adam that conceived our modern world on the blustery, barren steppes.

Although the word *steppe* eventually became synonymous with "wasteland" under the Soviet collectivized agricultural state, upon the domestication of the horse roughly 5,500 years ago, it was anything but. The Eurasian Steppe was a mosaic of cultural fusion and trade—a superhighway for the exchange of products, ideas, knowledge, pathogens, genetics, and language. It was a battleground, and at other times a buffer, between competing empires and nation-states. It was the center of a dominant and diffusing sphere unlocked by the domestication of horses upon its stretching savannahs that Russian playwright Anton

Chekhov described as "the music of the grass" in his 1887 short story "The Steppe." Prior to the horse and wheel, traversing the overland east-west Eurasian Steppe corridor from Europe to China—with temperatures soaring to 110°F (43°C) in summer and plummeting to -50°F (-46°C) in winter—was unfeasible on foot.

When humans harnessed the horse, the steppe took center stage, with the spotlight of global history focused squarely on its stretching and influential sway. "Horse domestication was an absolute lightning strike in human history, leading to incredible, widespread, and lasting social transformations all across the ancient world," stresses equine archaeologist at the University of Colorado William Taylor. "Horses were an order of magnitude faster than many of the transport systems of prehistoric Eurasia, allowing people to travel, communicate, trade, and raid across distances that would have previously been unthinkable. . . . [T]he domestication of horses made the steppes and prairies of the world into cultural centers, population hubs, and political powerhouses. Nearly everywhere they were introduced, from the steppes of Asia to the Great Plains or the Pampas of the Americas, they reshaped human societies immediately." Increased mobility promptly led to a thirst for territory, resources, riches, and power.

As the nineteenth-century German philosopher Friedrich Nietzsche hammered so eloquently, "This world is the will to power—and nothing besides!" Believing that this notion was an innate instinct fundamental to Charles Darwin's "preservation of the favoured races in the struggle for life," Nietzsche was incensed that his evolutionary concept had "been sheepishly put aside by Darwinists."* Horsepower, in all its forms, however, certainly adheres to his paradigm. With horses, *sapiens* now had the ability to continuously push the final frontiers, explore strange new worlds, and seek out new civilizations for conquest or cooperation more quickly and in greater numbers.

Framed under the shifting horizon of an unbroken sugared sky, the endless seas of waving steppe grassland became the conduit for trans-

---

* Nietzsche's initial mental breakdown in 1889 was purportedly caused by witnessing the flogging of a horse.

continental trade, migration, and cultural conversion and conversation. Four thousand years before the merging east-west traffic on the infamous Silk Roads, there was an equally significant and reverberating Steppe Road, or, more accurately, Steppe Roads.* "This enormous grassland was an effective barrier to the transmission of ideas and technologies for thousands of years," points out David Anthony in his widely heralded book *The Horse, the Wheel, and Language: How Bronze-Age Riders from the Eurasian Steppes Shaped the Modern World*. "Like the North American prairie, it was an unfriendly environment for people traveling on foot. And just like North America, the key that opened the grasslands was the horse. . . . The isolated prehistoric societies of China and Europe became dimly aware of the possibility of one another's existence only after the horse was domesticated. . . . The opening of the steppe—its transformation from a hostile ecological barrier to a corridor of transcontinental communication—forever changed the dynamics of Eurasian historical development." The mighty, reverberating impact of these first horses was felt immediately and shuttered unbridled across unassuming continents.

With the domestication of the horse, and the timely additions of the wheel and bronze, the steppe became the setting for a tsunami of change sweeping west to the gates leading into Europe between the pillars of Bucharest and Odessa and east to the redoubts and watchtowers of the Great Wall of China. What transpired at these bookends of Eurasia over the next three millennia was vastly different, but, nonetheless, molded the cultures and configurations of not only Europe and China but also our entire modern global order.

The four-thousand-year-old Chinese civilization was not shaped by its own horses as much as it was shaped by the foreign horses of the nomadic Xiongnu (she-ong-noo) equestrian raiders from the outside fringes trying to get in. To rebuff these mounting threats, the Chinese boldly ventured west into Central Asia to acquire "heavenly horses" and developed their own cavalry forces. They also erected a series of snaking

---

* The word *Seidenstraße*, or "Silk Road," was coined in 1877 by the German scientist and geographer Baron Ferdinand von Richthofen. He was the uncle of the celebrated German First World War flying ace Manfred von Richthofen, better known as the Red Baron.

walls, seeking refuge behind their safety. In this sense, horses defined not only the modern geographical, spiritual, and ethnic map of the larger region but also what it meant to be Chinese. The intricate and influential dynamics of the horse on these borderlands birthed the unique culture for which China is renowned.

Over 5,600 miles to the west, at the other end of the Eurasian landmass, another, perhaps even more momentous, encounter was unfolding. With the advantage of their domesticated horses, predatory Indo-European peoples spilled out of the steppe and confronted the relatively egalitarian and wall-less agricultural societies of "Old Europe"—which promptly vanished.

The entire cultural fabric of Neolithic Europe (and the Indian subcontinent) was rapidly, fully, and completely replaced with the hierarchy, customs, genetics, and language of these mounted marauders. This wholesale remodeling established a more militant, capitalist, male-dominated culture that has been both the bane and the blessing of Europe. As a relatively small, densely populated region without walls or even stable borders, Europe and its shifting states were vulnerable to invasion and, through successive generations, manufactured some of the bloodiest wars the world has ever witnessed.

CHAPTER 5

A Horse by Any Other Name
The Indo-European Domination of Eurasia

This was the farthest west these roving mounted scouts had ever pushed into Europe. They were almost eight hundred miles from their traditional homeland. In search of new pastures and tribute, the militant horsemen were the vanguard for a wagon train of wandering migrants and their herds of cattle, sheep, and horses.

Dismounting on the reverse slope of the foothills to conceal their approach, they methodically crawled up the incline to peek over the other side. What they spotted was a scene wholly foreign to their nomadic eyes, customs, and language. It was unspeakable. They had never seen anything like it.

This strange, densely populated village contained forty to fifty grand, multiroomed lodges. Constructed of sturdy timber and mud plaster, they were aligned in tight, orderly rows or in adjoined concentric rings to form a pseudo-defensive perimeter. There was, however, a conspicuous absence of tangible fortifications—no wall, no moat, and no palisade—and there did not appear to be any obvious palace, hall, or temple. These towering two-storied buildings and smaller but structurally sound cabins were pierced by purposefully sited and well-maintained streets.

Set on a rising mound, the massive enclave was elevated thirty to fifty feet above surrounding meadows teeming with cattle, pigs, and sheep, and cultivated fields traversed by oxen pulling rudimentary scratch plows called ards. In the distance, touching the horizon, a thick ring of high-crowned woodlands fenced in the dynamic community. For the scouts, it was startling to realize that there were no horses and no wagons. After all, it *was* 3400 BCE. What kind of people still didn't have horses?

As they looked closer and peered through the open doors at long range, they could easily make out swirling circular patterns painted on the walls, a design that also decorated multicolored ceramic pottery. Copper tools, decorative pieces, and an abundance of buxom female figurines, which seemed to hold great spiritual significance, were visibly scattered throughout all households.*

Continuing to survey the scene, they watched in disbelief: men and women apparently held the same social standing, and in some instances, women even seemed to give directions to their obedient husbands. What kind of place was this? A mysterious land with huge buildings, no defensive barricades, no wagons, and no horses, where women seemed to be on equal footing with men.

Having completed their reconnaissance, the bewildered scouts slowly snaked back down the slope to their patiently waiting steppe horses. As they turned their mounts from the fiery sunset, they began a slow trot back to their clan members, who were anxious to receive their report. There was just one hitch: they had no words in their language to describe what they had just witnessed.

The American anthropologist Edward Sapir, often considered one of the founding fathers of linguistics, recognized in the early twentieth century that "the complete vocabulary of a language may indeed be looked upon as a complex inventory of all the ideas, interests, and occupations that take up the attention of a community." This includes not only the words they possess but also those that they do not. The Nunavik Inuit of Canada, for example, have at least fifty-three words in their language describing various types of snow, while the Yupik of central Siberia have forty. Many languages do not have even one word for snow, since it is not part of their speakers' ecological and cultural worldview, just as these nomadic scouts had no term for *city* in their Proto-

---

* The global smelting of copper has its origins in Anatolia (Turkey) around 5000 BCE, followed by China (2800 BCE), the British Isles (2300 BCE), the Andes Mountains of Peru (2000 BCE), and West Africa (900 BCE).

Indo-European tongue. You cannot give name to something that simply does not exist.

I can safely say that the pre-Columbian Haudenosaunee (Five Nations Iroquois) Confederacy of upstate New York did not have words in their closely related languages for *steel, platypus, smallpox,* and *penguin,* just as the Aboriginal peoples of Australia had no words for *money, hyrax, measles,* and *horse,* and Europeans surely had no terms for *potato, kangaroo, koala,* and *corn.** People can be placed on a historical, geographic, and cultural map through their vocabulary.

Languages also evolve. Over time, root or source dialects become wholly foreign and unintelligible to modern speakers of that very same cognate language. To illustrate this point, I will borrow an example from the English language employed by David Anthony in his brilliant book *The Horse, the Wheel, and Language.* The standard opening line of the Lord's Prayer circulating in modern English around 1800 reads: "Our Father, who is in heaven, blessed be your name." In the Middle English of 1400 used by Geoffrey Chaucer in *The Canterbury Tales,* one would recite, *"Oure fadir that art in heuenes, halwid be thy name."* This is the antecedent limit of comprehension and phonetics for English speakers today.

Now step back another four hundred years to the Old English of 1000, around the time when its Anglo-Saxon speakers were defeated and replaced by the Norman duke William the Conqueror: *"Fæder ure þu þe eart on heofonum, si þin nama gehalgod."* An earlier conversation with the famed monk-author the Venerable Bede or the Anglo-Saxon scholar-king Alfred the Great would be unthinkable and unintelligible to modern English speakers. All is not lost, however.

Our ancestors, with the timely assistance of the first generations of domesticated horses, left us with a pivotal collective legacy. They bequeathed their language. Within *their* words, they buried clues to *our* past. Their immortal vocabulary presents a clear window into their daily lives and routines, while mapping the larger historical landscapes they traversed and shaped. Words are linguistic fossils, as important to the

---

* The Tuscarora joined as the sixth nation of the confederacy in 1722 (under Oneida sponsorship) after fleeing North Carolina following the Tuscarora War (1711–1715) against the British, colonial militias, and their Indigenous allies.

understanding of our past and our present as stones and bones, or books and DNA.

We know innately what words mean in the context of definitions related to our own specific experiences, cognitive environments, and the exact place and time we inhabit within the larger human chronology—or more formally since the publication of Samuel Johnson's *A Dictionary of the English Language* in 1755. According to the Merriam-Webster dictionary (the first American English version was published in 1806), snow is "precipitation in the form of small white ice crystals formed directly from the water vapor of the air at a temperature of less than 32°F (0°C)." A city is "an inhabited place of great size, population, or importance," and a horse is "a large solid-hoofed herbivorous ungulate mammal (*Equus caballus*, family Equidae, the horse family) domesticated since prehistoric times and used as a beast of burden, a draft animal, or for riding." I imagine that all of us would come up with something casually similar for our own rote descriptions of *snow, city,* and *horse*. But what do these dry definitions *actually* mean?

In other words, what do these terms and phrases tangibly represent and physically embody to diverse human populations globally, locally, and personally over vast stretches of space and time? What is the human story and historical significance *behind* these dictionary definitions? Like grass and the evolution of the horse, words themselves provide the answer. They explain our history, quite literally, and in the process reveal the influential power of the horse.

Words can pull back the curtain to unveil secrets of our past. For example, every child learning to speak English knows that we create past tense verbs through the addition of the suffix *-ed* or *-t* to the end of a word: *talk* becomes *talked,* and *mean* becomes *meant*. Another set of verbs requires a simple letter change in the middle of the word or stem, such as *sit* and *sat*.

What the child is not taught is that this latter method is the original form of past tense—*know* to *knew* or *understand* to *understood*—beginning about five thousand years ago in the Proto-Indo-European (PIE) language group to which English belongs. A vast catalog of languages can be traced to this shared PIE linguistic base spoken by the first humans in history to practice a dominant horse-based culture, including our bewildered scouts.

These first domesticated horses hastened the ability to get the word out, and PIE is now the ancestral language for roughly 3.7 billion people—or 46 percent of our global population.* All extant European languages, save Basque, Finnish, Magyar, and Estonian, belong to one of eight contemporary main branches of Indo-European: Albanian, Armenian, Balto-Slavic, Celtic, Germanic, Hellenic, Indo-Iranian, and Italic.†

For example, English, Yiddish, Dutch, Icelandic, and Swedish, among others, are all sub-tiers of the base Germanic language group. The same applies to the Latin-based Romance languages of Spanish, French, Italian, Portuguese, Romanian, Catalan, and Provençal within the Italic branch. Other Proto-Indo-European branches, such as Anatolian and Tocharian, are now extinct. According to *Ethnologue: Languages of the World*, an annual statistical linguistics publication, there are 445 living Indo-European languages, with 313 belonging to the Indo-Iranian branch.‡

In the late 1700s, Sir William Jones, a lawyer and civil servant from London with a fascination for everything India, began to uncover the clues left by our ancestors within PIE to unravel the secret meanings and enigmatic stories behind their words.§ In the process, the historical spotlight was focused on the founding generations of domesticated horses and their fierce riders, who trotted onto center stage, silhouetted against

---

* The runner up, Sino-Tibetan, has approximately 1.8 billion speakers (23 percent).

† In some Indo-European models, the closely related Baltic and Slavic are separate branch entries, as are Indo (Indic) and Iranian.

‡ The Indo (or Indic) languages, including Hindi/Urdu, Bengali, Punjabi, and Marathi, are confined to the Indian subcontinent except for immigrant diasporas around the world. The only exception is Romani, the language of the Romani people (known colloquially by the pejorative exonym *Gypsies*), who brought it with them into Europe during their migrations from northern India between the twelfth and fifteenth centuries.

§ Jones's father, also William Jones, was a mathematician and close friends with Sir Isaac Newton and Sir Edmond Halley. He is most recognized for his introductory use of the symbol π *(pi)* to represent the ratio of the circumference of a circle to its diameter.

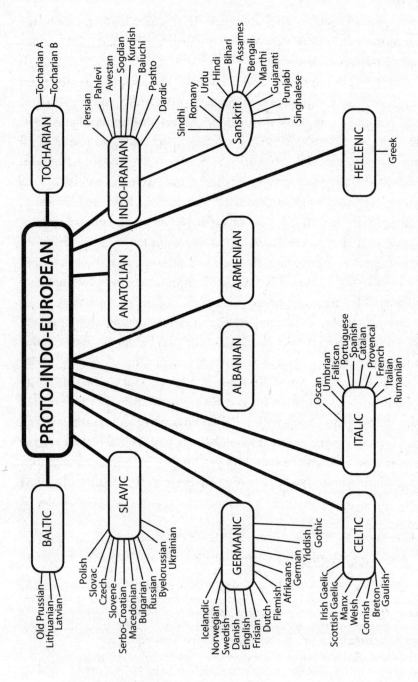

The Proto-Indo-European linguistic family tree. *(Jorde Matthews/Be Someone Design)*

the sprawling, transcontinental backdrop of the windswept Eurasian Steppe.

As chief justice of India, founder of the Royal Asiatic Society, and a respected scholar and philologist,* Jones pointed out astutely that there were too many similarities in vocabulary and syntax among Sanskrit, Latin, and Greek to be mere coincidence:† "A stronger affinity, both in the roots of verbs and in the form of grammar, than could have been produced by accident; so strong indeed, that no philologer could examine them all three without believing them to have sprung from some common source, which perhaps no longer exists."‡ His simple explanation was the presence of a common root "Indo-European" language at some point in the not-too-distant past. Upon further study, Jones concluded that the Germanic, Celtic, and Indo-Iranian languages also belonged to this shared ancestral tongue. The concept of Proto-Indo-European was born.§

More than 2,040 PIE root words and sounds have been isolated and reconstructed by linguists and anthropologists. Although Proto-Indo-European speakers left no written record, and it has not been spoken for roughly 4,500 years, in 1868 German linguist August Schleicher reassembled a story in PIE, *"Avis akvasas ka,"* translated to "The Sheep and the Horses," which also helped determine an approximate date of origin and *Urheimat* (proto-linguistic homeland) for PIE. Building on the models and comparative frameworks of Jones and others, Schleicher, who originally proposed and defined the Indo-European language family in 1863, also believed that contemporary European languages evolved from a core Proto-Indo-European tongue.

---

* Philology is the study of the structure, historical development, and relationships of a language or languages.

† His translation of ancient Persian poetry inspired the literary Romantic movement, including the English poets Lord Byron and Percy Shelley, during the first half of the nineteenth century.

‡ A simple comparison of the words for the numbers 1 through 10 or 100 across the branches of Indo-European languages reveals irrefutable similarities and unmistakable homologies (having the same or close relation, relative position, or structure within languages).

§ There is a hypothetical older root language called Nostratic, which includes Proto-Indo-European among other proto-languages as far afield as Koreanic, Japonic, Altaic, Dravidian, and, in some models, even the Eskaleut languages (Inuit-Yupik-Unangan/Aleut).

This hypothesis was bolstered by the application of the biological theories of Charles Darwin and the consonant principles of geological change proposed by his acolyte Charles Lyell. Schleicher argued correctly that languages followed the same principles of evolution espoused by his contemporaries, adhering to a progressive journey from a single point of phonological origin.

Linguists like Schleicher and Jones used comparative grammar, syntax, pronunciation, natural mouth movements, and sound systems to exploit similarities and differences to classify languages into base groupings. They then tracked and cataloged cognate words or roots, and other linguistic markers—but in reverse, much like Marsh and Huxley with their horse fossils—to resurrect PIE from the dead.

The translation of "The Sheep and the Horses," also known as "Schleicher's Fable," from the reconstructed PIE reads:

> A sheep that had no wool saw horses—one pulling a heavy wagon, another one a great load, and another swiftly carrying a man. The sheep said to the horses: "It hurts me seeing a man driving horses." The horses said to the sheep: "Listen, sheep, it hurts us seeing man, the master, making a warm garment for himself from the wool of a sheep, when the sheep has no wool for itself." On hearing this, the sheep fled into the plain.

In their own words, PIE speakers were obviously familiar with wool, horses, wagons, and riding. These marvels of early transportation explain the rapid success of their multipronged migrations and military campaigns, and the diffusion of their language, genetics, and culture.

Neither wool (or, more specifically, sheep with a genetic mutation producing longer fleece fibers capable of being spun into yarn), nor domesticated horses, nor wheels, nor wheeled vehicles existed before about 4000 to 3500 BCE, and yet all were common vocabulary to PIE speakers, as were words depicting the breadth of domesticated animals and their secondary products. Likewise, the intoxicating drink *medhu- (mead), made from honey, found service in the rituals of Indo-Europeans. Honeybees were not native east of the Ural Mountains.

Returning to our snow analogy, equally as important as existing words such as *horse, wheel,* and *wool* are those words that are absent from PIE. When looking at the varieties of fish found throughout Eurasia, for example, the two salmonids confirmed in PIE, in combination with other species that are *absent* from the lexis, point squarely to the water features intersecting the now-familiar Pontic-Caspian Steppe. The same process can be applied to the names and species of trees found—and not found—in PIE.

Modern archaeological evidence, comparative linguistic data, and the DNA sequencing of both humans and horses confirm that a shared Proto-Indo-European linguistic base was spoken some 5,500 years ago by a dominant horse-based culture on the Pontic-Caspian Steppe. "It is," writes linguist Thomas Olander at the University of Copenhagen, "the last station of the Indo-European language family before the speakers parted ways." In the process of their successive migrations, invasions, and "cultural eras," these nomadic Indo-European (Yamnaya) pastoralists quickly disseminated domestic horses throughout Eurasia, India, and the Middle East. "An important component of these movements," stresses J. P. Mallory, one of the foremost authorities of Indo-European, "is the spread of the domestic horse from the steppe."*

The root PIE word specifically denoting a domestic horse (**ek'wos*) traveled with them and is represented across the vast anthology of early Indo-European branch languages: *a-su-wa* (Luwian), *a-as-su-us-sa-an-ni* (borrowed into Mitanni for "horse trainer"), *as-va* (Sanskrit), *aspa-* (Avestan), *asa* (Old Persian), *áśwā* (Balto-Slavic), *asva* (Lithuanian), *hippos* (Greek), *i-qo* (Mycenaean), *equus* (Latin), *ekku* (Hittite), *eku-* (Venetic), *ehwaz* (Germanic), *aiƕs* (Gothic), *eoh* (Old English), *epo-* (Gaulish), *ech* (Old Irish), *es* (Armenian), and *yuk* and *yakwe* (Tocharian A and B respectively). In addition, the horse is the only animal to appear in the personal names of early Indo-Europeans.†

---

* Another recent theory posits an earlier origin for PIE in the southern Caucasus/northern Anatolia around eight thousand years ago. Subsequent migrations created branches of PIE, including that spoken by the Yamnaya, which quickly dominated the Eurasian Steppe.

† A brief list of examples, with the Indo-European term for *horse* in my italics: birid*aswa* (possessing great horses); satt*awaza* (he who won seven prizes at horse races); drv*aspa* (she who keeps horses healthy); pourus-*aspa* (he who has many horses); arbat*aspa* (master of

The expansions of Indo-Europeans and their horses across the steppe to Asia is evident in the Chinese, Mongolian, Japanese, and Korean words for *horse* adopted from the extinct Tocharian branch of PIE. Similarities can also be found for the words *wheel* and *plow*, revealing that these Eastern cultures did not have these technologies until they arrived from the steppe. They adopted the words along with the things they named.

From their homeland on the Pontic-Caspian Steppe, waves of aggressive PIE speakers ventured east toward China, southeast to Iran and India, south into Anatolia via the melting pot of the Caucasus, penetrated the Balkans by hugging the coast of the Black Sea, and, finally, trotted west into Europe. In the process, these highly mobile pastoralists conquered and absorbed local populations, replacing native tongues and genetics with evolving Indo-European dialects and DNA.

Indo-Europeans lived at a watershed moment and benefited from the first domesticated horses and wheeled vehicles. They were by no means more technologically sophisticated or socioculturally advanced than their European or Chinese neighbors. Evidence suggests that they were, in fact, less developed, characterized by the reaction of our scouts to the immense, bustling city they surveyed and scrutinized in "Old Europe." They did, however, possess horses, the single most lethal and innovative living machine that no contemporary cultures could counter or withstand.

The horse-powered world of Indo-Europeans was one of motion and militarism coordinated by a paternalistic social hierarchy of kin, clan, and tribe (the root word of which developed into the term *Aryan* in the Indo-Iranian branch). These ascending political tiers were all ruled by male chiefs and adhered to patriarchal lineage and inheritance within a patrilocal marriage arrangement where brides and a dowry moved to the husband.

They had no word for city and divided their possessions into two simple categories: transportable or fixed. They drove their domesticated cattle, sheep, and horses, and occasionally farmed grains. They spun wool

---

warhorses); hu*aspa* (having good horses); zari*aspa* (of golden horses); arima*spai* (of well-trained horses).

textiles, collected honey, and used wagons. Indo-Europeans sacrificed animals, including horses, in funerary rituals and as offerings to a vast array of temperamental and capricious sky deities. Some of the oldest myths and spiritual incarnations portray supernatural or divine horses and the horse-hero motif.* Massive kurgans (Russian for low burial mounds), usually reserved for elite males, although occasionally containing the bodies of female warriors, typify Indo-European culture.

The presence of kurgan pit graves, often including weapons, horses, wagons, chariots, and other burial wealth, are archaeological and cultural markers indicating the spread and routes of Indo-European invasions and migrations known as the kurgan hypothesis, or the steppe hypothesis. While their nomadic lifestyle renders their makeshift camps and transient settlements "archeologically invisible," these burial mounds leave an easily traceable connect-the-dots route map across Eurasia.

With their domestication of the horse, Indo-Europeans were the first people on the steppe to create a truly mobile migratory herding culture predicated on new or seasonal pastures. Horses and wagons permitted the transport of portable tent shelters, stockpiled foodstuffs, honey, dairy products, and potable water deep into the remote grasslands of the steppe.

In addition to reconnoitering suitable locations, mounted vanguards (like the scouts we met peering down on the hapless town) also preceded migrations or invasions to find soft targets to subdue and extort. Indo-Europeans practiced a parasitic patron-client reciprocity arrangement common among tribal societies with institutionalized tiers of prestige, power, and influence. Tributes and taxes were paid in decorative stone, bone, shell, and copper ornaments, precious metals, people, and livestock.

This equestrian lifestyle spread quickly across the steppe, and by 3400 BCE, localized boundary disputes and cyclical warfare shifted toward larger coalition war parties and heightened patron-client relationships. As tribal frontiers collided, steppe nomads pushed farther afield for grazing lands, trade, and tribute within an increasingly status-driven

---

* Examples stretching across the Indo-European expanse from Scandinavia to China include: Bellerophon and Pegasus, Hades and Alastor, Odin and Sleipnir, Celtic kelpies, Turkish Tulpars, Epona, and Tianma, the "heavenly horse" of China.

society. Correspondingly, environmental pressures led to a sharp decline in animal husbandry and wild game across the western steppe, requiring populations to uproot.

This exodus in search of greener pastures led successive waves of "Proto-Indo-Europeans emerging out of local communities in the forest-steppe of the Ukraine and south Russia," explains Mallory. "These intruded into southeastern Europe at a time when there was major restructuring of local societies." Their early monopoly on horses provided Indo-Europeans with an overpowering military-economic (and linguistic) dominance over surrounding sedentary farmers and pedestrian peoples with stiff frontiers and fixed sociocultural habits. "As strange as it may seem," notes anthropologist Anatoly Khazanov, "the nomads' first military advantage was connected with their underdeveloped division of labor and wide social participation."

Their horses and wagons established quicker delivery of an increasing variety of bulk goods, including a pastoral protein economy based on milk and secondary products, over extended distances and augmented routes. Evidence for milk drinking on the Pontic-Caspian Steppe (including Botai) also explodes around 3300 BCE. Roughly 94 percent of the fossilized dental tartar/calculus (calcified plaque) samples scraped from the teeth of people living at this place and time reveal milk proteins from cows, sheeps, goats, and horses, as compared with less than 10 percent before this Dairy Revolution. "The rapid onset of ubiquitous dairying at a point in time when steppe populations are known to have begun dispersing offers critical insight into a key catalyst of steppe mobility," summarizes bioanthropologist Shevan Wilkin. "The identification of horse milk proteins also indicates horse domestication by the Early Bronze Age, which provides support for its role in steppe dispersals."

Milk was only one instrument in a much larger toolbox that allowed these PIE-speaking steppe nomads to entrench their language and cultural attributes across an astounding geographical range. "Milk is a contributing factor, but not the only factor," Volker Heyd, an archaeologist at the University of Helsinki, asserts. "It's a new economy and a new way of life, and the origins are the invention of the wheel, horse riding, and dairying." This lethal combination unleashed unprecedented Indo-European expansion across Eurasia.

The isolated and disparate geographic spheres of Europe and Asia were separated by the vast, austere steppe intersected by lumbering rivers, forbidding mountains, and unforgiving deserts. With its anatomical and evolutionary designs, the horse offered Indo-Europeans a way to conquer this exactingly harsh, seemingly endless, and previously insurmountable corridor. As the linchpin, the horse single-handedly fused these two unique, formerly insulated geographical worlds into a united human whole through trans-civilizational exchange across Eurasia and beyond. History galloped forward and never looked back.

With their trailblazing horses, covered wagons, and an insatiable appetite for conquest, Indo-Europeans rapidly dominated the middle passages of trade from the Atlantic coast to the deserts of western China. This transcontinental interchange along the Steppe Roads initiated an international revolution in human encounters, language, genetics, transportation, farming, trade, metallurgy, warfare, fashion, and psychoactive drug use. By the beginning of the first millennium BCE, most of Eurasia had been overrun and completely "Indo-Europeanized." Neolithic, or "Old," Europe was the first target.

———

Between 6000 and 4000 BCE, Neolithic Europe urbanized into some of the most technologically advanced, culturally sophisticated, and socially complex agrarian settlements in the world. The largest communal super-sites, like that probed by our scouts, afforded shelter for upwards of fifteen thousand people.

By the time Indo-Europeans arrived from the steppe around 3400 BCE, however, Old Europe was reeling from failing agrarian yields, decreased cereal production, and a devastating plague—the earliest known forerunner to *Yersinia pestis*, the bacteria behind the cataclysmic Black Death of the fourteenth century CE—that cascaded across these densely populated centers. "The disease dynamic here may have been comparable to the European colonization process in America after

Christopher Columbus," relates archaeologist Kristian Kristiansen from the University of Gothenburg. "Perhaps Yamnaya brought plague to Europe and caused a massive collapse in the population." To make matters worse, temperatures plunged to levels not seen during the previous two thousand years, sending agricultural output into a tailspin. As this compounding survival crisis intensified, mounted Indo-European nomads and migrating warbands crossed the frontier.

Old Europe was ripe for rapid conquest and exploitation. Most communities had no defensive fortifications: no walls, no moats, and no palisades. When the Indo-European vanguard arrived, it found a scattered, starving, and weakened population that could offer little, if any, resistance. "The number of abandoned sites and the rapid termination of many long-standing traditions," explains David Anthony, "suggest not a gradual evolution but an abrupt and probably violent end. . . . It looks like the tell towns of Old Europe fell to warfare, and, somehow immigrants from the steppes were involved. . . . Simply put, they were the biggest human settlements in the world. And yet, instead of evolving into cities, they were abruptly abandoned."

There is widespread evidence of a sudden depopulation (including massacres) and vacated and burned villages. "A catastrophe of colossal scope," states Bulgarian archaeologist Henrieta Todorova, "a complete cultural caesura" brought about by steppe horsemen. Evgenij Chernykh, an archaeologist at the Russian Academy of Sciences, adds: "We are faced with the complete replacement of a culture."

It should be noted, however, that these marauding horsemen were not, as linguists Asya Pereltsvaig and Martin Lewis state fittingly in their book *The Indo-European Controversy: Facts and Fallacies in Historical Linguistics*, "pushing into new territories in a mindless manner, diffusing across an undifferentiated landscape in exactly the same manner as viruses spreading in an epidemic." Trade, prestige, power, pasture, and women were all corollary, if not parallel, impetuses for their influential militarized expansion across Eurasia. On the other hand, Indo-European riders were by no means modern cavalry.

Successive plundering raids by smaller clan-based war parties gradually penetrated deeper into Europe within a series of larger long-distance movements typical of mounted migrations. The initial horse-powered push into eastern Europe appears to have taken place around 3400 BCE, followed by the annexation of northern and western Europe between 2900 and 2700 BCE.

The relatively egalitarian, peaceful, and gender-equal societies of Old Europe were replaced by a male-dominated, warring, and capitalist culture that entrenched itself as the model for millennia. This "New Europe" laid the footing for the foundational hallmarks of Greater Europe and its (former) global empire to the present day.

While the inhabitants of Old Europe were not quite as inclusive or pacifist as often portrayed, they were nevertheless overrun quickly by successive waves of destructive Indo-Europeans who infused and cemented their language and DNA across Europe. "Although the desire to wish away the 'bloody swords' of the past is understandable, it is also naïve, as violence unfortunately pervades human history," write Pereltsvaig and Lewis. "But when it comes to premodern agricultural and pastoral societies, the evidence is overwhelming: enveloping violence was the norm. If one wants to rule out the possibility of bloody swords, one would be advised to examine something other than the human past."

The Y chromosomes of Indo-European men—who were comparatively stronger, taller, and healthier courtesy of their diets of fish, meat, and milk—quickly replaced those of native European and Indian men. "There is a heavy reduction of Neolithic DNA in temperate Europe, and a dramatic increase of the new Yamnaya genomic component that was only marginally present in Europe prior to 3000 BC," notes Eske Willerslev, a geneticist at the University of Cambridge. "Moreover, the apparent abruptness with which this change occurred indicates that it was a large-scale migration event, rather than a slow, periodic inflow of people." The replacement of DNA by the conquering Indo-Europeans was so extreme and absolute that Kristiansen has "become increasingly convinced there must have been a kind of genocide." With the power of the first domesticated horses in the hands of hostile and colonizing Indo-Europeans, the genetic composition of Europe was profoundly, and permanently, altered. As a result, the European population doubled to fourteen million between 2000 and 1500 BCE.

Within a few generations, Indo-European DNA accounted for an astounding 40 percent to 50 percent of the total genetic fingerprint across Europe. More telling is that almost all males have Y chromosomes from these steppe invaders, suggesting that only Indo-European men were producing children. This male lineage was largely absent across Europe and India prior to 3000 BCE. "The collision of these two populations was not a friendly one, not an equal one, but one where males from outside were displacing local males and did so almost completely," says pioneering geneticist at Harvard University David Reich. "Their descendants continued to spread."

Reich estimates that Indo-European genetic populations quickly replaced a full 40 percent of the Iberian Peninsula, roughly 50 percent of Scandinavia, 70 percent of present-day Germany, and 90 percent of Britain. More regionally, Indo-European DNA represents upwards of 50 percent across northern and central Europeans, roughly 30 percent across southern Europeans, and even 7 percent to 11 percent of the islanders of Sardinia and Sicily. "If you look at the Y chromosomes—the DNA that people get from their fathers—it was more or less 100 percent from the steppe," he explains. "What this means is that the males coming in from this Eastern group had completely displaced the local males." This male dominance was not limited to Europe.

Across the Indian subcontinent, the "genetic influx from Central Asia in the Bronze Age was strongly male driven, consistent with the patriarchal, patrilocal, and patrilineal social structure attributed to the inferred pastoralist early Indo-European society," states archaeogeneticist Marina Soares da Silva at the London-based Francis Crick Institute. "This was part of a much wider process of Indo-European expansion, with an ultimate source in the Pontic-Caspian region." As a result, between 60 percent and 90 percent of men now living in the Indian subcontinent, including the Indus River valley, can trace their patrilineal DNA to Indo-European horsemen. "Indigenous males seem to have been marginalized by the new arrivals much more than the women and were unable to have children to the same extent," reveals geneticist Martin Richards. "This seems unlikely to have been a wholly benign process."

In both Europe and India, significant portions of this ancestry are the result of "star clusters": the genes of one powerful man appear in millions of descendants.* Reich estimates that roughly 20 percent to 40 percent of Indian men and 30 percent to 50 percent of males across eastern Europe descend from a single Indo-European man who lived between 4800 and 2800 BCE.

There is also strong evidence that these horse-and-chariot-driven Indo-European invasions laid the foundation for the Indian caste system. "A strong clue comes from the fact that Aryan [Indo-Iranian speakers] genes register far more strongly in the higher castes," writes Namit Arora in *Indians: A Brief History of a Civilization*. "Further, DNA evidence has shown that endogamy first appeared and became the norm 'among upper castes and Indo-European speakers.' Indeed, as many scholars have long argued, the roots of the Indian caste system almost certainly trace back to the Aryan substrate."†

While Indo-European DNA quickly dominated huge portions of Eurasia, so too did their PIE language, which was creolized in different localities to produce the main Indo-European branches and bifurcated sublanguages spoken today. "There is now compelling evidence that the spread of the Yamnaya archeological culture was the vector that also spread all the Indo-European languages spoken today," contends Reich. "Genetics has made some important progress toward solving the more than two-hundred-year-old Indo-European language puzzle."

For example, in one of many references to horses and chariots in the Hindu Rig Veda, written around 1500 to 1300 BCE in northwest India, a prayer accompanying the sacrifice of a horse reads, "Let this racehorse bring us good cattle and good horses, male children and all-nourishing wealth."‡ Aside from demonstrating the patriarchal, pastoral, and

---

* Chinggis Khan is a fertile example of this phenomenon. Geneticists believe that 8 percent of people living in the former Eurasian heartland of the Mongol Empire are his direct descendants. To put this another way, roughly forty-five million to fifty million people currently on the planet are progeny of the fearsome thirteenth-century warrior-ruler.

† Endogamy is the cultural practice of marrying or mating only within a specific social, religious, caste, or ethnic group.

‡ The origin of the mythological unicorn dates to the Bronze Age Indus Valley Civilization (3300–1900 BCE).

commercial influences of Indo-Europeans, this spiritual collection of 1,028 hymns is the oldest Indo-Aryan (Sanskrit) text, save for borrowed words transcribed by the Mitanni of Syria.

The Mitanni were experienced charioteers and skilled horse managers. In the oldest horse-training manual in existence, the Mitanni "master horse trainer" Kikkuli used numerous Indo-Aryan numbers and terms to describe technical aspects of chariot horses and their conditioning and health. His 1,600-year-old text begins, "Thus speaks Kikkuli, the *assussanni* [horse trainer], from the land of Mitanni." To describe his profession, he reverts to the Indo-Aryan word. Given the pattern of domestication, Proto-Indo-European and its evolving regional tongues became the language of the horse.

―――

On the opposite end of the transcontinental Eurasian Steppe, China faced the same enemy, but its fate and future turned out markedly different. Initially lacking horses, with the help of natural *and* artificial barriers and a quick military response, China held out and rebuffed a complete Indo-European occupation. The what-if historical scenarios here are as boundless as the shifting desert sands and soaring mountain ranges that insulated China from these hostile Indo-European horsemen.

Prior to the domestication of the horse, China was comparatively isolated from the West by the interconnected depth of the Taklamakan, Gurbantunggut, and Gobi Deserts transected by the Tibetan plateau and Altai Mountains. This protective geographical perimeter was finally breached by horse-shuttled Indo-European traders and raiders using two main routes along the Steppe Roads and, later, the Silk Roads. As historian David Christian explains, "The Inner Eurasian steppelands were occupied, probably since the fourth millennium B.C.E., and certainly by 3000 B.C.E, by communities practicing extensive and mobile forms of horse pastoralism, which ensured that their contacts and influence would extend over large areas. Indeed, the emergence of mobile pastoralist lifeways should probably be regarded as the real explanation of the origin of the trans-Eurasian network of exchanges that the Silk Road came to symbolize."

From the Kazakh Steppe or the multicultural Central Asian trading centers sprinkled across the "Stans," including Merv, Bukhara, and Samarkand, the northern route pointed east through the Dzungarian Gate corridor. It continued through the mountain pass between the northern Altai Mountains and southern Tian Shan ranges before abutting the Gobi Desert to the Mongolian Steppe and the northern reaches of the Yellow River, eventually descending on Beijing.

Leaving these buzzing grand bazaars of Central Asia and the lush Fergana valley (Uzbekistan, Tajikistan, and Kyrgyzstan), the southern route penetrated the formidable Pamir Mountain pass to the commercial hub of Kashgar. Skirting the northern foothills of the Tibetan plateau, it pushed east through the Tarim Basin of the Taklamakan Desert of Xinjiang, the Jade Gate mountain pass, and the narrow Gansu Corridor on the southern edge of the Gobi to the dynastic capital of Xi'an.

Secondary routes, access tributaries, and snaking mountain passes—including the infamous Khyber and Bolan traversing the Hindu Kush of Afghanistan and Pakistan—fed into these two main east-west arteries, creating an interconnected Indo-European "world system" linking China to India, the Middle East, and Europe as early as 2000 BCE. "These pathways serve as the world's central nervous system, connecting peoples and places together," writes historian at the University of Oxford Peter Frankopan in *The Silk Roads: A New History of the World*. "This is the region where the world's great religions burst into life, where Judaism, Christianity, Islam, Buddhism, and Hinduism jostled with each other. It is the cauldron where language groups competed, where Indo-European, Semitic, and Sino-Tibetan tongues wagged alongside those speaking Altaic, Turkic, and Caucasian. This is where great empires rose and fell, where the after-effects of clashes between cultures and rivals were felt thousands of miles away."

Indo-European domination spread east to the gates of China through the northern route to its terminus in the harsh Dzungarian and Tarim Basins in the northwest province of Xinjiang. Beginning roughly five thousand years ago, this region served as the source for the broad-spectrum Indo-European cultural and commercial pipeline heading into the heartland of China. "The earliest accounts of the Tarim Basin depict a society whose linguistic and ethnic diversity," says Mallory, "rivals the type of complexity one might otherwise encounter in a modern transportation hub.... The existence of a vast extension of material culture, economy, ritual behavior, and physical type from the Pontic-Caspian eastwards... by about 3000 BCE."

The first horses and chariots, as well as bronze metallurgy, wheat, barley, goats, sheep, cattle, and intensified dairy pastoralism were all introduced to China through this region by peoples from the West with strikingly European biological features evidenced by the well-preserved, naturally freeze-dried collection of Tarim (and Dzungarian) mummies dating between 3000 and 1700 BCE.* This period corresponds with the earliest known association between horses and humans in China. The archaeological record, ancient shell and bone inscriptions, and the genomic analysis of horses all point to the wider dissemination of horses and chariots across China between 1500 and 1200 BCE during the Shang dynasty.

A series of recent DNA analyses on the mummies identifying "West Eurasian haplogroups" and "admixture from populations originating from both the West and the East" support the idea that these Indo-European communities represent the eastern limit line of successive migrations traced back to the Pontic-Caspian Steppe. "Our data indicate multiple population influences in the Tarim Basin during 4000–3500 yBP [years before present]," reports geneticist Chunxiang Li, "consistent mainly with the 'steppe hypothesis.'" Modeling Europe and the Indian subcontinent, Indo-European males also dominated the genetic profile of this region. Confirming their Western origins, DNA testing on the mummies

---

* This natural mummification was the result of a combination of factors: a hyperarid climate, frigid winters reaching -40°F (-40°C), and highly salinized soils, all of which create a hostile environment for corpse-decaying bacteria and result in the pristine preservation of organic materials such as human corpses, wood, leather, wool, animal and plant remains, and other human cultural materials. The mummies are so well preserved that archaeologists can clearly delineate biological features, including hair and skin color; eye, nose, lip, and cheek shape; and beautiful and intricate tattooing on the human canvas. Evidence suggests that tattooing was also passed into China from the steppe around 1200 BCE and subsequently became a stigma associated with barbarians. Additionally, the multitude of Bronze Age languages present across the Dzungarian and Tarim Basins have been remarkably preserved on a variety of mediums, among them stone, wood, leather, bone, and the Chinese invention of paper.

revealed a European paternal line and a mixed European and southern Siberian maternal admixture incorporated en route to the Tarim Basin.

Covering more than 642,000 square miles (approximately one-sixth) of modern China, Xinjiang Province is now home to a multicultural mosaic that includes the Turkic Uighurs, who arrived from the Mongolian Steppe in the ninth century CE. In addition to the dominant Han, China currently houses fifty-five minority groups with their own languages and scripts. More than eighteen of these languages, spoken across an astounding forty-seven distinct ethnic groups, can be heard in Xinjiang. "This was a region of incredible crossroads," remarks anthropologist Michael Frachetti of Washington University in St. Louis. "There was vibrant mixing of North, South, East, and West going back as far back as five thousand years." This cultural and linguistic diversity is the modern legacy of its prominence as a trading center on the ancient Steppe (and Silk) Roads opened by these eastern horse-shuttled Indo-European forays.

It is important to note that Indo-European expansion was not solely the result of violent exchanges. The Indo-European colonization of greater Eurasia adheres to the unbreakable bond between war and trade enshrined earlier by the Agricultural Revolution. "Indo-European was swept quickly forwards in the fifth millennium as the language of the colonizing," notes Barry Cunliffe. "By the mid-to-late fifth millennium BC, disparate communities, scattered across hundreds of kilometers, were linked, albeit loosely, in systems of reciprocal exchange through which commodities, and with them ideas and beliefs, could quickly spread." Mounted Indo-Europeans were the medium for long-distance trade trotting along the Steppe Roads from Europe to China.

English is currently the universal language of commerce because the capitalist powerhouses of imperial mercantilism, the industrial revolutions, and resultant military might—namely, the United Kingdom and the United States—spoke English. From the end of the global Seven Years' War in 1763, Pax Britannica ruled the high seas and the international markets, only to be replaced by the Americans following the unconditional surrenders of the Second World War. The adoption of

Proto-Indo-European and its evolutionary descendants across Eurasia was no different.

Carried across the steppe by horse and wagon, PIE became the language of economic enterprise and expansive trade. "Wealth, military power, and a more productive herding system probably brought prestige and power to the identities associated with Proto-Indo-European dialects after 3300 BCE," writes David Anthony. "All these factors taken together suggest that the spread of Proto-Indo-European probably was more like a franchising operation than an invasion. Although the initial penetration of a new region (or "market," in the franchising metaphor) often involved an actual migration from the steppes and military confrontations."

A similar diffusion of the Comanche language among the nation's sedentary Indigenous neighbors and Euro-American traders occurred in the early 1800s as these Horse Lords of the Plains expanded both their geographic empire and economic monopoly on bison hides.* The takeaway here is that if you need or want to enter the fray of free-market capitalism and conduct business with the dominant economic corporations during any time period, it is imperative to learn *their* language.

The flexible horse-fueled mobile pastoral economy of Indo-Europeans within their central geographic homeland was a stark contrast to the sedentary, alluvial, and horseless city-states of Europe, India, China, and the Middle East. While Indo-Europeans used their equine prowess to vanquish and plunder within an emerging "booty capitalism," these intrepid horsemen also utilized equally skilled commercial ventures and patron-client contracts to accumulate wealth. This lethal combination coupled to a monopoly on domesticated horses catapulted the infusion of their language across an extremely vast spectrum of terrains and tongues.

With their cherished horses, Indo-Europeans could travel six to ten times farther per day than all other people on the planet. The typical trail horse can cover roughly sixty miles per day, while those properly trained and conditioned for endurance can push upwards of one hundred miles.

---

* Comanche is an Uto-Aztecan language related to the Aztec Nahuatl tongue.

On horseback at the leisurely pace of fifteen to twenty miles per day, the entire 5,600-mile journey across the Steppe Road trading routes from eastern Europe to western China could be accomplished in less than a year. By continuously changing out for fresh horses, the elite Mongol dispatch "arrow riders" of the thirteenth century CE could cover the 4,300-mile distance from the Mongol capital of Karakorum to the Hungarian Plain in about a month!

Up to this point in human history, not one person or collective had ever traveled so far so quickly. It was the dawn of a new age of rapid movement and the lightning transmission of ideas, materials, innovation, trade goods, human migrations, and conquering armies. According to Yale University anthropologist William Honeychurch:

> The first long-distance contacts across East and Inner Asia began in earnest around the middle of the second millennium BC with transfers of bronze technologies, artifact styles, wheeled vehicles, belief systems, and especially, animals . . . when secure and competent horseback riding became widespread. This novel "transport technology" promoted the building of relationships far beyond the local area, and impacts from this transformation were felt from . . . the functional use of horse riding for greater mobility and its interactive importance for extending and projecting relations, alliances, and conflicts.

Mounted and mobile Indo-Europeans controlled the middle passages of trade between Europe and the fringes of Asia along the Steppe Roads. These horse riders were the FedEx, UPS, or Amazon Prime of their age, and their revolution in transportation allowed them to control the ebb and flow and distributable quantity and supply of valuable and exotic goods across the now-interconnected landmass of Eurasia. For instance, they acted as intermediaries, swapping their livestock and secondary products for gold, silver, copper, and tin from the mines of Afghanistan, the Caucasus, and the Balkans. They then shuttled these goods north and east, back-dealing for the precious metals of the south.

Indo-European conquests, migrations, and trade did more than just cement their language, genes, and male-dominated ethos across Eurasia. This transformative, history-defining "Indo-Europeanization" also disseminated domestic horses, wool sheep, wagons, chariotry, bronze, and other profound cultural alterations, including pot and pants, across a significant portion of the planet, with ripple effects that have not yet subsided.

As a component of the "chariot package deal," domesticated horses show up in substantial numbers across Europe, including Italy, Britain, and Greece, between 2200 and 1800 BCE. Greece was occupied by drifting, and perhaps invading, groups of Indo-Europeans from Anatolia and Armenia. These newcomers ushered in the Mycenaean era (1750–1050 BCE), representing the first developed and distinctively male, warrior-based "Greek" civilization on mainland Greece. This introduction of horses also gave rise to the fabled centaur.

There are stories across Indo-European cultures involving erotic ceremonies of women, usually of royal pedigree, fornicating with horses. In Greek mythology, Centaurus, a deformed son of Ixion, the sun god, mated with mares from the region of Thessaly to produce the human-horse centaur hybrids. The main job of centaurs in Greek lore is to undermine and sabotage the sanctity of marriage. Apparently, as the original wedding crashers, they get roaring drunk and attempt to seduce or abduct the bride. In Juliet Clutton-Brock's view, however, "A more likely explanation is that a group of people from Thessaly were seen from a distance riding on horses, and this unfamiliar scene was perceived as proof that creatures existed which were half human and half horse."

It is hard for us to imagine a time or a world where this familiar "centaur scene" of a person on horseback, which is so cemented in our collective human consciousness, was completely foreign. It would literally be like watching aliens exit their UFO at the end of your driveway, presumably offering diplomatic greetings or patron-client demands in a branch language of Proto-Indo-European.

In short, between 3000 and 2000 BCE, there is an unmistakable and marked increase in the frequency of horse (and wagon) remains across greater Eurasia. But additional clues lie deeper than just the archaeological record. In the embryonic city-states of the Middle East—Ashur, Babylon, Ur, and Uruk, among others—the arrival of horses required the addition of new words to the local lexicon.

Written evidence from Ur dating to 2100 BCE documents the creation of two new Sumerian cuneiform characters from Indo-European loan words devised for these strange new beasts: *anse.kur.ra* ("ass from the mountains") and *anse.zi.zi* ("speedy ass"). There are also rare references to "mounted messengers," although we cannot be sure that the equid being ridden is a horse or an ass. Prior to the introduction of horses, donkeys were carrying the load in Egypt as early as 3500 BCE and in Mesopotamia by 3000 BCE. While the donkey remained an important beast of burden to Middle Eastern populations, horses were a highly prized and sought-after supplement.

A healthy donkey can carry a pack load of a hundred pounds for a workable distance. Although they can hit short bursts of thirty miles per hour, the evolutionary structure of their leg bones and hooves prevents donkeys from maintaining any sustained speed. Unlike horses, they have only two speeds: these brief breakneck spurts and a walking pace slower than the average human. Horses possess twice the traction force of donkeys, twice the weight-bearing capacity, three times the horsepower, and almost twice the average work speed. By their very nature, mules are

Standard of Ur: The *War Panel* depicting Sumerian battle wagons drawn by equids (either asses or onagers), circa 2600 BCE. *(Wikimedia Creative Commons)*

obviously faster and stronger than donkeys but are clearly not comparable to horses. Like the horse and donkey exchange in the Middle East, a similar swapping of animals was occurring throughout Europe, marking a pivotal changing of the guard.

By 2000 BCE, the horse was gradually replacing the ox as a much stronger and faster means of pulling carts and wagons. "The ox cart," the late British archaeologist Stuart Piggott proclaimed, "creaks and groans its way into bucolic oblivion." When horses replaced oxen to pull vehicles, the average daily distance more than doubled from 15.5 to 37.3 miles. With their superiority in speed, horses can generate more than double the foot-pounds per second of oxen. For hauling capacity, the hourly workload of a horse in miles per hour is 8.7, compared with 2.5 for an ox. A horse hitched to a wheeled vehicle can pull twice its own body weight; consequently, the standard thousand-pound horse could move two thousand pounds of cargo for trade, war, and travel. Along with horses and PIE, by the end of the second millennium BCE, carts and wagons can be archaeologically confirmed across Eurasia, India, and China, followed closely by the chariot. The same can be said for bronze.

Bronze alloy was first made from copper and 2 percent to 8 percent arsenic (a natural grayish metalloid element poisonous to humans and most animals), which was replaced with the rarer but far superior addition of tin in slightly higher percentages. The oldest bronze artifacts and smelting technology dating to 3700–3500 BCE are found in the north Caucasus and the Pontic-Caspian Steppe before spreading west on the backs of Indo-European horses to Europe between 3200 and 2400 BCE and east to China through the Tarim Basin around 2000 BCE. Chinese metallurgists eventually perfected the lost-wax molding method of casting bronze into extremely intricate and ornate products, including vessels, musical instruments, weapons, and chariot fittings.

Copper-tin alloyed bronze provided a stronger, harder substance than all other metals of the time, with obvious ramifications for weaponry, armor, and military and agricultural machinery. Unsurprisingly, tin became one of the most valuable commodities across the ancient world.

Voluminous tin prospection and extraction in Afghanistan, Uzbekistan, and Tajikistan began around 2000 BCE. This tin was then carried along existing trade routes that were already shuttling lapis lazuli, turquoise, and precious metals from Central Asia to Europe, the Middle East, and Egypt. Over a fifty-year period during the early second millennium BCE, for example, more than eighty tons of tin were transported by horses and donkeys from Afghanistan to Anatolia. Seafaring Phoenician and Mycenaean traders from the Mediterranean may have even exploited the rich tin mines of Cornwall and Devon in southern England.

While the lethal combination of horses, chariots, and bronze fashioned by Indo-Europeans forever altered *and* accelerated the art of war, they also peddled other commodities with a different kind of potency. Indo-Europeans are thought to have introduced cannabis culture and the smoking of psychoactive marijuana to Europe. "Narcotics in the form of *Cannabis*," says Anthony, "were one of the important exports from the steppes." Archaeological evidence of marijuana use in Europe is chronicled as early as 3500 BCE in a kurgan burial mound in Romania.

Linguistically, during the third millennium BCE, descendant cognates from the PIE root *kanna* spread widely across Indo-European languages, correlating the spread of mounted PIE speakers and marijuana use. The lexical cognates of *kanna* include: Greek *kannabis*, Sanskrit *sana*, Germanic *hanipa* (Old Prussian *knapios*, German *hanf*, English *hemp*), Russian *konoplja*, Albanian *kanep*, and Armenian *kanap*. According to Stanford University historian Adrienne Mayor, "cannabis/hashish [were] both readily available."

Writing in the fifth century BCE, Greek historian Herodotus mentions the cannabis practices of the nomadic horse warrior Scythians of the Pontic-Caspian Steppe. Among other vibrant commentary about the Indo-European Scythians, which makes up the bulk of book 4 of his *The Histories*, Herodotus reports, "They have also discovered a kind of plant whose fruit they use when they meet in groups. They light a bonfire, sit around it, throw this fruit [cannabis buds] on the fire, and sniff the smoke rising from the burning fruit they have thrown on to the fire. The fruit is the equivalent there to wine in Greece; they get intoxicated from the smoke, and then they throw more fruit on to the fire and get even more intoxicated, until they eventually stand up and dance, and burst into song."

He goes on to write that "the Scythians take cannabis seeds [buds containing seeds], crawl in under the felt blankets, and throw the seeds on to the glowing stones. The seeds then emit dense smoke and fumes, much more than any vapour-bath in Greece. The Scythians shriek with delight at the fumes." In modern cannabis practices, this marijuana sauna is known as hotboxing.

Excavations of cannabis paraphernalia at Scythian sites, including male and female burials containing personal cannabis kits, support Herodotus's account. The burial chamber of a well-preserved fifth-century BCE tattooed Scythian mummy known as the Siberian Ice Maiden—nestled in the southern Siberian Altai Mountains (at the four corners of Russia, China, Kazakhstan, and Mongolia)—contained, among other artifacts, six horses in full harness, cannabis and accompanying paraphernalia, and opium. It is theorized that she used cannabis, traces of which were also found in her hair, to relieve the considerable pain of terminal breast cancer (mirroring a modern use for medicinal marijuana).

Like today, Indo-European hemp was also utilized for making rope, twine, bags, and clothing. As recorded by Herodotus: "Now, there is a plant growing in their country called cannabis, which closely resembles flax, except that cannabis is thicker-stemmed and taller. In Scythia, in fact, it is far taller. It grows wild, but is also cultivated, and the Thracians use it, as well as flax, for making clothes. These clothes are so similar to ones made out of flax that it would take a real expert to tell the difference between the two materials. Anyone unfamiliar with cannabis would suppose that the clothes were linen." Hemp clothing was customary among Chinese peasantry as early as 1600 BCE, while luxurious silk, still relatively unknown outside of China, was reserved for privileged aristocrats.

Marijuana was not, however, the only psychoactive drug in the Indo-European medicine chest. Archaeologist Andrew Sherratt, best known for his theory of the Secondary Products Revolution, mentioned earlier, reports in the book *Consuming Habits: Drugs in History and Anthropology* that analyses of "ash repositories from sacred fires, and preparation rooms where vats and strainers for liquid were found . . . identified traces of *Ephedra, Cannabis,* and *Papaver* [the genus of the opium poppy]."

Mystery surrounds the identification of the Indo-European ritual psychotropic intoxicant known as soma in the Rig Veda and haoma in the Avesta, the sacred text of the ancient Persian religion Zoroastrianism. The Persians referred to the eastern Scythians (the region of the seven "Stans") as Śakā haumavarga or "haoma consuming Scythians." Sherratt, and others, have revisited William Jones's 1794 suggestion that soma could be the harmel plant, or Syrian rue, which possesses potent hallucinatory properties the Nazis utilized as a "truth drug" during interrogations. Although numerous plants and fungi have been suggested, the modern consensus is that soma was most likely the plant-based stimulant derived from several species of *Ephedra*.

Aldous Huxley, the grandson of our eccentric cartoon-drawing horse evolutionist Thomas Henry Huxley, firmly entrenched soma in the lexicon and public imagination of the Western world. In his 1932 dystopian novel *Brave New World*, the drug "soma" is used as a mechanism of social control and brainwashing. It is freely distributed to all citizens of the World State, making them slaves to industrial society and its future.

With this revelation between Indo-European horses and the dissemination of recreational drugs, the term "drug mule" takes on a whole new (or in this case, old) meaning.* While blazing their way across Eurasia, mounted Indo-Europeans also kicked off a functional gender-fluid fashion revolution.

There is a direct correlation between horseback riding and the origin and widespread adoption of pants. "The design was an innovation that facilitated riding on horseback," explains Adrienne Mayor. "In other words, trousers were the world's first 'tailored' clothing. Trousers did not just happen but had to be invented." Prior to the horse-fabricated stitching of pants, people wore gowns, robes, skirts, togas, tunics, leggings, and kilts. Certainly not ideal riding garments. "From now on," stresses J. D. Hill, the head of research at the British Museum, "most would wear trousers."

The oldest recovered pair of purposefully fashioned pants, dating to 1300 BCE, belong to two horsemen unearthed in the Tarim Basin. These

---

* A kurgan at Maykop in the northern Caucasus dating to 3500 BCE contained what is believed to be the oldest recovered "beer bongs." The eight decorated and finely crafted gold and silver "beer straws" are roughly a yard long and almost a half inch in diameter.

trendsetting warriors are part of the larger group of Indo-European mummies mentioned earlier, some of whom are also buried with hemp, ephedra, cheese, and other horse-related artifacts. The German-led archaeological team responsible for the find confirmed in 2014 that "the invention of bifurcated lower body garments is related to the new epoch of horseback riding, mounted warfare, and greater mobility."

With a straight-leg taper, wide inseam crotch, and a string-drawn waist, the ancient woolen pants (complete with intricate woven designs) allow for both freedom of movement and protection of the legs and genitals. "Definitely supports the idea that trousers were invented for horse riding by mobile pastoralists," remarks linguist and sinologist at the University of Pennsylvania Victor Mair, "and that trousers were brought to the Tarim Basin by horse-riding peoples."* Progressively, for those on horseback, which was increasingly becoming the rule rather than the exception, pants were all the rage. These peripatetic pot-smoking Indo-European fashion junkies and their horses created one of the most enduring and universal items of clothing in human history.

These original pants and their Tocharian-speaking Indo-European owners resemble the Dothraki nomadic warrior horsemen who patrolled the vast grassland steppes in the Emmy-winning television series *Game of Thrones*. In fact, almost the entire fictional horse culture of the Dothraki, from their clothes and koumiss to their khalasar hordes, is borrowed from numerous equestrian societies, including these original horse lords of the steppe. Writer George R. R. Martin, from whose epic fantasy-novel series A Song of Ice and Fire the long-running *Thrones* was adapted, openly acknowledged this, saying, "The Dothraki were actually fashioned on an amalgam of a number of steppe and plains cultures," including the Scythians, Huns, Mongols, Sioux, and Cheyenne. History is usually stranger than fiction.

While the original riders from the steppes might have spread PIE, horses, wagons, chariots, bronze, pants, pot, and other cultural attributes across Eurasia, the propagation of Indo-European languages was

---

* Sinology is the study of Chinese language, history, and culture.

far from over. "If we can draw any lessons about language expansion," summarizes David Anthony, "an initial expansion can make later expansions easier . . . and that language generally follows military and economic power." As we will see, global waves of militaristic European imperialism disseminated Indo-European languages, particularly English, Spanish, French, and Portuguese, across the vast colonial expanse to the fringes of empire and the four corners of the Earth.

Within the cultural upheaval of the Columbian Exchange, Indigenous peoples the world over—from the Iroquois Confederacy to the Aboriginal Australians—were quickly compelled to create words in their own tongues for *horse, wagon, pants,* and anthropologist Jared Diamond's conquering trifecta of *guns, germs,* and *steel*. "Indo-European languages," affirms Brian Joseph, a historical linguist at Ohio State University, "are official state—that is to say, national—languages on all continents except Antarctica." Over a period of roughly five thousand years, thanks to the horse, the progenies of PIE—the language of our mounted scouts who had no word for *city*—are currently spoken by almost half the people on our planet.

By 2000 BCE, from their last refuge and homeland on the Pontic-Caspian Steppe, horses and their Indo-European masters fused two previously distinct geographical worlds—Europe and China—as the bookends of the Eurasian Steppe. As philologist at Indiana University Christopher Beckwith writes in *Empires of the Silk Road,* "Central Eurasia was the home of the Indo-Europeans, who expanded across Eurasia from sea to sea and established the foundations of what has become world civilization." Multifarious trading routes crisscrossing the transcontinental Steppe Roads cultivated broadly comparable Indo-European cultures across this extensive region.

This horse-powered expansion and conduit of exchange created an immeasurably smaller world, instigating a commercial revolution and an insatiable demand for horses, domesticated animals, and their secondary products, in addition to slaves, tin, copper, precious metals, knowledge, and, above all else, power. "The steppe is the most remarkable natural corridor in the world," says Barry Cunliffe. "It is here that the first horses were domesticated and ridden, where the first two-wheeled chariot was invented, and where riders first learnt to work together as cavalry with world-shattering effects."

The quick-step succession of the Agricultural and Equine Revolutions gave birth to the first true empires emboldened by a new era of imperial capitalism and state-sponsored conflict. "The most direct contribution of plant and animal domestication to wars of conquest was from Eurasia's horses," declares Diamond, "whose military role made them the jeeps and Sherman tanks of ancient warfare on that continent."

As trade accelerated, and wealth accumulated, so too did the ancient world's appetite for war. The horse revolutionized combat in the ancient world, and by 2000 BCE, avaricious kings in command of burgeoning city-states and embryonic empires began to strike out and clash in a cyclical game of thrones. While the Akkadians, Assyrians, Egyptians, Israelites, Hittites, Medes, and Scythians, among others, vied for political power and economic control of the greater Middle East, the very existence of the nascent unified Chinese state was threatened by militaristic mounted raiders from the northern steppe.

The poetic author of the biblical book of Job, writing in the sixth century BCE, clearly understood the combat value of horses: "Did you give horses their strength and the flowing hair along their necks? Did you make them able to jump like grasshoppers or to frighten people with their snorting? Before horses are ridden into battle, they paw at the ground, proud of their strength. Laughing at fear, they rush toward the fighting, while the weapons of their riders rattle and flash in the sun. Unable to stand still, they gallop eagerly into battle when trumpets blast. Stirred by the distant smells and sounds of war, they snort in reply to the trumpet." During this age of empires, history hung in the balance, and was spurred forward by the warhorse.

PART II
FORGE OF EMPIRES

CHAPTER 6

Behold a Pale Horse
Apocalyptic Chariots and Imperial Ambitions

In his acclaimed 1905 work *The Life of Reason: The Phases of Human Progress*, philosopher George Santayana penned his immortal phrase "Those who cannot remember the past are condemned to repeat it." In some historical instances, however, the opposite can be true: "Those who remember the past can successfully repeat it." During the First World War, General Edmund Allenby and his vaunted British cavalry did just this.

In the autumn of 1918, four-plus years into the war, the British campaign snaking through the biblical lands of the Middle East to secure the Levant and the prized petroleum of Mesopotamia from the crumbling Ottoman Empire and its German benefactor was entering its final phase. British troops were pushing up the Tigris River from Baghdad toward the bountiful oil fields of Mosul. Farther west in Palestine, however, Allenby's seventy-five thousand soldiers—not only British but also Indian, Australian and New Zealand (ANZAC), and Arab—were bottlenecked by narrow mountain passes on the Megiddo Plains.

Nicknamed "the Bull" for his towering six-foot-two, barrel-chested frame and fiery temper, Allenby had a reputation for getting things done. Prior to the outbreak of war in August 1914, he had served as inspector general of cavalry. Allenby always demanded that his officers follow his golden rule: "Look after the horses first, then the men, then yourself."

With the advance of his troops stalled on the Plains of Megiddo (Jezreel valley), Allenby drew up plans to push his renowned horsemen through the tight mountain canyons and dangerous passages, or defiles.

He outfoxed his opponents using a carefully coordinated strategy of deception, surprise, and the concentration and swift deployment of his twelve thousand cavalrymen.

Dummy camps were erected with mock stables housing fifteen thousand canvas horses to conceal the actual movements of his cavalry. British newspapers leaked spurious dispatches and spread disinformation about erroneous British and allied Arab forces amassing along the Jordan River valley with a view to assaulting the commercial railway hub of Amman (Jordan). To authenticate this ruse, small detachments of British infantry, the ANZAC Mounted Division, and the Imperial Camel Corps were openly marched east toward the valley each morning under the gaze of enemy spies. These same troops were then secretly shuttled back each night, to repeat their march the next day. Additional misleading troop movements were orchestrated to mimic the dust clouds of spearhead cavalry units.

These tactics, as Allenby was aware, had been used on the ancient battlefield of Chengpu 2,550 years earlier. Fought between the rival Chinese states of Jin and Chu in 632 BCE, during the engagement the Jin commander outmaneuvered his opponent by luring him into a kill zone behind the screen of a dust cloud generated by chariots dragging trees. Employing a classic pincer movement, Jin infantry and chariots quickly surrounded the Chu troops and scored a resounding victory while confiscating "400 four-horse teams of chariot horses."

With his deceptions and feint toward Amman in motion, Allenby launched his main combined infantry-cavalry surprise assault on the western coastal mountain passes on September 19. In less than a day and a half, his cavalry covered an astonishing sixty-two miles. This lightning-fast line of march surprised not only Allenby, who exploited the pace of his cavalry, but also the German commander, General Otto Liman von Sanders, who was directing the Ottoman defenses. So quick and complete was the Allied encirclement of the German-Ottoman positions that Liman was forced to flee in his pajamas. Allenby's mounted forces tore through enemy positions and, in quick succession, captured Amman, Damascus, and Beirut.

Following the war, in March 1919, as the acrimonious Paris Peace

Viscount of Megiddo: General Edmund Allenby and his vaunted cavalry enter Jerusalem through the Jaffa Gate, December 1917. *(Library of Congress)*

Conference entered its third month, recently promoted Field Marshal Allenby assumed his post as the high commissioner for Egypt and Sudan. While chatting with famed American archaeologist and Egyptologist James Breasted, Allenby revealed that he had drawn his inspiration for the deployment and punching power of his cavalry force at Megiddo not only from the ancient Chinese but also from a commander much closer to home: the warrior-pharaoh Thutmose III, who had himself fought at Megiddo 3,400 years earlier.*

According to Allenby, his advance was "exactly old Thutmose's experience in meeting an outpost of the enemy and disposing of them at the top of the Pass leading to Megiddo! You see, I had been reading your

---

* Breasted founded the prestigious Oriental Institute at the University of Chicago in 1919 with a fifty-thousand-dollar grant from Standard Oil tycoon John D. Rockefeller.

book," he told Breasted, who had translated the firsthand Egyptian account into English, "and I knew what had taken place there." The resounding victories of both the Egyptian pharaoh circa 1479 BCE and the British general in 1918 owed their successes to the agility, mobility, and speed of the horse. Debouching from the mountain defiles undetected, their chariots and cavalry were able to exploit the element of surprise and the terrain to quickly surround and overtake enemy positions.

Like Allenby, Thutmose sought to drive his enemies north after the recent expulsion of the foreign Hyksos from Egypt, ending at last their 125-year occupation. The pharaoh realized the need to create a buffer for Egypt by taking control of the Levant. Spearheaded by a force of ten thousand to twenty thousand soldiers and a thousand chariots (including three thousand to four thousand horses, which would have represented the entire trained stock in Egypt), Thutmose's territorial expansion was challenged at Megiddo by a coalition of ten thousand to fifteen thousand Canaanites, including roughly a thousand chariots, led by the king of Kadesh, who had revolted against Egyptian rule. The ensuing clash at Megiddo is the first battle in history to be recorded by an eyewitness with reliable detail. The inscriptions on the Temple of Amun at Karnak are based on the daily journal of Tjaneni, an Egyptian scribe and the first known embedded combat reporter.

In a logistical feat made possible by the horse, mirroring Allenby's rapid movement and concealment, Thutmose marched his army 250 miles from Egypt into Gaza in nine days—undetected. According to Tjaneni, "[Every man] was made aware of his order of march, horse following horse, while [his majesty] was at the head of his army." Like Allenby in 1918, Thutmose prevailed at Megiddo by using two simple but supreme fundamentals of military doctrine: the element of surprise and seizing the initiative with quick-strike mobility. Following a crafty line of advance, he maneuvered his concentrated forces through one of the same passes utilized later by Allenby.

Leading from the center ranks, Thutmose "went forth in his chariot of electrum adorned with his weapons of war, like Horus armed with talons, the Lord of might, like Mentu of Thebes, his father Amen-Ra strengthening his arms." The warrior-pharaoh and his troops struck quickly, folding the Canaanite flanks and forcing a general retreat to the walled city of Megiddo. In a scene reminiscent of the German general Limon von Sanders hastily absconding in his pajamas, "When the rebels saw His Majesty prevailing over them, they fled straight away to Megiddo with faces of fear. They abandoned their horses and their chariots of gold and silver, so that they could reach safety," reports the Egyptian eyewitness. "Now the people of Megiddo had shut the city gates, but they let down garments in order to hoist them over the walls"—the old bedsheet ladder trick.

After a protracted siege, the exhausted and starving Canaanites surrendered. Our scribe Tjaneni writes that the Egyptians took home an impressive cache of loot: 340 prisoners, 2,238 horses, 924 chariots, 200 suits of armor, 503 bows, 1,929 cattle, 22,500 sheep, an assortment of other military hardware, and 426 pounds of precious metals and gems.

The Egyptian victory under Thutmose has perhaps left an even greater legacy. The Greek translation for the Hebrew name *har megiddo* is Armageddon. In one of the great ironies of human history or imagination, Megiddo represents the site of the first recorded battle and, according to the book of Revelation, also the last. Megiddo/Armageddon is the portended gathering place of the armies of good and evil during the long-awaited end times.

When John of Patmos penned his ominous Revelation sometime between 81 and 96 CE, he was sure to include the pale horse of annihilation, the last of the Four Horsemen of the apocalypse, which self-proclaimed oracles have been prophesying and promising ever since: "And I looked, and behold a pale horse: and his name that sat on him was Death, and Hell followed with him. And power was given unto them over the fourth part of the earth, to kill with sword, and with hunger, and with death, and with the beasts of the earth." Edmund Allenby, however, spurned these auguring soothsayers: his horse had a bay-brown coat.

As a reward for his brilliant service, the British general was awarded (and chose) the title 1st Viscount of Megiddo and Felixstowe. "They wanted me to add 'Armageddon' to the title, but I refused to do that," he recounted. "It was much too sensational and would have given endless opportunity to all the cranks in Christendom, so I merely took Megiddo."

Allenby, who used the term "the Fields of Armageddon" in his dispatches to the War Office, became irrevocably known as "Allenby of Armageddon"—the last of a line of Christian crusaders (although he was an atheist)—a title that he publicly received with embarrassment and privately renounced with derision.

Historically, however, proficient cavalry like Allenby's (or the apocalypse) was still on the distant horizon when the first Battle of Armageddon was fought. At this time in our story, chariots, not cavalry, were the elite equine combat arm across ancient battlefields, as witnessed by Thutmose's victory. "The adoption of the war chariot and the imposition of the power of charioteers throughout the centres of Eurasian civilisation," states acclaimed military historian John Keegan in *A History of Warfare*, "is one of the most extraordinary episodes in world history."

Eventually, however, after a millennium of dominance, the chariot was cast aside around 900 BCE by the Assyrians, who raised the first true cavalry to conquer their neighbors and construct an impressive empire. From Megiddo to Megiddo, the horse dominated the art of war for almost four thousand years. Although equines have disappeared from modern battlefields, the legacy of their combat prowess still shapes our everyday lives.

The current tradition of driving a carriage from the right side, driving a car from the left side, and mounting a horse from its left side all have to do with chariotry, cavalry, and the fact that 90 percent of people are right-handed. A charioteer drove from the right side so that his sword or spear weapon could be utilized by his dominant hand on the right side of the vehicle, while he held the reins with his left hand. The archer fought from the left side of the chariot. Carriages simply copied the chariot position of the driver and are still operated from the right side.

Following this established design, early automobiles were also originally driven from the right side. In 1908, with the rollout of his mass-produced, world-changing Model T, Henry Ford moved the steering wheel from the right side to the left side so that the driver could engage the controls in the center console with their dominant right hand. This modification also made it easier for right-handed people to climb into the passenger seat on the vehicle's right side. When Cadillac finally made the switch in 1915, all American manufacturers produced cars driven from the left.

Horses are also always mounted from their left side. Traditionally, the sheath was worn on the left hip so that the sword could be pulled and presented quickly by the right hand. This placement necessitated a soldier's mounting his noble steed from the left (near) side to avoid the sheath impeding his leg and interrupting his climb onto the back of the horse. Similarly, if horses are hitched in tandem, or postilion hitch, the driver rides the left, nearside horse, so that the right hand is free with the whip on the off (right) horse. We can thank chariotry and cavalry for these modern attributes of civilization. While the technology has advanced and changed from animal to mechanical, and from chariot to horseless carriage, the premise remains the same.

Based on archaeological evidence, Hittite cuneiform inscriptions, and Indo-European timber grave burials, the chariot first appears around 2000 BCE on the steppe of southern Russia and Kazakhstan. "The chariot was such a sophisticated, highly tuned machine, it was extremely expensive to build or buy, to train its horses and drivers, and to maintain. Its users had to be experts," says Christopher Beckwith. "Indo-European peoples of the second wave became the world's first experts in the maintenance and use of chariots and chariot horses, and they were the first to use them successfully in war." These Proto-Indo-Iranian–speaking peoples disseminated chariot technology as part of the larger Indo-European exchanges across Eurasia. According to sinologist Edward Shaughnessy at the University of Chicago, "If we now compare the technical characteristics of the Chinese and Trans-Caucasian chariots, I think there can be no doubt as to their typological similarity, or even identity."

Surviving correspondence between ancient rulers often includes requests or demands for chariots and horses. Horses were difficult and expensive to acquire, train, and sustain, while chariots were costly big-ticket items in both construction and maintenance. Although most two-wheeled chariots pulled by two to four horses were built exclusively of laminated wood and durable leather (some with ornamental metals), the construction, model types, and effectiveness varied across time and place.

Transporting a skilled driver and any combination of one to three specialized archers, swordsmen, and spearmen, chariots served as both long-range and close-order fighting platforms. Unlike wagons or earlier battle carts, lighter, spoke-wheeled chariots were engineered for reduced weight and speed behind expressly trained horses. The Mitanni horse trainer Kikkuli, mentioned earlier, wrote his manual (c. 1400 BCE) specifically for training and conditioning Hittite chariot horses.

Equine enthusiast and historian Ann Hyland, who has also trained horses professionally, replicated Kikkuli's seven-month exercise, drill, and dietary program with modern Arabian stallions. The result was healthy horses possessing significantly increased speed and endurance—vital characteristics of chariot horses.

While speed is largely correlated to leg length as evidenced by modern Thoroughbred horses, weight-carrying capacity and endurance are proportionately independent of height. The overall build of a horse rather than strictly height determines the occupation for which it is best suited. Horses of the late Bronze Age, standing fourteen and fifteen hands, would have been plenty big enough to offer invaluable military service over expanded areas of operation. Charioteers could travel upwards of thirty to forty miles a day.

For aspiring city-states and conquering empires, the application of Kikkuli's equine training regimen was instrumental in forging formidable militaries. "Three critical components combined to create the war chariot, the first effective horse-based weapon system: lethality with the composite bow, speed and endurance in the trained modern horse, and a stable and maneuverable two-wheeled platform with improved harness," explains military historian Louis DiMarco in *War Horse: A History of the Military Horse and Rider*. "This system revolutionized the nature of warfare. Chariot-based armies dominated the centers of the civilized world in the Middle East from around 1800 BC to approximately 900 BC." By harnessing the horse directly for war, humanity was pulled into a new era. Organized violence entered new territory, as did those rulers steering economic aggression across the ancient world. For the first time, trade, war, *and* peace became international commodities.

With horse-fueled transport, chariotry, and eventually cavalry, kings and their military strategists poured over increasingly larger maps, continuously projecting their power into previously uncharted terrain. Strangers entered strange lands to ally with or conquer distant peoples. Mysterious raiders came from across the seas and landed on foreign shores to rape, plunder, and raze unsuspecting populations—only to vanish as quickly as they had appeared. Fierce, nomadic barbarians from the northern steppe supplied the insatiable demand for horses among other exotic trade goods, but were also objects of fear, loathing, and curiosity.

By 2500 BCE, multifarious networks of trade and economic tentacles fanned out from the first densely populated urban centers, creating intense commercial competition that led to fickle alliances and frequent war. Strategic conflict extended beyond immediate rivalries and neighborly disputes to include distant city-states and competing empires ruled by dynasties of divinely inspired kings and pharaohs. As the horse quickened the marching steps of modern civilization, waves of trade and war flooded outward from the great rivers of the Middle East and Egypt. The horse was invariably caught up in these sweeping currents of change it helped create.

Throughout history, legendary empires expanded through imperialism, conquest, and political or economic leverage. Each was defeated in time and replaced by another, continuing a cyclical rise and fall of ancient kingdoms. "The period from 4000 to 539 BC was marked by the almost constant occurrence of large-scale war," writes military historian Richard Gabriel in *The Culture of War: Invention and Early Development*, "in which human slaughter and the destruction of societies were commonplace. Horses, chariots, wagons, and increasingly accurate and lethal projectile weaponry propelled humanity into the modern military age." Prior to the arrival of the horse and chariot, however, the geopolitical designs of Egypt and the Middle East had already been established largely through trade and conflict.

Wielding the first professional military with a standing army 5,500-strong and utilizing donkey-drawn carts, around 2300 BCE Sargon of Akkad conquered and subsequently unified the feuding principalities of Mesopotamia and northern Syria into the earliest true empire in history. Trade and agriculture flourished under the coordination of a centralized bureaucracy. His reign marked the first time that a large body politic incorporating more than a million subjects was governed by a single ruler, who was, also for the first time, viewed and worshipped as a god-king.

Sargon ruled for an astounding fifty-five years and set in motion the precedents for empire, divine rule, and single-state governance over multiethnic populations in the Middle East and beyond. From its capital of Akkad along the Tigris River, for the next 170 years, the Akkadian Empire dominated the region, extending its economic, political, and military interests through trade, diplomacy, and war into the Levant, Anatolia, and Persia. These geopolitical configurations and social conventions became hallmarks of Western civilization.

By 2193 BCE, however, drought, compounded by a decline in trade for essential materials, thrust the Akkadian Empire into disarray. It eventually buckled under the combined weight of famine, rebellion, and chaos wrought by outside raiders. So complete was its collapse that the city of Akkad sank into the sands of time. It has yet to be found. Uncovering its hiding place, vast treasures, and ancient secrets represents something of a holy grail for archaeologists.* By pushing the limits of imperial power, Sargon and his successors not only rewrote the geopolitical and military doctrine for Western civilization but also rubbed elbows with other blossoming empires, including Egypt.

Agricultural pursuits, mirroring those emerging in the first Sumerian city-states in Mesopotamia around 4000 BCE, allowed a relatively isolated Egypt to flourish on the banks of the Nile River. Egyptian unification and agricultural expansion began around 3000 BCE, at which time Egypt had an estimated population of one million clustered around the northern delta of the Nile.

Given its geographic isolation and austere desert surroundings, Egypt was a minor player in the higher echelons of external geopolitical affairs

---

* It is believed to rest somewhere along an eighty-mile section of the Tigris River stretching north from Baghdad.

and had been relatively insulated from the events transpiring to its northeast. While the Egyptians ventured sporadically into the biblical lands of the Levant, bringing them into conflict with the Canaanites, Israelites, Hittites, and Assyrians, they never secured a lasting foothold in the region.

Egypt was essentially an empire unto itself, reaching its territorial and cultural zenith during the era known as the New Kingdom, from 1550 to 1070 BCE, noted for some of the best-known pharaohs, including Akhenaten and his wife, Nefertiti; Ramses II; and Tutankhamun. The death of eighteen-year-old King Tut in 1323 BCE marked the beginning of the end of Egyptian imperial power and cultural achievement.* Never again was it an esteemed international player. Egypt ultimately became a vassal state for a series of conquering empires, beginning with the Libyans around 1000 BCE, followed by the Assyrians, Persians, Macedonians, and Romans.

A comprehensive and diverse collection of sources and mediums provide for a relatively complete, if imperfect, insight into the political and military affairs of the ancient world. As Juliet Clutton-Brock explains, "From this time onward, throughout the second and first millennium BC, the period of history covered by the Old Testament, the chariot horse played an integral part in the endless wars fought between the peoples of western Asia. The Israelites came to depend on a supply of horses and chariots from Egypt to defend themselves against Syria and the Assyrians." The scarcity and status of horses is evident in their prohibitive cost. A single horse was worth, respectively: 7 bulls, 10 donkeys, 30 slaves, 500 sheep, or 5.3 pounds of silver. The first mention of horses in the Bible (Genesis 47:17), depicting Egypt and Joseph during the great famine around 1700 BCE, is one of the rare references outside the context of war.

For the people of the great civilizations of the Old Testament,

---

* It has been suggested that King Tut was born of an incestuous brother-sister relationship, causing numerous congenital deformities. It was common for Egyptian nobility to marry siblings and even their own children. Queen Cleopatra, for example, was married to each of her two adolescent brothers, Ptolemy XIII and Ptolemy XIV, with whom she co-ruled Egypt. Of the fifteen marriages of Ptolemaic (Macedonian) Egyptian rule, ten were between brother and sister and two were with a niece or a cousin.

including the Egyptians, Canaanites, Hittites, Babylonians, Israelites/Judeans, and Assyrians, the horse swept down from the northern steppe fully domesticated as an animal purposefully adapted and harnessed for war. Herodotus recorded an old Hittite folk memory referencing the fact that "horses were beasts of war and not native to the country." While chariot warfare was in its infancy across the Mediterranean world, Egypt was the first to feel the full wrath of this innovative quick-strike technology. Lacking horses and chariots, it was vulnerable to foreign armies outfitted with these overpowering cutting-edge weapons.

Initiating their invasion from their homeland in the northern Levant around 1675 BCE, the Hyksos subdued the majority of Lower (northern) Egypt through their proficient use of horses, chariots, and composite bows. A thousand years after the construction of the Great Pyramids, they established the Hyksos dynasty.* "Warriors in chariots drove into the country like arrows shot from a bow, endless columns of them," recorded the Egyptians. "Day and night horses' hooves thundered past the frontier posts."

Compounding this advantage, the Hyksos bow exceeded the range of its Egyptian counterpart by one hundred to two hundred yards. "A composite bow could kill at three hundred yards," explains Gabriel, "and, when fired in salvos along a parabolic arc, could strike targets at twice that range with killing force. The composite bow allowed armies to engage and kill each other across greater distances than ever before." In other words, soldiers (and the states they served) had the ability to project power and reach out and touch someone to a far greater extent than ever before. The swift Hyksos southern thrust was a campaign of maneuver and firepower to which the Egyptians had no answer.

Archaeological and linguistic evidence reveals that the Hyksos introduced the horse, chariot, and metal bits (far superior to their customary organic counterparts) to Egypt during their roughly 125-year rule. The earliest vocabulary related to horse skeletons (exhibiting bit wear) and chariots corresponds with the vanguard of the Hyksos invasion. "Hundreds of Canaanite words turn up in New Kingdom documents,

---

* Hyksos: "foreign rulers" or "rulers from the mountains."

laboriously transcribed by Egyptian scribes into hieroglyphs," observes Egyptologist Donald Redford in his award-winning book *Egypt, Canaan, and Israel in Ancient Times*. "As one might expect, fully one-quarter of those terms that can be identified have to do with the military. Technical expressions describing the chariot, its parts, and accoutrements account for half of these." While the horse, and chariot, spread slowly and in limited numbers from the Nile delta, across the inhospitable northern deserts, to Morocco by the ninth century BCE and to the Horn of Africa by the fourth century BCE, it did not find a home in the southern climes of the continent for quite some time.

The earliest unequivocal evidence of horses in sub-Saharan Africa appears around 900 CE, although as historian at the University of Sterling Robin Law points out in his hidden gem *The Horse in West African History: The Role of the Horse in the Societies of Pre-Colonial West Africa*, "[I]t is likely that horses in fact reached West Africa much earlier, probably during the first millennium BC."

Taxing ecological conditions and a cocktail of tropical pathogens, however, prevented the horse from establishing any consequential permanent hoofprints. Sweltering temperatures and drenching humidity, an inventory of parasites, the midge-borne virus that causes deadly African horse sickness, and, above all else, the tsetse fly vectoring trypanosomiasis made sustaining any sizable horse populations unviable. As a result, horses were nothing more than an exotic status symbol for African elites. The first Portuguese traders and slavers to venture into the area in the mid-1400s noted that there were "very few" horses.

The Hyksos in Egypt, however, set the stage for their eventual downfall by introducing the horse and chariot. As modern Western governments have found out the hard way in Iran, Afghanistan, Somalia, and Iraq over the last half century or so, arming fleeting friends who may one day become enemies is a dangerous game. The eviction of the Hyksos from Egypt, like their conquest, was steered by the chariot.

The wars of expulsion commenced around 1555 BCE. After a series of Theban victories, the pharaoh Ahmose I, who prided himself on his chariotry, presided over the final defeat of the Hyksos, including the razing of their capital. "As for Avaris on the Two Rivers," he tells us in an inscription, "I laid it waste without inhabitants; I destroyed their towns and burned their homes to reddened ruin-heaps forever, because of the destruction they had wrought in the midst of Egypt." His pyramid temple is adorned with striking reliefs portraying Ahmose as a towering warrior routing his opponents from his horse-drawn chariots.

While the Hyksos failed to "bite and hold" Egypt, they left behind a far greater legacy than their brief occupation and incomplete reign: horses, chariots, and the composite bow. Egyptians quickly mastered the art of chariotry, which proved advantageous in the decisive battles to come.

Although the Old and Middle Kingdoms of Egypt had practiced a policy of isolation, following the exodus of the Hyksos, Ahmose and his successor, Thutmose I, campaigned into the Levant, as far as the Euphrates River in northern Syria. The Hyksos invasions had sounded the alarm that Egypt needed a defensive buffer zone. These campaigns were intended to break the ability of the Hyksos, and other suitors with lusting eyes trained on Egypt, to mount any meaningful military operations. After hunting elephants in Syria to celebrate his successes and wanton destruction of the Levant, Thutmose I and his armies returned to Egypt.

He regaled his royal subjects with strange tales of his adventures, including "that inverted water which flows upstream when it ought to be flowing down stream." Egyptian rivers flow from south (upstream) to north (downstream), and the Euphrates River was the first they had ever seen flowing from the north, which was "downstream" on the Nile. Subsequently, the Euphrates became known simply as "inverted water." His son and successor, Thutmose II, solidified Egyptian control over the southern Levant to the Syrian border.

Upon his death in 1479 BCE, Thutmose II's half sister and principal wife, Hatshepsut, broke the tradition of patriarchal monarchy and became Egypt's most successful female ruler, although she was not the hereditary and lawful heir to the throne. This claim belonged to her two-year-old stepson, Thutmose III (whom we met previously at Megiddo), born to one of her husband's concubines or secondary wives.

To solidify her claim and validate her reign, Hatshepsut adopted the regalia, fashion, and symbolism of traditional male pharaonic office, including the *khat* and uraeus headwear, the *shendyt* kilt, and the false beard. She donned body armor to compress and conceal her breasts, and altered the ending of her name to the masculine form and became "His Majesty, Hatshep*su*" rather than the feminine "Hatshep*sut*." She also claimed to be "the most beautiful woman in the world."

During her twenty-one-year reign, she led a relatively peaceful kingdom marked by diplomatic trade and cordial external relations. Focusing on Egypt itself, rather than foreign conquest and military expenditures, she commissioned massive infrastructure and building projects. Hatshepsut is celebrated, along with Nefertiti and Cleopatra, as one of the most illustrious and influential women in Egyptian history. Until recently, however, she was entirely expunged and absent from the annals of our past. But unlike the lost city of Akkad, thankfully she was found.

When she died, her name and likeness were purged and erased from monuments and registers to maintain the legitimacy of a male pharaoh and prevent future women from emulating her power. It worked. Hatshepsut vanished from the historical record and was completely forgotten. It was as if she had never existed. It took more than three thousand years for her life and legacy to finally be resurrected from this *damnatio memoriae* by determined Egyptologists.*

Following Hatshepsut's death in 1458 BCE, her stepson Thutmose III ruled Egypt for an astounding fifty-four years. As a true Renaissance man, he was an acclaimed historian, author, botanist, architect, athlete, and, in stark contrast to his stepmother, became Egypt's greatest and most celebrated warrior-pharaoh. His legendary exploits, including those at Megiddo, earned him the modern nickname "Napoléon of Egypt." Leading his troops into battle, Thutmose fought more than seventeen major campaigns in twenty years, and—unlike Napoléon and more like Alexander the Great—went undefeated.† He established Egypt as a first-rate military power to contend with those of the Middle East, including the Canaanites, Hittites, Mitanni, Assyrians, and Babylonians.

---

* A Latin phrase meaning "condemnation of memory"—the erasing of all official record and heritage of a person.
† Napoléon's own Egyptian campaign between 1798 and 1801 was a miserable failure.

Thutmose completely modernized the Egyptian military based on the weapons, equipment, and tactics of the Hyksos. He also engineered improvements to the design and combat application of chariots, including platform-mounted archers. With long-range composite bows, these archers screened the advance of the infantry before launching swift frontal chariot assaults. In military terms, chariots were the first mounted or mechanized infantry. They were the armored personnel carriers (APC) and light armored vehicles (LAV) of the ancient world, and Thutmose used them to great effect.

In the wake of his reverberating victory over the Canaanite coalition at Megiddo around 1479 BCE, Thutmose directed his army north to establish Egyptian hegemony over the entire Levant and to demonstrate his strength of force to would-be rebels. Following these lucrative campaigns, the annual tribute from vassal states flowing into Egypt reached unprecedented levels. Thutmose used these taxes to subsidize enhancing and expanding his armies and to unveil the greatest construction and engineering achievements of Egyptian civilization save the Sphinx and Pyramids. The string of successful military actions orchestrated by Thutmose III, courtesy of the horse and chariot, propelled Egypt to its zenith of cultural and territorial power.

Empires are hard to take, and even harder to hold. Although Egypt maintained a fragile grasp on foreign territory, two hundred years after Thutmose carved out the Egyptian Empire, Ramses II would be engulfed in conflict to maintain its borders. From the thirteenth century BCE onward, Egypt was challenged on all fronts by a series of invaders who all either severely weakened or exercised outright dominion over the fading civilization. In the immediate flickering afterlight of the golden reign of Thutmose III, the contest for commanding the greater Middle East had begun and would be decided by the horse. The Hittites would strike first.

The Hittites were Indo-European speakers who drifted from the steppe to Anatolia sometime between 3000 and 2200 BCE, displacing or absorbing the resident Hattian and Hurrian populations. Although the Hittites were the first to forge iron weapons around 1300 BCE, this specialized knowledge was not the primary advantage in their efficient war machine. Their military prowess rested on their mastery of heavy chariots, stocks of highly trained horses, and advanced siege tactics.*

The secret craft of Hittite iron weapons production was replicated and quickly adopted across the ancient armies of Eurasia, reaching India and the Horn of Africa by 1000 BCE and China by 700 BCE. As with bronze, the Chinese took iron fabrication to the next level, mass-producing a catalog of cast-iron wares—including pots, sickles, shovels, and weapons—a thousand years before any other culture. They were also tempering iron with coal, a technique not replicated anywhere in the world for another two thousand years!

The importance of iron did not rest with its strength or its ability to hold an edge but rather with its abundant availability and its relatively easy extraction from its carrier ore. The mass production of iron weapons, which was widespread by the rise of the Assyrian Empire around 900 BCE, was cheap and easy.

The financial strength buttressing the Hittite military was its control of long-distance trading routes from the Mediterranean and Black Seas. Shipwrecks from this period reveal diverse cargoes from across the known world. The strategic Hittite geographical home base in Anatolia also enabled them to dominate the north-south market for raw materials, precious metals, and prized horses across the Pontic-Caspian Steppe, the Caucasus, the Middle East, and Egypt. Others, however, wanted in on this lucrative exchange.

Hittite expansion was resisted by the Egyptians, who were promoting their own commercial position and interests in the Levant. Although the Hittite king, Suppiluliuma I (whom Kikkuli served as horse master), wrote to the Egyptian pharaoh Akhenaten around 1350 BCE that "I wish good friendship to exist between you and me," he took advantage of the tumultuous reign of Akhenaten and Nefertiti. The royal couple had triggered domestic turmoil by abandoning traditional Egyptian polytheism in place of monotheistic Atenism.

Exploiting this internal friction, Suppiluliuma initiated his invasion

---

* The Hittite military expanded to include a formal navy, which, in the first recorded naval battles, secured the island of Cyprus between 1275 and 1205 BCE.

of Egyptian holdings in Syria and Canaan, informing Akhenaten that "as soon as my chariots are ready to carry the cloth [battle standard], I shall send it." These preliminary Hittite provocations elicited the desired effect of fomenting insurgency among numerous Egyptian vassals. These initial events and opening salvoes set the stage for protracted conflict between the two vying imperialist powers, including the legendary clash of Egyptian and Hittite chariots at Kadesh around 1274 BCE in western Syria.

The Battle of Kadesh is believed to be the largest chariot battle ever fought, with an estimated 6,000 machines taking the field. Like Megiddo, we have reliable documentation of troop movements, strengths, and tactics. In a remarkable feat of logistics made possible by the horse, in roughly a month, the shameless, self-promoting pharaoh Ramses II marched his massive procession of 20,000 troops, 2,500 chariots, and supply wagons over five hundred miles to the battlefield at Kadesh, where he would face off against King Muwatalli II, commanding a Hittite force of 17,000 troops and 3,500 chariots.

The Koller Papyrus, written by the Egyptian army scribe Amenope shortly before the departure of Ramses II's "Syrian Expedition" to Kadesh, provides explicit instructions for the sophisticated care, grooming, and feeding of military horses and the preparation of chariots and weapons for combat: "Take good heed to make ready the array of horses which is bound for Syria, together with their stable-men, and likewise their grooms; their coats [?] and filled with provender and straw, rubbed down twice over. . . . Their chariots are of bry-wood filled with all kinds of weapons of warfare." Planning and preparation, however, cannot always overcome incompetent leadership.

At the onset of battle, Ramses II's forces blundered into a Hittite trap. The Egyptians were saved from annihilation only by the greed of their enemies. Hittite soldiers were more interested in looting and plundering than in pressing their military advantage. Despite his bumbling, Ramses spun Kadesh as a resounding victory, won single-handedly by his own heroics. "I shot on my right and captured with my left," he claimed dubiously. "I found 2,500 chariots, in whose midst I was, sprawling before my horse. Not one of them found his hand to fight. . . . I caused them to plunge into the water even as crocodiles plunge, fallen

Ramses II at the Battle of Kadesh: This self-aggrandizing scene depicting his heroics adorns his Great Temple, Abu Simbel, Egypt, circa 1260 BCE. *(Robert Harding Images)*

upon their faces one upon the other. I killed among them according as I willed." Although Ramses would have us believe otherwise, Kadesh ended in a draw, forcing the two war-weary empires to negotiate the earliest-known deliberated peace treaty sixteen years later.

Ratified in 1258 BCE, both the Egyptian and Hittite copies of the armistice survived, marking the first verified example of a negotiated nation-state accord. This relatively stable cease-fire and delineated spheres of influence secured a balance of power in the region for the next few decades. This atypical tranquility was short-lived, however, proving to be the calm before the storm. By 1200 BCE, the entire Mediterranean world was thrust headlong into an orgy of violence, and its occupants faced a far greater threat than traditional enemies and fluctuating allies.

Enigmatic and powerful coalitions of sundry maritime raiders, collectively known as the Sea People, cast a dark shadow of fire and ash over Egypt and the Middle East for the next two hundred years. According to an Egyptian report from 1186 BCE, "No country could stand before these arms." These plundering raids, amid dire drought, famine, and a series of earthquakes and tsunamis, set off a sweeping chain of destabilizing apocalyptic events across the greater Middle East. Fragile transnational trade routes were severed, farms were razed, livestock slaughtered, and refugees wandered and looted the charred landscape. The ensuing, and exponentially increasing, humanitarian crisis led to the collapse of long-established city-states and empires, plunging the region into the ancient Dark Ages.

The infamous (and perhaps fictional or at least misrepresented) Trojan War is set against this historical backdrop. In the storybook version, an army from Greece crosses the Aegean Sea to ransack and pillage the city of Troy in northwestern Asia Minor, giving rise to one of the most fabled horses in history: the mythical Trojan horse.

Although known as Sea People, their ferocious attacks were not confined to coastal cities such as Troy. Roving bands of marauders pushed inland, penetrating and plundering the heartland of Anatolia and Mesopotamia, bringing the Hittite and Babylonian Empires to their knees. Others, such as the Egyptians and the Assyrians, staved off their attackers by buckling and folding inward to ensure survival.

When the smoke cleared, and the mysterious invaders vanished from both the landscape and the literary record, the power vacuum was filled by a revitalized empire riding a new and far more lethal military machine. The Assyrians discarded the cumbersome chariot in favor of the first true cavalry. "The old polities of Egypt, Assyria, Babylonia, and Elam still maintained a degree of internal cohesion," writes Barry Cunliffe, "Egypt in grand isolation, the other three in a state of constant conflict. Out of this mêlée would arise the world's first empire": the Neo-Assyrian Empire.

With the freedom of action afforded by their cherished warhorses, merciless Assyrian riders launched the Cavalry Revolution completed a half millennium later by Alexander the Great. Cavalry propelled the world into the modern era. It was a game-changing concept and decisive weapon system that steered the fate of battle for the next 2,800 years, from the vicious expansion of the Assyrian Empire to General Edmund Allenby's victories of the First World War.

CHAPTER 7

Riders on the Storm
Cavalry, Assyrians, Libraries, and Scythians

After crushing an Aramaean uprising in 883 BCE, King Ashurnasirpal II of Assyria tells us in his own words what happened to these rebels of the city of Tela along the Euphrates in Syria:

> I built a pillar over against the city gate and I flayed all the chiefs who had revolted and I covered the pillar with their skins. Some I impaled upon the pillar on stakes and others I bound to stakes round the pillar. I cut the limbs off the officers who had rebelled. Many captives I burned with fire and many I took as living captives. From some I cut off their noses, their ears, and their fingers, of many I put out their eyes. I made one pillar of the living and another of heads and I bound their heads to tree trunks round about the city. Their young men and maidens I consumed with fire. The rest of their warriors I consumed with thirst in the desert of the Euphrates.

Not to be outdone, two centuries later, another Assyrian king, Sennacherib, describes his campaign against the Elamites in 691 BCE in equally gory detail: "I cut their throats like lambs, cut off their precious lives as one cuts string. Like the many waters of a storm, I made the contents of their gullets and entrails run down upon the wide earth. My prancing steeds, harnessed for my riding, plunged into the streams of their blood as into a river. . . . With the bodies of their warriors I filled the plain, like grass. Their testicles I cut off and tore their privates like the seeds of cucumbers in June."

Between 900 and 640 BCE, the Assyrians, who honored their god through conquest and expansion, fought 108 wars of varying intensity to acquire territory or to unleash unforgiving punitive expeditions against insolent neighbors. "Neo-Assyrian royal inscriptions describe hundreds of military campaigns, often in great detail," writes Assyriologist at Yale University Eckart Frahm in *Assyria: The Rise and Fall of the World's First Empire*. "Few civilizations of the ancient world have left richer sources for a careful study of military history than the Neo-Assyrian age, vividly illustrating how the Assyrian war machine actually worked." While the Assyrians revolutionized warfare with the first use of mobile siege towers, wall-scaling ladders, battering rams, and tunnelers, cavalry was their war-winning weapon and, like the pale horse, their messenger of death.

Originally a city-state dissected by the Tigris-Euphrates valley and important trade routes, Assyria seemingly always attracted the lusting gaze of the major imperial powers of the second millennium BCE: the Egyptians, Hittites, and Mittani. As vassals of the Mitanni, the Assyrians acquired a chariot tradition and then used it, in conjunction with adopted Egyptian battle tactics, to defeat the Babylonians, Hittites, Mittani, and smaller polities between 1300 and 1230 BCE.

Assyria ruthlessly filled the power vacuum created by the Sea People and the collapse of the Hittite Empire. Around 900 BCE, the resurrection and consolidation of the traditional Assyrian homeland centered on the three major cities of Nimrud, Nineveh (Mosul), and Ashur—all situated on the Tigris in northern Mesopotamia—began to take shape.

With the aid of the cavalry horse, Assyria rose from the ashes as the phoenix of the Iron Age to restore its former glory—and then some. Under the direction of adroit kings, the Assyrians carved out an impressive empire unprecedented in size, scope, and sophistication to obtain vital raw materials (notably wood and iron) and pastures for their horses, the heartbeat of their unrivaled military. Assyrian society was based on military might and imperial conquest to feed its insatiable appetite for the spoils of war. The stunning martial, academic, and engineering exploits of the vaunted Assyrians and their eminent kings have gained historical renown through a substantial and varied collection of ancient inscriptions and texts, including the Old Testament.

The imperial period of the Neo-Assyrian Empire witnessed the reign of six influential monarchs: Ashurnasirpal II (883–859 BCE); his son Shalmaneser III (858–824 BCE); Tiglath-pileser III (744–727 BCE), followed by the greatest of all Assyrian kings, Sargon II (721–705 BCE).* He was succeeded by his son Sennacherib (704–681 BCE) and then by his great-grandson Ashurbanipal (668–631 BCE). At the height of its power under Ashurbanipal, the Assyrian Empire stretched some 540,500 square miles, from southern Anatolia in the north, down the biblical rivers to the Persian Gulf in the south, and from the Egyptian Nile delta in the west to Persia in the east.

The Assyrians unabashedly built their power base and economic dominance through fear, violence, and brutality. They utilized deportation, slavery, forced marches, rape, mutilation, and gruesome torture. Ruthless secret police and elite militias—fed information and targets by an extremely efficient intelligence system—conducted methodical and coordinated terror campaigns. Even by ancient standards of warfare and cruelty, the Assyrians were especially sadistic. Their ultimate power, however, rested with their warhorses.

While there is preceding evidence of horseback riding in a military context, the earliest known true cavalry formations belonged to the Assyrian king Tukulti-Ninurta II (890–884 BCE). These small, mounted detachments were initially tasked with acting as the supporting "eyes and ears" for the main infantry forces through patrolling and reconnaissance. His initial operations northward into the horse-rich lands of Urartu, centered in Anatolia and the southern Caucasus, supplied his cavalry with superior mounts standing roughly fifteen hands. These imported horses, most likely bred and shuttled south by the nomads of the steppe, were unparalleled in the southern reaches.

The conventional cavalry practiced by the Assyrians was a sophisticated enhancement of the individualistic and fluid mounted raiding tradition established beyond the Assyrian gaze by the horsemen of Urartu

---

* Chosen in homage to Sargon of Akkad.

The ascent of cavalry: Assyrian relief at Nimrud depicting horses and riders, circa 728 BCE. *(British Museum/Robert Harding Images)*

and their northern nomadic Indo-European Cimmerian and Scythian neighbors. "Hard-learned strategic lessons gradually passed southward from the steppes into chariotry circles," notes archaeologist Brian Fagan. "Once convinced of the value of mounted warfare, the Assyrians learned quickly." So did others.

To counter Assyrian cavalry incursions, Urartu quickly raised formal mounted units of its own. Following a raid against Urartu in 714 BCE, Sargon II conceded, "The people who live in that province of Urartu are all very able in matters of cavalry, and there are none equal to them." Well, none other than the Assyrians, that is.

The reasons for this initial contrast between the casual, organic, pseudo-cavalry of the steppe and the calculated, regimented cavalry of the Assyrians was due largely to environmental differences. Farming practices in the river valleys of the Middle East demanded hydraulic and irrigation technologies associated with higher population densities. Regulating the masses to maximize agrarian production required stringent and complex sociopolitical structures, including centralized bureaucracies, sophisticated transportation networks, and professionalized militaries. These rigid attributes of sedentary society were far more inflexible and draconian than what was needed on the open forage pastures and nomadic hunting grounds of the Eurasian Steppe patrolled by relatively small, free-ranging populations.

The disciplined organizational scaffolding that permeated Assyrian culture was applied to all components of the military. "Altogether, a radical rationality seems to have pervaded Assyrian military administration, making their armies the most formidable and best disciplined the world had yet seen," states acclaimed historian William H. McNeill. "Yet it seems no exaggeration to say that the fundamental administrative devices for the exercise of imperial power which remained standard in most of the civilized world until the nineteenth century AD first achieved unambiguous definition under the Assyrians." When measured against these benchmarks, the modern blueprint for centralized governance and empire building was assembled by Assyrian warhorses.

The Assyrian conversion from chariotry to cavalry, however, was gradual. Like all else, this transition was guided by trial-and-error evolution, practice, and the training of both horse and rider. The horse and weapon are only as effective as their human handler. "Cavalry expertise improved from the two-man team of warrior and horse-handler, both seated precariously behind the horse's motion, to the cavalry of Sargon II and later kings which were seated independently in a more balanced and secure fashion," explains Ann Hyland. "Using this balanced seat cavalry was now wholly a strike force, with each rider a combat trooper. It lessened the need for horses per unit, or alternatively allowed the strike force to be doubled."

Initially, Assyrian cavalry fought in pairs. The archer or swordsman always rode the horse on the right, allowing for freedom of movement with his right weapon hand, while his companion on the left held a shield and the reins of both horses. Early Assyrian cavalry was certainly not the dashing and debonaire "light cavalry" of imagination.

The arrival of an effectively independent and highly mobile Assyrian mounted warrior was tied directly to the dramatic augmentation in Assyrian horse stocks acquired through tribute and raids. Sargon, for example, demanded that his subjects deliver their equine excises to a specific collection point on the first of the month. "Should even one day pass by, you will die," he wrote of the literal deadline. "Whoever is late will be impaled in his own home, and his sons and daughters too will be slaughtered." Cavalry horses were a serious, life-or-death business.

Exemplified by the forays of Tukulti-Ninurta and Sargon, the Assyrians specifically targeted steppe horses from distant fields beyond the Zagros Mountains of northern Anatolia and Persia. These "mighty steeds" were legendary and referred to by Herodotus as the "sacred 'Nisean' horses, named after the huge Nisaean Plain in Media which produces these tall horses." Although the exact location of the Nisaean Plain remains a mystery, it can be safely said that these horses, the most valuable south of the steppes, came from the lands of Urartu and Media.

The grazing fields south of the Median capital of Ecbatana (now Hamadan, Iran) were awash with alfalfa, also called lucerne. This clover relative, known locally as Median grass, contains upwards of 13 percent digestible protein, roughly double that of most hays. It is also rich in vitamins and minerals, including high levels of iron and calcium. This exceptionally nutritious and widely available fodder explains the muscular, robust, and athletic descriptions of Nisean horses, which averaged between fifteen and sixteen hands.

When Alexander the Great visited the renowned Nisean herds in 324 BCE upon his return from India, he counted roughly 50,000 to 60,000 horses grazing on the plain. This represented a fraction of the estimated herd peak of 150,000 during the high point of the Persian Empire. Now thought to be an extinct breed, these horses, known to the Chinese as *Tianma*, or "Heavenly Horse," were the elite cavalry mounts of the ancient world.

Given their ubiquitous presence across our epic battle-scarred human saga, we intrinsically associate horses with great exploits of military history. We must remember, however, that horses do not come born, built, broken, and trained for war. They are instinctively temperamental and skittish animals, evolved for *flight*, not fight! Training a horse to charge headlong into a whirlwind of destruction, smoke, noise, crowds, and chaos is no easy feat. The innate evolutionary follow-the-leader attribute alluded to earlier goes only so far. Tutoring and drilling a horse to become a fearless, finely tuned weapon of war takes time, effort, and the guidance of a trained hand. Kikkuli and his ancient contemporaries certainly understood this challenge.

Assyrian texts reveal the existence of horse masters like Kikkuli who were skilled in the arts of breeding and training. This must have also been of paramount concern for all cultures seeking to establish professional cavalry. In his seminal manual *On Horsemanship*, written around 365 BCE, the Greek soldier and author Xenophon proffered guidelines on how to train a horse to navigate the swirling maelstrom of battle, advising: "You must also tell your groom to take the colt through crowds and to familiarize it with all kinds of sights and sounds; and if the colt finds any of this alarming, the groom must not lose his temper with it, but should calm it down and gently teach it that there is nothing to be afraid of." Horses, like most humans throughout our long history of violence, had no choice in the matter.

Securing horses through plunder and tribute from conquered territory was a primary Assyrian war aim. There was a constant need for horses to expand the cavalry branch and to replace those lost through combat, disease, and natural decay. These levies and plunders, particularly against northern kingdoms, netted thousands of horses. Tiglath-pileser III captured five thousand Nisean horses in a single campaign against the Medes, while acquiring another two thousand in disciplinary fines from a subjugated king whose only crime was that, apparently, he "was indifferent towards Assyria's achievements." While conducting more than fifty military operations, Tigleth-pileser also resettled or exiled over six hundred thousand deportees.

An independent ministry of "horse recruitment officers" called *musarkisus* was established to obtain and manage horses (and mules and camels) and provide for the needs of the cavalry.* In their prestigious

---

* It is thought that the Assyrians were the first to use the camel for military purposes. The two-humped Bactrian camel, domesticated around 4,500 years ago in Afghanistan/Turkmenistan, is an evolutionary marvel, able to withstand temperature extremes from -40°F (-40°C) to 109°F (43°C). In frigid cold and snow, it is protected by its thick, woolly coat, it can "drink" snow, and it can fuel itself from the fat stored in its humps. A 1,300-pound camel can drink 50 gallons, or 417 pounds (190 liters) of water in three minutes! They can lose up

position, the musarkisus were responsible for maintaining the imperial system of stables, corrals, and feed. Assyrian writings also hint at rudimentary equine veterinary services.

The "Horse Reports" from Nineveh, a series of twenty letters spanning a three-month period during the reign of King Esarhaddon, circa 680 BCE, register 2,654 horses entering the city through tributes, trades, and raids. These imported horses were then handpicked and branded into specific roles by the musarkisus: 1,840 chariot horses, 787 cavalry mounts, and 27 breeding studs. As demonstrated by this horse specialization, the battle-hardened Assyrian military had matured into the most sophisticated, formidable, and effective fighting force of its time. It was also massive.

The professional Assyrian army under Sargon II fluctuated between one hundred thousand and two hundred thousand soldiers (possibly upwards of three hundred thousand with the total mobilization of conscripts and foreign mercenaries) and twenty thousand horses across various military occupations. It was the largest the world had seen, roughly doubling that of runners-up the Egyptians under Ramses II, 550 years earlier. Assyrian soldiers were also completely outfitted with iron weapons, helmets, and armor, making them the first "iron army" in history.

The specially forged weapons room in the armory at Sargon's newly constructed royal complex at Dur-Sharrukin contained more than two hundred tons of iron weaponry. During his sixteen-year reign, Sargon II conducted at least ten major campaigns. The biblical "weeping prophet" Jeremiah describes these sweeping offensives with chilling accuracy:

---

to 25 percent of their body weight to dehydration before potential negative effects set in, compared with 12 percent for most mammals. In blistering summer climates, a camel can go for months without water, and seal its nostrils with a skin flap during frequent desert sandstorms, while a third eyelid and two sets of eyelashes dislodge sand and dust from each eye. Its feet are made of durable, tough soles and can splay widely to accommodate both sand and rock, allowing it to navigate the desert floor with ease, stability, and impressive, albeit clumsy, dexterity. Camels can carry loads of more than 550 pounds over distances of twenty-five to thirty miles a day. The dromedary, or Arabian camel, possesses one hump and was domesticated (independent of the Bactrian camel) farther south in Somalia or the Arabian desert between four thousand and five thousand years ago.

"Behold, he shall come up like clouds, And his chariots like a whirlwind. His horses are swifter than eagles. Woe to us, for we are plundered!"

While Assyrian kings were the destroyers of worlds, they were also great builders of infrastructure and sprawling royal complexes, palaces, and temples, interconnected by an extensive network of roads (including the main Royal Road) and way stations utilized by efficient mounted couriers. It is estimated that it took no more than five days to deliver a message across the farthest reaches (some 450 miles) of the Assyrian Empire. "The speed and frequency with which messages were sent back and forth throughout their empire," recounts Frahm, "was nothing short of revolutionary in its time."

Numerous kings relocated the capital to their own custom-built cities. When Ashurnasirpal II, the sadistic flaying king mentioned earlier, crafted his own nine-hundred-acre capital complex enclosed by five miles of defensive wall at Nimrud, it took fifteen years to complete. "The palace of cedar, cypress, juniper, boxwood, mulberry, pistachio wood, and tamarisk, for my royal dwelling and for my lordly pleasure for all time, I founded therein. Beasts of the mountains and of the seas, of white limestone and alabaster I fashioned and set them up on its gates," he boasted at the grand opening in 879 BCE. "Silver, gold, lead, copper, and iron, the spoil of my hand from the lands which I had brought under my sway, in great quantities I took and placed therein." To share in his commemorative celebration and parade his wealth and power, the king invited 69,574 of his closest friends and family to indulge in a ten-day housewarming party featuring an all-you-can-eat smorgasbord buffet and a fully stocked open bar.

The elaborate complex at Nimrud remained the capital of Assyria for 150 years, until Sargon II erected his own capital at Dur-Sharrukin. Following his death in 705 BCE while leading a cavalry charge in Anatolia, his son and successor, Sennacherib, immediately refurbished and expanded the ancient city of Nineveh, including the construction of an archival library, to serve as the royal seat. To irrigate his lavish, verdant garden at Nineveh (most likely the actual Hanging Gardens of Babylon, one of the Seven Wonders of the Ancient World), Sennacherib invented the water screw (Archimedes' screw) technology using a pioneering technique for bronze metal casting.* Of course, the bills for such extravagant construction projects were all covered by the spoils of war, courtesy of the cavalry horse.

Sennacherib is arguably one of the most notorious Assyrian kings, achieving biblical immortality for his protracted military offensive against Jerusalem in 701 BCE and for brutally destroying the pesky city of Babylon twelve years later. "With their corpses I filled the city squares. . . . The city and its houses, from its foundation to its top, I destroyed, I devastated, I burnt with fire," Sennacherib recorded proudly. "Through the midst of that city I dug canals, I flooded its site with water, and the very foundations thereof I destroyed. I made its destruction more complete than by a flood. That in days to come the site of that city and its temples and gods might not be remembered, I completely blotted it out with floods of water and made it like a meadow."

During his earlier cordon of Jerusalem, Sennacherib attempted to badger Hezekiah, the king of Judah, and his subjects into battle: "The king of Assyria wants to make a bet with you people! He will give you two thousand horses, if you have enough troops to ride them. How could you even defeat our lowest ranking officer, when you have to depend on Egypt for chariots and cavalry? Don't forget it was the Lord who sent me here with orders to destroy your nation! . . . These people, like you, will soon have to eat their own excrement and drink their own urine." Sennacherib and his futile siege of Jerusalem were further consecrated in Lord Byron's "galloping," rhythmic 1815 poem "The Destruction of Sennacherib," narrated from the Hebrew perspective, with its famous opening line, "The Assyrian came down like the wolf on the fold."

The legend of Sennacherib was sealed when he was assassinated by two of his sons in 681 BCE, shortly after his renovation of Nineveh. His grandson, Ashurbanipal, is generally regarded as the last great king of the Assyrian Empire before its downfall and the sack of Nineveh in 612 BCE. Over the course of three decades, he launched more than thirty

---

* One of the earliest hydraulic machines used to transport water up a gradient to higher elevation.

Riders on the storm: Ashurbanipal on horseback, from a relief at Nineveh, circa 645 BCE. *(Zev Radovan/Alamy Stock Photo)*

military campaigns to put the finishing strokes on restoring traditional Assyrian territory.

Like his grandfather, Ashurbanipal was also immersed in academia and invention. "I solved complicated mathematical problems that have not even been understood before," he inscribed. "I read the artfully written texts in which the Sumerian version was obscure and the Akkadian version for clarifying too difficult. I am enjoying the cuneiform wedges [writing] on stones from before the flood." In keeping with his thirst for knowledge, the scholarly king oversaw the creation of the first systematically organized library in history, known as the Royal Library of Ashurbanipal.

The Assyrian monarch regarded his remarkable library at Nineveh, which contained more than thirty thousand clay tablets from across the empire, including *The Epic of Gilgamesh* (a foundational religious text written as a heroic saga, and perhaps the oldest surviving literary work on earth), as his crowning achievement and understood the historical magnitude and eternal moment of his immense archive: "I, Ashurbanipal, king of the universe, on whom the gods have bestowed intelligence, who has acquired penetrating acumen for the most recondite details of scholarly erudition (none of my predecessors having any comprehension of such matters), I have placed these tablets for the future in the library at Nineveh for my life and for the well-being of my soul, to sustain the foundations of my royal name." Although it appears his motivations were not entirely altruistic, as he was also attempting to secure his legacy, we nevertheless owe Ashurbanipal a debt of gratitude for his progressive and enlightened thinking.

The value of his universal library to future generations cannot be overstated. It contained texts on everything from medicine, politics, astronomy, and botany, to warfare, engineering, spirituality, and trade. In addition to tablets, Ashurbanipal also collected scholars from across his domain to stock and oversee his magnificent library. In his 1920 book *The Outline of History*, author H. G. Wells christened it "the most precious source of historical material in the world." The vast anthology of human history and the living echoes of ancient voices contained in this extraordinary collection were acquired across the far reaches of the Assyrian Empire conquered by the first horse-charged cavalry. Not only did the horse propagate language, but also it preserved history.

I suppose the library stands as small penance for the otherwise ruthless Assyrian era. No empire lasts forever, and Ashurbanipal might have realized when he was reciting his earnest library dedication that the end was near. It turned out to be a presaging requiem of sorts. Like the Roman Empire almost a thousand years later, maintaining a centralized seat of power over a strained multiethnic empire proved too much for the Assyrians. As their influence waned, compounding pressures hastened their collapse.

After Ashurbanipal's death in 631 BCE, the Assyrian Empire faced escalating threats on three simultaneous fronts. Palace intrigue, domestic strife, and tenuous power struggles between competing political parties erupted into a series of bloody conflicts of succession that spilled over into broader civil wars. The draconian rule of the Assyrian regime over its subjugated peoples eventually ignited a string of popular uprisings. The eastern borders of the Assyrian heartland were breached by the Medes and Persians; wild Cimmerian and Scythian horse warriors from the frontier steppe roamed freely across the former empire, pillaging as they pleased. Eventually the walls of Nineveh, and its library, came crumbling down. Twenty years after his death, Ashurbanipal's great library was buried under the charred and smoking rubble of his burning palace at Nineveh, lost to history for 2,500 years.*

The vengeance wrought upon Nineveh in 612 BCE by a motley coalition of Medes, Babylonians, Persians, Cimmerians, and Scythians was captured in vivid prose by the biblical prophet Nahum:

> Doom to the crime capital! Nineveh, city of murder and treachery, here is your fate—cracking whips, churning wheels; galloping horses, roaring chariots; cavalry attacking, swords and spears flashing, soldiers stumbling over piles of dead bodies. You were nothing more than a prostitute using your magical charms and witchcraft to attract and trap nations. But I, the Lord All-Powerful, am now your enemy. I will pull up your skirt and let nations and kingdoms stare at your nakedness. I will cover you with garbage, treat you like trash, and rub you in the dirt. Everyone who sees you will turn away and shout, "Nineveh is done for! Is anyone willing to mourn or to give her comfort?"

Assyrian brutality and sadism were apparently met in kind. "So complete was the destruction that two centuries later, Xenophon and his Greek mercenary army of ten thousand men passed the ruins of Nineveh unaware of what they were passing," reports Richard Gabriel. "Not a single vestige of the Assyrian power remained. A people who had lived on the Tigris for more than two thousand years had literally disappeared from the face of the earth." The ruin of Nineveh was the turning point

---

* The first remnants of the library were uncovered in 1850. The vast collection of tablets is now housed in the British Museum. Unfortunately, many of the excavated, standing Assyrian sites in Iraq, including Nineveh, were systematically destroyed by the terrorist organization ISIS (Islamic State of Iraq and Syria), in 2015. To fund its militant campaign, ISIS also illicitly sold priceless Assyrian artifacts to unscrupulous private collectors.

in the unmitigated collapse of Assyria, with the mantle of power passing to the Scythians and Persians.

By pressing their borders and prowling northward to procure horses for their daunting cavalry, the Assyrians had collided with the advance guard of nomadic Indo-European Cimmerian and Scythian riders driving south from the steppe. By 700 BCE, the northern kingdom of Urartu, a favorite target for Assyrian horse raids, lost its frontier lands to the Cimmerians, who had encircled the Black Sea, breached the Caucasus, and usurped large portions of Anatolia.

During the early seventh century BCE, the Cimmerians were quickly chased off by semi-cohesive bands of nomadic predatory Scythians, who charged southward through the Caucasus from the Pontic-Caspian Steppe in search of pastures, plunder, and adventure. These skilled mounted warriors entered the fray with lightning speed. "The Scythians prefer mares as chargers," wrote the Roman historian and philosopher Pliny the Elder, "because they can urinate without checking their gallop." After a brief pause in the fertile grazing grounds in central Anatolia, brazen groups of Scythians continued southwest across the biblical rivers, growing fat off the decaying carcass of the Assyrian Empire en route to the Levant and Egypt.

From their homeland on the Pontic-Caspian Steppe, smaller columns of Scythians also trotted west toward Europe. They penetrated the Carpathian Mountains and punched through Romanian Transylvania to pasture on the Hungarian Plain. In his comprehensive account *The Scythians: Nomad Warriors of the Steppe*, Barry Cunliffe portrays these roving groups as "a pioneer force of Scythians riding westward . . . following routes already opened up by horse riders from the steppe two or three centuries before." Scythian-style military artifacts from the sixth and fifth centuries BCE have been found as far west as Austria and Germany.

While seldom mentioned in popular narratives or textbooks, the Scythians, and their multidimensional impact on the historical and cultural trajectory of Eurasia from Europe through China, deserve better. With the wonders of their culture and through their hobby of war, they changed the world from the back of a horse. "In fact," stresses

Christopher Beckwith in *The Scythian Empire: Central Eurasia and the Birth of the Classical Age from Persia to China*, "the Scythian Empire is one of the least known but most influential realms in all of world history." While they are relatively absent from our modern literary record, they did inspire the imagination and ink of the ancients.

According to Greek mythology, the Scythians were the descendants of the hero Scythes. When the demigod adventurer Hercules was driving cattle through what later became known as Scythia (Pontic-Caspian Steppe), his horses went missing. While searching a cave, he found them in the possession of an *echidna* (she-viper), a creature with the upper portion of a woman and the lower portion of a snake. She promised to return the horses upon his completing one simple request: Hercules had to fornicate with her. Of course, he consented, and their serpentine passion produced three sons.

When asked what should become of them, Hercules handed her a bow and a belt. When they came of age, he explained, whichever son could string and draw the bow and wear the belt should be raised, while the other two were to be banished.* The youngest son, Scythes, fulfilled the prophecy and became the founder and eponym of the Scythians. "From Scythes, the son of Heracles," declares Herodotus, "the kings of Scythia descend." The mythical story is telling for its inclusion of the two central pillars of Scythian culture and military power: the combination of the horse and the bow.

Prior to Herodotus, earlier Greek historians were fascinated by the mysterious Scythians. In his book *Journey Around the Earth*, Hecataeus (c. 550–476 BCE), the first known Greek historian, described both the fierce Celts of southeastern France and the barbarian Scythian "mare-milking nomads" of the steppe beyond the Black Sea, as did Hellanicus of Lesbos (c. 490–405 BCE) in the chapter on Skythika in his volume *Barbarika Nomina*.† As Barry Cunliffe explains, "Scythian archers were

---

* According to medieval Irish chroniclers, the exiled sons gave rise to the Picts of Scotland.
† The term *lesbian* is derived from the Greek island Lesbos, home to the sixth-century BCE poet Sappho, whose name survives in the English word *sapphic*. Her poetry often depicted the lives, rituals, and beauty of women as well as her admiration, love, and attraction toward females.

frequently depicted on Attic Black-Figured pottery, and historians like Herodotus recorded stories from their history exploring, with undisguised delight, their unusual behavior and beliefs.... Yet to the Greeks, it was just this that made them so fascinating. And rightly so."

Herodotus possessed a relatively knowledgeable understanding of the homeland of the Scythians, describing it as "level, well-watered, and abounding with pasture." It was intersected by eight major rivers, which he names, all flowing into the Black Sea. Traditional Scythian territory extended from the Danube River in the west, across the Pontic-Caspian Steppe, to the Don River in the east.

In southern Siberia, at the international crossroads of Russia, Mongolia, Kazakhstan, and China, the multi-tomb kurgan complexes of Arzhan and Pazyryk reveal the geographical extent, cross-cultural steppe trade, and intricacies of Scythian society shaped by the evolving revolution in equine transport and mounted warfare. These earthen and timber vaults, spanning roughly five hundred years, run like braille dots across the map of the Eurasian Steppe, extending west through Ukraine into Romania.

Scythian trading caravans also trundled and clattered east toward Mongolia, and, *seemingly*, vanished into the misty void for the next two hundred years. They reemerged around 300 BCE as the Xiongnu, a revamped nomadic civilization based on expert mounted raiding and, as we will see, introduced themselves to the neighborhood by climbing the steppes and violently knocking on the door of northern China.

These Scythian burial mounds are literal gold mines, containing sizable collections of stunningly beautiful and artistically crafted gold artifacts, from belt buckles, hair combs, and vessels, to plaques, sculptures, and weapon ornaments. Many of these relics were salvaged in the wake of Tzar Peter the Great's imperial expansion into the sprawling region of historical Scythia between 1682 and 1725.

Although Russian penetration into these areas began during the mid-1500s under the legendary reign of Ivan the Terrible, it was Peter who consolidated Russian rule over these fringe outposts of empire. The kurgans and rotunda burial mounds stimulated the curiosity (and grave robbing) of Russian migrants and rambling explorers. For example,

Nicolaes Witsen, the Dutch ambassador to Tzar Alexis I (Peter's father), amassed a stunning collection—and a fortune—of golden Scythian grave goods. "How civilized must have been the people who buried these rarities," he remarked in 1664. "The gold objects are so artfully and sensibly ornamented that I do not think European craftsmen could have managed better."

Peter was an intellectual, erudite man who sought to refashion Russia into a modern empire through a series of reforms, including the establishment of a formidable navy. As an educated man and a rabid student of the Enlightenment and the Scientific Revolution, Peter was also an avid and eclectic collector of art, curiosities, and ancient artifacts—notably those crafted by the Scythians from his newly annexed territories.

To curry favor, politicians and wealthy patrons flooded Peter's inbox with golden Scythian and Siberian treasures, and he quickly amassed a striking collection. He also passed edicts to protect and secure Scythian relics, including a death sentence for anyone "searching for golden stirrups and cups." He ordered all regional governors and administrators to

Dying hero with horses under the tree of life: A Scythian gold belt buckle from the collection of Peter the Great, fourth century BCE. Notice the bow case/quiver hanging from the tree. *(State Hermitage Museum, Saint Petersburg, Russia/Alamy Stock Photo)*

Scythian warrior: Gold plaque from the Kul'-Oba kurgan, Ukraine, fourth century BCE. *(State Hermitage Museum, Saint Petersburg, Russia/Alamy Stock Photo)*

collect "from earth and water . . . old inscriptions, ancient weapons, dishes, and everything old and unusual."

After a series of homes, since 1860 his trove of over 250 Scythian objects has been displayed as the Siberian Collection of Peter the Great in the Gold Room of the Hermitage Museum in Saint Petersburg, the city he established in 1703 to serve as Russia's cultural capital. Peter's contribution, along with several excavated Scythian graves—among them those at Arzhan and Pazyryk—provides invaluable historical context to the events unfolding across Eurasia and the Middle East in the centuries after the fall of the Assyrian Empire.

The Arzhan valley contains more than three hundred kurgans of various sizes arranged in parallel chains, styling it as the Siberian "Valley of Kings" to rival that of Luxor (Thebes), Egypt. The largest, 360 feet in diameter and over 13 feet high, was constructed at the turn of the ninth and eighth centuries BCE (roughly a century before Homer wrote the *Iliad* and the *Odyssey*) with more than six thousand larch logs. These trees were roughly a hundred years old when felled and were carefully shaped and trimmed. It would have taken three hundred people over a month just to process this timber! Within a labyrinth of seventy chambers, the central grave contained the remains of a man and a younger woman, presumably a sacrificed consort or wife, positioned on a bedding of twenty horse tails.

This massive tomb also contained the remains of fifteen attendants and 160 horses. These sacrificed companions were saddled, bridled, and outfitted in full regalia, with a dazzling assortment of antlered masks and gold, bronze, and turquoise ornaments. The style and cultural attributes of the horse gear and other grave goods suggest that they were tributes from considerable distances, indicating that the king commanded sway and allegiance over a vast domain. Another three hundred horses (along with cattle, sheep, and goats) were eaten in a lavish feast for a large, diverse contingent of mourners from the far reaches of the Scytho-Siberian realm.

A slightly smaller kurgan dating to the mid-seventh century BCE held the bodies of a royal couple adorned in ornate clothing decorated with golden beads, cuffs, and panther figurines. The sprawling, multi-chambered barrow contained more than 9,300 objects—of which 5,700 were made of gold weighing a combined forty-five pounds. Other grave goods were made of turquoise, garnet, malachite, and more than 400 beads fashioned from Baltic amber originating 2,500 miles to the west. The interment also contained sacrificed horses and servants, customary to Scythian burials—including those farther west at Pazyryk, where the breadth of horse-powered international trade and cultural fusion is on full display.

Straddling the fifth to third centuries BCE, the twenty-five kurgans at Pazyryk contain an impressive array of imported artifacts and goods. Inside were: silk, cloth, and a striking canopied carriage from China; mirrors and cotton fabrics from India; masks and a harp of Greek origin

Pazyryk horseman: A felt wall hanging from a kurgan burial depicting an elegant pants-wearing man riding a well-groomed horse, circa 300 BCE. *(State Hermitage Museum, Saint Petersburg, Russia/Alamy Stock Photo)*

or design; and finely crafted carpets, tapestries, and felt hangings from Persia and Armenia, including the often-depicted, pants-wearing Pazyryk Horseman and the handmade Pazyryk Carpet.*

These kurgans were initially excavated by Soviet archaeologist Sergei Rudenko between 1947 and 1949. According to M. W. Thompson, the translator of Rudenko's book *Frozen Tombs of Siberia: The Pazyryk Burials of Iron Age Horsemen*, "the Persian and Chinese textiles at Pazyryk are older than any surviving examples in Persia or China. . . . [T]he horses which had been buried adjoining them in the same shaft, together with their saddlery, were in all cases undisturbed, and thus constitute far and away the richest archeological remains throwing light on the early history of horse-riding."

---

* Measuring six by six feet, it is woven with images and stylized designs, including warriors on horses. Hand stitched around 400 BCE, it is likely the oldest surviving pile-style carpet in the world, with knot density higher than most modern luxury productions.

The cluster of kurgans at Pazyryk, including that of the tattooed Ice Maiden mentioned earlier, also contain cannabis seeds and smoking paraphernalia. An additional five bodies, all wearing pants, are also tattooed to varying degrees. The "Pazyryk Chief" is inked across his entire body with intricate circular shapes and elaborately drawn animals both real and imaginary. Personal possessions uncovered include a pair of leopard-skin boots and stringed musical instruments!

The magical mystery tour of Scythian song and dance, which Herodotus associated with marijuana use, was not welcomed by everyone. One Scythian king grumbled that he greatly preferred the neighing, nickering, and snorting of his horse to the strumming music and stoned caterwauling of his companions.

The numerous entombed Pazyryk horses are represented by two types, suggesting that many had been acquired, like those at Arzhan, through trade and tribute from a wide berth. Some are stocky local horses (resembling the Przewalski's horse) standing twelve to fourteen hands. Others are the prized, long-legged Turkoman horses standing fifteen to sixteen hands.

Possessing a shiny coat with a glistening, metallic sheen, giving them the nickname "Golden Horses," Turkomans are tough and hardy mounts, gaining a reputation for both speed *and* endurance. Russian records reveal that a Turkoman horse covered sixty-six miles of desert terrain in 4 hours and 30 minutes, while another traveled thirty-three miles in 1 hour and 58 minutes. With careful selective breeding, the Turkoman horse has influenced several modern breeds, including the Akhal-Teke and the Thoroughbred. This extraordinary international mosaic of horses and items found in Scythian kurgans on the remote steppe of Siberia was corollary to the widespread shift from chariotry to cavalry.

The increasing demand for steppe horses by the imperial armies and urbanized centers of China and the Middle East generated extensive transcontinental trade. Contemporary records indicate that at the height of the horse trade, between fifty thousand and one hundred thousand animals per year were being exported from the steppe to China and India, and through the Nisaean Plain of Persia to the Middle East. The tomb of Duke Jing of Qi (roughly modern-day Shandong Province) in

China, dating to 490 BCE, for instance, contains the sacrificial remains of more than six hundred imported horses. This lavish burial is easily the largest horse sacrifice site discovered in China and represents what would have been a vast fortune in equine capital. Steppe horses were shipped and swapped across Eurasia in exchange for exotic goods. The modern world was beginning to take shape.

The natural by-product of heightened trade was war, which was progressively being carried out by men on horseback, or, in the case of the Scythians, men *and* women on horseback. "The horse was the great equalizer of males and females on the steppes, probably one of the chief reasons for the nomads' noteworthy gender equality," writes Adrienne Mayor in *The Amazons: Lives and Legends of Warrior Women Across the Ancient World*. "A skilled archer horsewoman could hold her own against men in battle. Riding horses liberated women. . . . [T]he steppe culture was the perfect environment for women to become mounted hunters and fighters." Accordingly, Scythian graves are also occupied by esteemed female warriors.

A revealing Scythian crypt from 500 BCE in southwest Russia contained four women with weapons, golden headdresses, and other jewelry representing three generations of female warriors. The youngest was twelve to thirteen; two were in their prime, between twenty-five and thirty-five; and the fourth was between forty and fifty years old, a "respectable age" for the time. "Such conditions of relative female empowerment existed," explain Indo-European linguists Asya Pereltsvaig and Martin Lewis. "It is clear that we cannot simply assume overwhelming male domination based on pastoralism and military prowess."

Roughly 30 percent of all excavated Scythian female burials, which number more than three hundred, contained horses and various weapons. Several women bore unmistakable battle wounds, including battle-axe cleaves and sword slashes. The Scythians "were notable for the great prominence of women in general and especially for the presence among them of women warriors," says Beckwith. "The unusual status of women . . . was noticed by Herodotus and has received solid confirmation by archeology."

These Scythian women are the famed "Amazon" horse warriors that captivated—and perhaps confused—Homer, Herodotus, and their fellow Greeks among other ancient scribes and artisans. As the Greek orator Lysias reminded the Athenians in 392 BCE, "Amazons were the first people to ride horses."* A mid-fifth-century BCE Greek medical textbook, *On Airs, Waters, and Places,* often attributed to Hippocrates but likely written by a conversant contemporary, reveals:

> In Europe there is a Scythian race... different from all other races. Their women mount on horseback, use the bow, and throw the javelin from their horses, and fight with their enemies as long as they are virgins; and they do not lay aside their virginity until they kill three of their enemies, nor have any connection with men until they perform the sacrifices according to law. Whoever takes to herself a husband, gives up riding on horseback unless the necessity of a general expedition obliges her. They have no right breast; for while still of a tender age their mothers heat strongly a copper instrument constructed for this very purpose, and apply it to the right breast, which is burnt up, and its development being arrested, all the strength and fullness are determined to the right shoulder and arm.†

Herodotus calls these women Amazons (*a-mazos* in Greek means "without breast") but warns that in the Scythian tongue they are known as *Oiorpata* ("man-slayers"). As we will see, Cyrus the Great, founder of the Persian Empire, would eventually suffer the fatal wrath and fury of a Scythian warrior-queen. There is no evidence, however, that they mutilated their breasts in any way. Nevertheless, 2,500 years later, this myth persists as fact in cocktail-party conversations.

---

* Many of the Greek names assigned to Amazons contain the roots *hipp-* or *-ippe* for horse: Alkippe (Powerful Horse), Ainippe (Swift Horse), Melanippe (Black Horse), Lysippe (Lets Loose the Horses), Philippis (Loves Horses), Hipponike (Victory Horse), Hippomache (Horse Warrior), Hippothoe (Mighty Mare), and Hippolyte (Releases the Horses). Other Amazon names reflect gender equality: Antibrote (Equal of Man), Isocrateia (Equal Power), Atalanta (Equal Balance), Antianeira (Man's Match), Andromache (Manly Fighter), and Polemusa (War Woman).

† Hippocrates: *Hippos* (horse) and *Kratos* (power).

The ancient Greek historian also tells us that many of these women died unmarried at an old age, suggesting that they were not willing to forsake their lifestyle and assume more traditional female gender roles within Scythian society. According to Mayor:

> In a nutshell: Amazons, the women warriors who fought Heracles and other heroes in Greek myth, were long assumed to be an imaginative invention. But Amazon-like women were real—although of course the myths were made up. Archaeological discoveries of battle-scarred female skeletons buried with weapons prove that warlike women really did exist among nomads of the Scythian steppes of Eurasia. So Amazons were Scythian women—and the Greeks understood this long before modern archaeology. And the Greeks were not the only ones to spin tales about Amazons. Thrilling adventures of warrior heroines of the steppes were told in many ancient cultures besides Greece. . . . Dramatic excavations of tombs, bodies, and artifacts illuminate the links between the women called Amazons and the warlike horsewomen archers of the Scythian steppes.*

While the self-righteous, patronizing Greeks pretty much labeled everyone as barbarians, including the Amazons, four groups were held in higher barbaric esteem than the remainder of the great unwashed masses: Celts, Persians, Libyans, and, of course, the Scythians.

There was also a sharp distinction between sedentary civilization and nomadic barbarism. As William Honeychurch explains:

> This dichotomy of cultures, often referred to as "steppe versus sown," is a powerful and implicit concept even today. . . . These stereotypes still support divisive and shortsighted

---

* Female warriors were not limited to the Scythians. Fu Hao was a top-ranking general in the army of her husband, Wu Ding, emperor of the Chinese Shang dynasty from 1250 to 1192 BCE. In command of more than thirteen thousand soldiers, including six hundred women, she was responsible for numerous military campaigns against neighboring mounted nomads.

government policies in traditional nomadic regions.... It is important to note, however, that both in the present and past, these ideas reflect not just cultural bias but real political concerns about the military power of nomadic groups in what were often politically important borderlands. If for nothing else, nomadic people are known for military capabilities that centralized states and empires, even in the twenty-first century, have found difficult to contend with.... For combatants acculturated in a tradition of one-on-one warfare, the nomadic use of hit-and-run tactics was considered devious, deceitful, and even cowardly—as well as highly effective.*

Mirroring the colonizing European attitude toward nomadic Indigenous peoples in the Americas, in the minds of Greeks, Persians, and Chinese, civilization (and its perceived advantages) was tied to fixed agriculture and static city-states. Peripatetic drifters and mounted nomads equated to barbarism.

Ancient Greeks also stereotyped the Scythians as drunkards—as well as obese, bow-legged, cannibals, eunuchs, impotent, and infertile.† In addition to their habit of smoking cannabis, Scythians preferred to drink their wine undiluted. In Greek culture, this was uncouth and distasteful behavior. This distinction between civilized Greeks and

---

* Recent NATO, American-led coalition, and Russian campaigns in Vietnam, Somalia, Afghanistan, Iraq, and Ukraine highlight the inability of superior technology and fighting strengths to combat nomadic asymmetric warfare fought by small units of largely independent guerilla forces and decentralized quick-strike "Scythian-style" insurgents. Inexpensive, homemade improvised explosive devices (IED) and cheap, readily available rocket-propelled grenades (RPG) are extremely effective weapons against the most sophisticated adversary and armaments.

† There may be some truth to these stereotypes. Routine and prolonged horseback riding can cause the testes to overheat, lowering sperm count. Recent studies have also shown that men who use marijuana more than once a week have a 30 percent reduction in sperm count, significantly less seminal fluid, and abnormalities in sperm production and performance through testicular atrophy. Chronic marijuana use can also cause erectile dysfunction, including impotency and flaccidity. It also increases food cravings by stimulating appetite receptors in the brain as well as triggering hunger hormones, leading to weight gain. The Scythians also rode horses or were pulled in traditional covered wagons instead of walking and had a high milk- and meat-based diet. It is also known that certain groups of Scythians practiced ritual cannibalism of their deceased male elders in an elaborate spiritual banquet ceremony.

uncivilized Scythians found its way into common lyrical prose written by the poet Anacreon (c. 582–485 BCE), who gained renown through his popular drinking songs and erotic passages: "Let's not fall into riot and disorder with our wine like the Scythians. But let us drink in moderation listening to lovely hymns." While these depictions of the savage, drunken Scythian were the product of Greek xenophobia, Scythians were, in fact, present across Greece in the flesh.

Small numbers of Scythian slaves trickled into Greece by the late sixth century, supplied to Black Sea emporiums by Scythian elites as a compelling reminder to their own people of their ability to wield power. Rogue Scythians were hired as mercenary mounted archers to augment Athenian infantry. Distinguished Scythian veterans were then mustered into special urban police detachments courtesy of the Athenian taxpayer.

Within Athens, Scythian officers also acted as capital guards to the popular assembly, or ecclesia. One story has these constables using ropes to corral reluctant citizens to vote. Scythian police also make an appearance in the dramatist Aristophanes's comedy *Lysistrata*, debuting in 411 BCE at the height of the Peloponnesian War, where their cowardice and drunkenness in the face of an angry mob of women yields easy laughs.

While the Scythians were scorned and mocked, the Greeks were savvy enough to afford them a begrudging martial respect. Thucydides makes it perfectly clear in his *History of the Peloponnesian War* (c. 410 BCE) that they were, in fact, no joke: "The Scythians, with whom indeed no people in Europe can bear comparison, there not being even in Asia any nation singly a match for them if unanimous, though of course they are not on a level with other races in general intelligence and the arts of civilized life."

Herodotus illustrates in blood-curdling detail the dangers of underestimating the Scythians:

> Here is how they conduct themselves in war. When a Scythian kills his first man, he drinks some of his blood. He presents the king with the heads of those he kills in battle, because his reward for doing so is a share of the spoils they have taken in the battle, but no head means no spoils. The

way a Scythian skins a head is as follows: he makes a circular cut around the head at the level of the ears and then he picks it up and shakes the scalp off the skull . . . which he proudly fastens to the bridle of the horse he is riding. The reason for his pride is that the more of these skin rags a man has, the braver he is counted. Many of them make coats to wear by sewing the scalps together into a patchwork leather garment like leather coats. . . . As for the actual skulls—the skulls of their enemies, that is, not all skulls—they saw off the bottom part . . . he gilds the inside and then uses it as a cup. . . . They also often skin the whole of a corpse and stretch the skin on a wooden frame, which they then carry around on their horses.*

As the Scythians encountered the Cimmerians, Assyrians, Babylonians, Medes, Israelites, and Egyptians, they quite literally carved out a ferocious reputation. The Persian armies of Cyrus and Darius I and the Macedonian forces under Philip II and his son Alexander the Great would soon observe the reality of this martial status at the sharp end of battle.

The Old Testament prophet Jeremiah issued a direct warning (c. 600 BCE) about the viciousness of these mounted warriors just as they were beginning their push through the Levant toward Egypt: "Behold, a people shall come from the north, And a great nation and many kings Shall be raised up from the ends of the earth. They shall hold the bow and the lance; They are cruel and shall not show mercy. Their voice shall roar like the sea; They shall ride on horses, Set in array, like a man for the battle." It did not take long for mounted Scythian archers to fulfill the prophecy. "Their insolence and oppression spread ruin on every side," declares Herodotus. While their quick-strike forays and whirlwind of chaos

---

* This account has been corroborated by later Chinese sources and through a substantial collection of archaeological evidence. Serving a similar cultural function to that of the Scythians, scalping existed among many pre-Columbian Indigenous nations of the Americas. For example, of the five hundred bodies of the Crow Creek massacre along the Missouri River in South Dakota dating to 1325 CE, roughly 90 percent of the skulls show evidence of scalping.

across the biblical landscape were fleeting, within the long view of history, their shock waves and influence were not.

Using their hit-and-run tactics honed on the steppe, they quickly obliterated the Cimmerians as an identifiable people, completed the annihilation of Urartu, and arrived in full command at the gates of Assyria.* Egypt saved itself for the time being with large tribute payments, Media was overrun after weakening itself against Assyria, and the Scythians marched on the Assyrian capital of Nineveh in 612 BCE.

Two years later, the vassal king of Media, Cyaxares (625–585 BCE), lured his overlord Scythian leaders to an extravagant banquet, where he wined and dined them, and, after they were properly drunk, murdered them, initiating a full-scale rebellion against Scythian rule. In his view, the Scythians got their just deserts for previously killing and serving up a local boy on his dining table after he'd ridiculed them for returning empty-handed from a hunt.

With a lack of central authority to coalesce the various factions and clans, and having already conquered and pillaged, the bulk of the grizzled Scythian veterans rode north over the mountains and plains, returning to their steppe homeland. While they left a trail of wreckage in their wake, including the destruction of the Assyrian Empire, they also left behind their expert knowledge of horsemanship and mounted warfare—including a piece of technology that would change the world: the saddle.

The earliest structured saddles are found in Scythian tombs dating to the early fourth century BCE. Prior to this, blankets and animal skins were used to reduce chafing and shield the rider from horse sweat, but these lacked any scaffolding or support. Intricately decorated early Scythian saddles consisted of two leather cushions stuffed with horse or deer hair, reinforced with thick "sweat padding" felt, attached to both sides of a wooden frame.

These innovations provided increased comfort and safety to both rider and horse, and a more even distribution of weight across the dorsal

---

* The Cimmerians, however, left a modern legacy in place names such as Cimmeria (the Crimea).

muscles and ribs, alleviating wear and tear on the spine. The saddle greatly improved the stability and security of riders as they delivered (or received) more powerful blows with handheld weapons, while boosting velocity, accuracy, and distance with projectile weapons, including spears and composite bows.

This rudimentary but effective saddle also provided the ability to execute what later came to be known as the legendary "Parthian shot." Essentially, while performing a real or feigned retreat at full gallop, the mounted archer would turn their body around with the wind at their back to shoot at the pursuing enemy. "In fact, because of the direction of the propulsive forces," explains Juliet Clutton-Brock, "it may be easier to stay on a galloping horse's back, without stirrups, when shooting an arrow backwards rather than forwards."

With both hands engaged with the bow and arrow, this contortion and twisting of the body while maintaining control of the horse without reins required dazzling equestrian skills and tremendous lower-body and core strength. The horse was controlled by squeezing pressure from the legs as stirrups were not yet invented. Over time, the term "Parthian Shot" was bastardized and corrupted into the phrase "parting shot" meaning "an insult or barbed comment issued as the speaker departs, walks away, or the conversation comes to an end."

The Scythians are also credited with the tactical invention of the wedge or V battle formation. This tight configuration was beautifully suited to the shock action of Scythian cavalry because it covers a narrow but piercing front with concentrated striking power and is easier to maneuver in close combat at top charging speed. The cavalry wedge punches through the center or anchor of the enemy line, widening the breach for follow-on forces while preparing the ground for flanking and enveloping pincer movements.

This indispensable alignment crafted by the Scythians on the Eurasian Steppe was perfected by Alexander's elite Companion Cavalry during his quest to realize immortality by reaching the eastern "ends of the world," and remains a standard military configuration. It is also applied by civilian police for "snatch squad" riot control. This Scythian battle formation even found its way onto your Sunday living-room

Parthian shot: Statue of a female Amazon warrior adorning an Etruscan cinerary urn, circa 500 BCE. *(Smith Archive/Alamy Stock Photo)*

television screens, from the grasses of the steppe to the turf of the gridiron stadium. Applying this military principle to contact sports, wedge blocking was used in American football before its recent ban under revised concussion protocols.

In addition to modern sports, the mounted legacy of the Scythians would be taken up by the Persians and subsequently by Alexander, both of whom had run-ins with the Scythians during their imperial ventures. "The Scythians, who were influential in the downfall of the Assyrians, rose to prominence in their place," says military historian at the US Army Command and Staff College Louis DiMarco. "In fact, they became the first of the steppe nomadic horse peoples to organize their significant military skills to undertake offensive operations against the sedentary empires of Europe, the Middle East, and Asia. Though they were the first, they would not be the last. . . . Cavalry would prove to be a crucial element in these historical confrontations."

In the wake of Assyria's downfall, western Eurasia and the greater Middle East were divided into three competing spheres of power: the Greeks/Macedonians, the Persians, and the Scythians. Farther east, the Warring States of China faced a troubling threat from the highly skilled mounted Xiongnu nomads from the northern steppe. As Xiongnu raids and tributary payments escalated, the Chinese, like the Greeks in the face of a Persian onslaught, were forced to put aside their differences to turn and face a more dangerous enemy with a united national front. The frontiers of Eurasia from Greece to China became decisive horse-dominated battlegrounds that forever changed history.

CHAPTER 8

The Education of Alexander
Academia and Empires

When Alexander set out in 334 BCE to conquer the Persian Empire and touch the ends of the Earth, he brought with him a highly trained and inspired army of forty thousand troops and three of his most treasured personal possessions: his faithful horse and friend Bucephalus, and his two favorite books, gifted to him by his private tutor, the Greek polymath Aristotle.

The first was a copy of Homer's *Iliad* annotated by Aristotle. The narrative epic recounts the heroic exploits of Achilles during the Trojan War, a much earlier Greek invasion of Persia. The archetypal warrior, Achilles was the golden child of the gods, and the masculine prototype Alexander sought to emulate during his own invasion almost a millennium later. He was obsessed with the heroic saga, and, as a boy, memorized lengthy passages and recited them to spellbound audiences. During his eleven years campaigning across twenty-two thousand miles of distant and exotic lands, Alexander kept his cherished copy under his bed—beside a dagger.

The second book, *Anabasis*, commonly called *The Persian Expedition*, was far more applicable. Written around 370 BCE by the Athenian military leader, philosopher, and historian Xenophon, it recounts the adventures of ten thousand Greek mercenaries, led by Xenophon, who were fighting for the Persians. The description includes minute details of Persian culture, governance, and military training, composition, and tactics.

Alexander's acquisition of Bucephalus, as described by the Greek historian and biographer Plutarch in *The Life of Alexander*, illustrates his keen observation, intelligence, courage, and horsemanship skills at a young age. "The story is plausible in itself," posits renowned historian and author Adrian Goldsworthy in *Philip and Alexander: Kings and Conquerors*, "It is a story that could be true, and anyone of sentiment will feel it ought to be true." Alexander and Bucephalus are arguably the most distinguished and venerated human-horse dyad in history.

When the prince was twelve years old, a frustrated trader abandoned a massive Thessalian horse to roam the streets of the Macedonian capital of Pella. The muscular, raven-black stallion marked by a menacing white star on its brow and one penetratingly piercing blue eye, refused to be mounted and chased off any attempt to be corralled. Surveying the spectacle, Alexander pleaded with his father, King Philip II, to purchase the horse. Originally interested in obtaining the magnificent creature, Philip quickly rescinded his offer after witnessing the ferocity of the wild animal.

The one-eyed king had no use for an unruly, insubordinate steed that quickly drew a swelling audience of curious onlookers. "What a horse they are losing," protested Alexander, "and all because they do not know how to handle him, or dare not try!" Much to his son's disappointment, Philip could not be swayed.

"Are you finding fault with your elders because you think you know more than they do or can manage a horse better?" the king snapped impatiently.

"At least I could manage this one better!" replied a defiant Alexander, as he shrugged off his cloak and crept toward the panic-stricken horse amid jeers from the raucous crowd.

Already renowned for his equestrian abilities, Alexander summoned the teachings of Xenophon to handle horses kindly by touching, caressing, and talking to them. "As the result of this treatment," he concluded, "necessarily the young horse will acquire—not fondness merely, but an absolute craving for human beings." Sensing that the horse was afraid of its own shadow, Alexander stunned the now-silent spectators. He clutched the dangling reins and turned the horse toward the sun to shroud its silhouette as he gently stroked its nose and whispered soothing words. He had tamed the savage beast.

As Alexander galloped away, a tearfully delighted Philip paid the thirteen-talent asking price, enough to sustain a Greek laborer for a century. "O my son," Philip proclaimed, "look thee out a kingdom equal to and worthy of thyself, for Macedon is too little for thee."

Eventually, the warhorse and loyal companion, which the prince named Bucephalus (ox-head), would carry his master across the known and unknown worlds as far as India, the eastern limit of his vast empire and one of the largest kingdoms in history. "The future was spread out before them like a giant canvas on which they could paint their own glorious deeds," writes Alex Rowson in *The Young Alexander: The Making of Alexander the Great*. "Together they would see more of the world than any before them. . . . They took with them their language, traditions, beliefs, and ideas on art and architecture, a distinctive Greco-Macedonian mix that would prove to be one of history's most enduring cultural exports." With that tender moment between a headstrong boy and a stubborn horse on the streets of Pella, together Alexander and Bucephalus would profoundly change the world.

Plutarch also relates that after Alexander had tamed Bucephalus, Philip "would not wholly entrust the direction and training of the boy to the ordinary teachers of poetry and the formal studies, feeling that it was a matter of too great importance." Alexander required an exceptional tutor to rival his exceptional horse. King Philip, who was busy transforming the Macedonian military with his own often overlooked genius, summoned Aristotle to instruct his child prodigy.

Alexander's formal education was the product of a powerful academic pedigree. Aristotle studied for twenty years under his mentor, the philosopher Plato, who founded the Academy in Athens, the first true institution of higher learning. Plato and Xenophon were both students of the infamous "Athenian gadfly" Socrates. This torch of progress eventually found its way from Aristotle into the ambitious grip of a young prince in the northern wilds of Macedon.

Aristotle left his mark on every modern academic field. He wrote more than 150 treatises on everything from metaphysics, politics, and geology to psychology, the arts, and anatomy. He combined detailed investigation and the scientific method with biological reasoning and empiricism. In his work *On the Generation of Animals*, Aristotle examined the reproductive cycle, heredity, and breeding of horses with inquisitive accuracy. In his eight-volume *Politics*, Aristotle commented that cavalry was the preeminent and decisive military arm, a teaching point he no doubt passed on to his young apprentice.

Under Aristotle's personal guidance, the adolescent Alexander was an avid reader and an astute student of history, and exhibited an early fascination with the Persian Empire. In addition to *Anabasis*, Alexander would have read Xenophon's impressive catalog of books, including *Cyropaedia* (a half-truth biography of Cyrus the Great, another of Alexander's childhood heroes), and his two treatises on horses.

On the Cavalry Commander specifies the duties and obligations of a cavalry officer, while the more renowned and studied *On Horsemanship* is an erudite discourse on the selection, training, equipment, riding, and maintenance of horses, including their military application. "The greatest generals of the ancient world such as Julius Caesar, Hannibal, and, above all, Alexander, would not have won their most famous victories without the proper appreciation of the battle-winning potential of cavalry," stresses military historian Philip Sidnell in *Warhorse: Cavalry in Ancient Warfare*. The definitive guidelines and humane standards espoused by Xenophon were disseminated widely across Eurasia courtesy of Alexander's exploits and endured well into the Middle Ages.

In addition to the permanent residence of Aristotle, Philip's court at Pella was, to borrow a phrase from Xenophon, "a workshop of war." Soldiers, politicians, academics, and adventurers visited from across the known world, forming a melting pot of concepts and conversation. Alexander once greeted the official Persian envoys by bombarding them with questions about their road system, cavalry horses, military training, and, of course, his hero Cyrus. The ambassadors were captivated by the young prince, and, according to Plutarch, "regarded the much-talked-of ability of Philip as nothing compared with his son's eager disposition to do great things." With this exceptionally syncretic and stimulating upbringing, Alexander was the culmination of all these brilliant minds.*

---

* According to later sources, Marsyas of Pella (brother of the acclaimed general Antigonus) wrote a work called *The Education of Alexander*. No copies survive.

Given that the premise of this book is the influence of the horse on human history, this chapter depicting the rise and supremacy of cavalry has been hitched squarely to Alexander and Bucephalus. "Even more important for our purposes," writes Jeremiah McCall in his detailed investigation *The Cavalry of the Roman Republic*, "scholars have agreed that the Macedonian cavalry serving under Alexander the Great represented the pinnacle of ancient cavalry." While the horse was a vital part of the armies of Hannibal (and his skilled Roman adversary Scipio) and Caesar, and was instrumental in the rise, and eventual fall, of the Roman Empire, it did not have the shock power and shattering impact it did under Alexander. On the heels of his father, Philip, he was the pioneering trailblazer. In the millennia that followed, their shock cavalry and combined arms warfare became the norm rather than the exception.

Alexander's broader intellectual and historical base proffered by Aristotle and others would have picked up where our story left off, with the rise of cavalry and the power vacuum created by the collapse of the Assyrian Empire around 600 BCE. An appreciation of Alexander's preeminent position in military history and his unrivaled use of cavalry must be contextualized within the eras of war that preceded him. "An unexamined life," professed Plato, "is not worth living."

To examine the life and legacy of Alexander, we must first take a brief step back into the affairs that created the atmosphere for his meteoric rise and enduring imprint on the modern world—specifically the Greco-Persian and Peloponnesian Wars. While there can be no denying that he possessed an uncanny innate intellect for the art of war, accompanied by reckless, charismatic courage, Alexander also learned from, and was the product of, the past.

At this time, our well-traversed greater Eurasian Steppe and its tributaries were divided into three theaters of war: Greece at the western gate, Persia occupying the central stage, and the warring states of China and their nomadic neighbors the Xiongnu trading barbs in the east. For all the major players of the period, the horse entered its age of ascendancy.

The Assyrians had shown the value of cavalry as a powerful tool in the construction of empire. The complete collapse of Assyria, however, set the stage for a new superpower of the ancient world. While the majority of the Middle East was rebuilding from the wreckage left by the Sea People and Assyrians, a new power quietly emerged from the shadows in the East.

Having secured the Persian throne in 559 BCE, Cyrus immediately set out consolidating his rule and conquering what would become his massive Achaemenid Empire. Initially, however, the small mountainous Persian enclave centered on the city of Pasargadae lacked suitable pasturelands for what was quickly becoming a life-or-death commodity: cavalry horses.

The early Persian annexation of the rich pasturelands and horses of the Nisaean Plain and Anatolia altered its equine fortunes. "Persian kings and their armies obtained horses from many sources," reports Ann Hyland. "What is remarkable is how rapidly the Persians became thoroughly proficient horsemen without a tradition of early equestrianism." When combined with their unique composite bow, incorporating both Assyrian and Scythian elements, and cast-bronze tree-leaf arrowheads, Persian cavalry was a potent tool in the Achaemenid military arsenal.

Cyrus the Great founded his empire through decisive military campaigns, skilled diplomacy, and, above all else, a visionary human rights policy the United Nations would applaud. His affirmation of cultural and religious tolerance and the repatriation of enslaved peoples to their homelands was immortalized by the egalitarian inscriptions on the small, clay Cyrus Cylinder. He is mentioned twenty-three times in the Bible and was anointed as messiah by the Judean Israelites, whom he freed from Babylonian captivity.

The empire he created embraced the coastal Greek Ionian city-states dotting western Anatolia, all the former imperial states of the Middle East, and stretched into the southern Caucasus and steppes of Central Asia. Approaching the Aral Sea in the autumn of 530 BCE, Cyrus encountered the matriarchal Massagetae branch of the Scythians (whom the Persians called Saka), led by the warrior-queen Tomyris. "Cyrus was eager to bring the Massagetae under his rule. They are said to be a large tribe, with a reputation of being warlike," wrote Herodotus. "The Massagetae resemble the Scythians in both their clothing and their lifestyle. In battle they may or may not be on horseback, since they rely on both methods, and as well as using bows and spears." Appreciating the disadvantages his army faced fighting the skilled armored cavalry of the Massagetae on their home turf, Cyrus wisely visited diplomacy first.

He sent Tomyris an offer of marriage, which she laughed off and flatly rejected. "King of Persia, abandon your zeal for this enterprise. You cannot know if in the end it will come out right for you. Stop and rule your own people, and put up with the sight of me ruling mine," she chastised. "But no: you are hardly going to take this advice." She then cordially invited him to battle. Cyrus should have heeded her suggestion. He was not, however, immediately willing to accept her entreaty to war. He preferred treachery.

Cyrus feigned retreat and left behind a tabled feast, stockpiles of wine, and a token party of expendable soldiers, who were quickly routed by a force led by Tomyris's son, Spargapises. While feasting and drinking, the Massagetae were taken by surprise and slaughtered by hidden Persian troops. A disgraced Spargapises died by suicide. Tomyris immediately mustered the full might of her military. "You bloodthirsty man, Cyrus! What you have done should give you no cause for celebration," she admonished the Persian emperor. "I swear by the sun who is the lord of the Massagetae that for all your insatiability I will quench your thirst for blood."

Tomyris was true to her word. In what Herodotus considered "the fiercest battle between non-Greeks there has ever been," the Persians suffered massive casualties, and Cyrus was killed. His body was brought before Tomyris, who hacked off his head as a trophy of war. As she plunged it into a cask filled with blood drained from the Persian dead, she screamed, "I warned you that I would quench your thirst for blood, and so I shall!" Based on Scythian ritual, Cyrus's skull likely ended up as a great gilded drinking cup and one of Tomyris's prized possessions. What was left of him made its way back to his beloved capital of Pasargadae and was interred beneath a modest limestone tomb, now suitably recognized as a United Nations World Heritage Site. Cyrus is regarded as one of the most influential and illustrious figures in history and is genuinely deserving of his "Great" suffix.

After a brief civil war, Darius, a distant cousin of Cyrus, eventually took the throne. Another (likely apocryphal) account has Darius and five other claimants deciding the fate of the Persian Empire through a contest of horse snorts. The man riding the steed who neighed initially at the sight of first light would become emperor. At the first shimmers of dawn, Darius's servant Oebares placed his hands, which he had rubbed across the genitals of a mare, around the snout of his immediately aroused horse. Darius allegedly erected a statue of himself on his horse to commemorate his coronation in 522 BCE with the following inscription: "Darius, son of Hystaspes, obtained the sovereignty of Persia by the sagacity of his horse and the ingenious contrivance of Oebares, his groom."

Darius inherited a vast and diverse empire. "This country Persia which I hold," he proclaimed, "is possessed of good horses, of good men." Both were vital cogs in the Persian military machine as he campaigned on the fringes of control from Egypt and Central Asia to the edges of India, including futile punitive expeditions against thorny Scythians. Although Darius boasted that "As a horseman I am a good warrior. As a bowman I am a good bowman, both afoot and on horseback. As a spearman I am a good spearman both afoot and on horseback," in his offensive against the Scythian "Saka beyond the Sea," he was outmatched and outmaneuvered.

Darius reigned over the empire at its territorial zenith. His cultural mosaic of subjects totaled an astounding fifty million people—almost half the global population—encompassing no fewer than forty-seven nations. Across the thriving and prosperous empire, both Cyrus and Darius promoted cultural, technological, and religious reciprocation and nurtured artistic, engineering, and scientific innovation. Darius introduced a universal coin currency through a state banking system to regulate international trade and taxation.

Like the Assyrian kings and Cyrus before him, Darius was also a brilliant mind and an impressive builder. To facilitate commerce and ensure the quick passage of troops and lines of communication across the empire, he constructed a novel network of infrastructure that included roads, bridges, irrigation systems, canals, way stations, post offices, and other travel services. The earth-packed roads, which served as the "eyes and ears of the king," could accommodate sixteen oxen or horses lashed four abreast hauling siege towers weighing upwards of seven tons. The most renowned components of this sprawling transportation system were the thirty-five-mile-long technological marvel the Darius Canal (forerunner to the Suez Canal), connecting the Nile to the Red Sea, and the legendary Royal Road, which was fed by smaller intersecting routes.

The Royal Road, which had Assyrian antecedents, extended roughly 1,600 miles across the empire, from the city of Sardis in western Anatolia to Susa in the heart of the Persian homeland. More than a hundred rest stops, complete with caravanserais, stables, bed quarters, a post office, a military garrison, and other amenities, were erected one day's ride apart, or roughly every fifteen miles.

What would have taken three to four months on foot, using a relay exchange of fresh horses, military personnel and mounted messengers of the Angarium postal institution could cover the entire distance in seven to fourteen days. "There is nothing mortal that is faster than the system the Persians have devised for sending messages," praises Herodotus. "And neither snow nor rain nor heat nor dark of night keeps them from completing their appointed course as swiftly as possible." This horse-shuttled Persian model served as the precedent for the informal US Postal Service creed, the short-lived Pony Express during the settlement of the American West, and the National System of Interstate and Defense Highways established in 1956 under President Dwight Eisenhower.

Persian economic reforms and landmark infrastructure projects created the first fully integrated society. Innovative horse transportation along the royal roads and secondary thoroughfares connecting the empire to vibrant Steppe Road trading centers allowed for the creation of multicultural "think tanks." For example, when Darius was building his custom capital at Persepolis, he imported skilled tradesmen from across the far reaches of empire. As a result, science, engineering, spirituality, and the arts flourished under Persian rule, and its premier cities of Pasargadae, Persepolis, Susa, Ecbatana, and Babylon became thriving academic centers.

The expansion of Persian primacy, however, led to a legendary showdown with another young and aspiring power. Herodotus wrote that Cyrus brought together "every nation without exception." The glaring exception, however, was Greece itself. The sparring Greek city-states were drawn into a life-or-death struggle against the Persians instigated by their Ionian brethren, who, with support from Athens, revolted against Persian rule in 499 BCE. After quickly subduing the rebellion, Darius vowed to punish Athens for its insolence. His full-scale invasion of Greece seven years later ignited the Greco-Persian Wars.

At this point, Greece, like China, was an assemblage of warring states. Sparta and Athens emerged as the dominant polities and sought allies to bolster their braided military and economic agendas. Also, as in China, these rival city-states were forced to put their mutual hostility on hold to withstand the onslaught of a far more lethal threat. For China, as we will see, it was plundering Xiongnu raiders from the steppe, while for the Greeks it was imperialistic Persians. At the onset of hostilities, however, the Greeks and the Chinese faced a decisive disadvantage: they both lacked any noteworthy cavalry tradition.

Taken as an ecological whole, the majority of Greece lacked suitable grasslands for the sustainment of any truly sizable horse populations. South of Macedon and Thrace, the Greek city-states, apart from Thessaly, were slow to adopt competent cavalry to complement heavy hoplite infantry. These troops fought shoulder to shoulder and shield to shield in deep ranks within the basic combat formation of the phalanx. Aside from inadequate pasture, there are additional reasons why the Greeks were discouraged, or dissuaded themselves, from establishing an earlier horse-based or cavalry culture.

Greece was dominated by a mountainous craggy landscape of cities and surrounding clusters of homestead farms inherently incompatible with raising horses. Within a semifeudal pastoral system, any arable land was intensively cultivated with crops and livestock, primarily goats and sheep, and their secondary products. The high population densities of

Greece could not afford to waste viable, and valuable, land on pasture for luxury animals such as horses. Moreover, unlike the stretching royal roads of the Persian Empire or the Eurasian Steppe, the relatively short distances between the crowded urban centers of Greece did not necessitate horse-based commercial or conversational transport. Most people were simply not acquainted with horses.

They were also unreasonably expensive. At the onset of the Greco-Persian Wars, the average price for a Greek horse standing fourteen to fifteen hands was around 500 drachmas, with the most precious military mounts costing as much as 1,200 drachmas. By comparison, a goat or sheep commanded 10 to 15 drachmas, with a cow fetching around 50. Human slaves ranged from 140 drachmas for a donkey cart driver to 360 for a master goldsmith. The daily wage for a typical Greek laborer in the year 400 BCE was a single drachma, equivalent to one salted fish, a gallon of table wine, or five pounds of wheat.

Additionally, training both horse and rider demanded free time, something the average Greek workhand or farmer did not enjoy. The diminutive cavalry cadre remained a status symbol reserved for the aristocracy, who played at war by galloping on the fringes of battle. By default, heavy hoplite infantry dominated Greek military doctrine.*

The Greek victories over the Persians at the battles of Marathon (490 BCE), Salamis (480 BCE), and Plataea (479 BCE) extinguished any future Persian designs toward Europe and propelled the balance of power west to Greece. The ensuing golden age of Greece would be the substance from which modern Western civilization was built. As military historian Paul Davis points out:

> Many commentators point to Salamis and Plataea as the turning point in all of European history, the point at which Europe became a culture based on Greek civilization and not a vassal of Eastern emperors. . . . Thus, the basis of western political

---

* While environmental and cultural barriers limited Greek cavalry, during the Greco-Persian Wars, both Darius and his son Xerxes intentionally reduced their cavalry component for their futile invasions of Greece. They based their decision on logistical considerations: water, fodder, and suitable pasture would be in short supply.

institutions, philosophies, and sciences comes from Greece; little is done today, or even conceived of, that the Greeks did not ponder upon more than two millennia past. Had the Persians prevailed, they might well have spread their empire deep into Europe.... No European population had the organization to mass against them; even the previously successful Scythians may have failed against a reinforced Persian military.... The world, indeed, could have been completely different.

As it was, the Greek triumph fostered the education of Alexander, fomenting his revolutionary employment of cavalry. He firmly entrenched the warhorse as an overriding, history-wielding combat arm that was here to stay. In the process, as both legendary conquerors and trailblazing explorers, Alexander and Bucephalus broadcast the Hellenic intellectual and cultural podium to the world.

Although the eastern invaders were repulsed, the Greco-Persian Wars proved to be a wake-up call for Greek military minds. Greece was saved as much by disease, luck, and Persian blundering as by Spartan bravery, home field advantage, and crafty Athenian strategy. Their decisive lack of cavalry could no longer be ignored. The inadequacies and shortcomings of their one-dimensional infantry system were wholly exposed, forcing Greeks, and Macedonians, to reevaluate their overall military configuration. After all, there was still the lingering question of hegemony within Greece itself. The mutual odium and common defense against Persia were the only bonds holding together the arranged marriage of Greek city-states.

Their bitter and spiteful divorce was finally settled by the reciprocal slaughter of the Peloponnesian War, derided by Aristophanes in *Lysistrata* and documented by Thucydides. Despite this warfare, raging intermittently from 460 to 404 BCE, or perhaps because of it, Greeks of the fifth century BCE (primarily Athenians) whose names hang in the hallways of immortality fashioned their most acclaimed academic innovations in architecture, science, medicine, philosophy, and the arts. Socrates, Plato, and Thucydides, for example, all fought for Athens during the Peloponnesian War.

These wars sounded the abrupt and grim denouement of the golden age of Greece. Fifty-six years of intermittent conflict had left both Athens and Sparta, and their subordinate allies, shattered and vulnerable. The weary, war-torn Greek world was in financial decay and political disarray. The Peloponnesian War prevented Athenian/Greek authority and influence from unfurling across the broader Mediterranean world toward the aspiring, rival states of Rome and Carthage. The expansive empire of Alexander and his pursuit of glory looked east, not west, to the ends of the Earth.

For those who were paying attention and reading up on their Thucydides and other accounts of the Greco-Persian and Peloponnesian Wars—among them Philip and Alexander—the untapped potential of cavalry came into sharp focus. In his authoritative book *Cavalry Operations in the Ancient Greek World,* historian Robert Gaebel emphasizes, "[F]or any bright young military leaders who were capable of profiting from past mistakes, there were some lessons to be learned about the use of cavalry. . . . This, together with a professionalism fostered by almost incessant warfare, set the stage for the remarkable military achievements of the fourth century [BCE] that reached their culmination in Alexander."

By studying and coalescing the evolving martial principles and experimentation of the previous century, both Philip and Alexander were ready to challenge the military status quo from horseback. While the rise of the backwater kingdom of Macedon to superpower status was no accident, it was also by no means inevitable, and rested more on brains than brawn.

From the smoldering cinders of the Greco-Persian and Peloponnesian Wars, a new power arose, and a young horse-whispering prodigy would lead it beyond the summits of supremacy to gaze from his unimaginable heights toward the horizons at the ends of the Earth. Although he secured the prestigious titles of basileus of Macedon, hegemon of the Hellenic League, pharaoh of Egypt, shah of Persia, and lord of Asia during his meteoric rise to conquering celebrity, he is now known simply as Alexander the Great.

Rarely does one individual change the course of history and single-handedly carve his or her name indelibly on the pages of immortality.

There are also only a handful of people who shaped not only their own contemporary world but also every future that followed. Alexander's history-altering achievements and undefeated 20–0 battlecard, however, were made possible by the combination of the cavalry horse and his own heroic abilities. It is also worth remembering just how young Alexander was (twenty years old), and the relatively small size of his military forces (forty thousand soldiers), when he marched on Persia and challenged Darius III to a risky game of world domination.

Notwithstanding the ceaseless and futile intellectual squabbling over his character and motives, there can be no questioning the pure, raw genius that was Alexander. "If in reality he never wept because there were no more worlds to conquer or sheathed his sword for lack of argument," contends Adrian Goldsworthy, "it does not diminish what he did in such a short time. Only in the modern mechanized era have a few armies managed to advance so fast for such a long period as Alexander's men.... The sheer scale of what Philip and Alexander achieved is staggering.... It is easier to consider the impact these two men had on their world, although even here it is impossible to know what would have happened if they had not done what they did."

In less than forty years, Philip and Alexander forged a backwoods vassal region into a mighty kingdom that conquered the battle-scarred Greek city-states and the entire Persian Empire, and even ventured into uncharted territory, claiming more than two million square miles (roughly 4 percent of the planet) in the process. By all comparisons, Alexander fielded one of the greatest armies the world has ever seen. Although he had proficient infantry, archers, and other specialized troops, it was his vaunted cavalry that delivered his string of victories and carried the complete Hellenistic cultural and academic package to Eurasia, and eventually, the world.

While the adolescent Alexander, whom Plutarch described as "ambitious and serious," immersed himself in the teachings of Aristotle, his father, Philip, also a student of history, instigated a military revolution. "The internal workings of the new army were given birth to by Greek intellect, and one cannot help but think this represented the collective cerebral effort of many ruminating and scientifically inclined minds," stresses David Karunanithy in *The Macedonian War Machine, 359–281 BC*. "The Macedonians under Philip and Alexander became adept at experimenting with, absorbing and improving upon the most progressive military thought of their antecedents, contemporaries, and enemies. Their activities acted as catalyst for a virtual explosion of intellectual activity on things military."

When Philip assumed the throne of Macedon in 359 BCE at the age of twenty-two, he immediately set out to transform the Macedonian army through sweeping alterations to weaponry, tactics, and the combined arms deployment of infantry and cavalry. "Over the next twenty years," notes Sidnell, "Philip campaigned almost continuously, all the while honing his military machine and its tactics on the whetstone of experience, so that it effectively became a permanent professional army."

Philip also mandated an unprecedented regimen of intense training and discipline—a mindset drilled into his son. Alexander later joked to his soldiers that, as a child, he was served "night marches for his breakfast, and for his dinner his frugal breakfast." Writing in the second century BCE, the Greek historian Polybius relates that "it is universally acknowledged that from his childhood he was well versed and trained in the art of war."

The elite heavy Companion Cavalry commanded directly by Alexander in all his major battles was positioned on the right flank. Light cavalry, including the crack Thessalians, occupied the left flank. A special training school was set up by Philip for potential Companions. At age fourteen, the same age that Alexander began his studies under Aristotle, the leading sons of Macedon and other worthy applicants received a formal education at the king's expense. Many of Alexander's Companion cavalrymen were also his childhood friends and classmates at this exclusive military academy, adding to the esprit de corps of the elite unit.

The syllabus placed a high priority on horsemanship. Xenophon advised that a prospective cavalry mount be "tested in all particulars in which he is tested by war. These include springing across ditches, leaping over walls, rushing up banks, jumping down banks." Unlike the Persians, the heavily armored Greek cavalry was based primarily on the sword and spear and not the bow. Macedonian cavalrymen were trained rigorously for close-quarter combat, prompting the ancient Greek historian Arrian to muse that Alexander's mounted troops fought like hoplites on horseback.*

Philip and Alexander understood fully the value and versatility of cavalry. "Given a strong cavalry force, there were three uses you could make of it," summarizes historian Sir William Tarn. "You could merely fight with the enemy's cavalry; or you could take his infantry in the flank or rear; or you could break through his line. Alexander used all three methods, but merely to defeat the enemy cavalry was clearly to him the least important." Using impeccably timed, coordinated, and decisive cavalry actions to strike exposed enemy flanks or vulnerable spots in the main line became hallmarks of Alexander's victories.

Given the pivotal role of cavalry within Macedonian military doctrine, procuring, breeding, and training quality mounts was a priority. Like the Mittani and Assyrians before them, the Macedonians developed a series of stud farms administered by the Office of the Secretary of Horses. Numerous Macedonian entries won at the Olympics. In 356 BCE, the same year Alexander was born, Philip's racehorse won at the 106th Olympiad. His chariot teams were victorious at the following two Olympic Games.

Although the first recorded Olympic Games were held in 776 BCE, horse and chariot races were not introduced until 680 BCE. Fittingly, the first recorded winner in horse racing was Crauxidas, from Thessaly, in 648 BCE. By the third century BCE, shortly after Philip's victories, the Olympic equestrian program contained six events. As a testament to the horse's venerated status, to this day it is still the only nonhuman participant at the Olympic Games and is considered as much an athlete as its rider. Equestrian is the only individual sport in which men and women compete head-to-head on equal terms.

Through trade and conquest, both Philip and Alexander continuously augmented their herds. After the Battle of Issus, for example, seven

---

* Although Arrian wrote in the mid-second century CE, he is considered the most reliable source on Alexander and his campaigns due to his reliance on, and specific reference to, eyewitness accounts and earlier secondary accounts, many of which have since disappeared from the literary record.

thousand Persian mounts were added to the Macedonian count.* The Nisaean Plain in Persia provided Alexander with fifty thousand of the highest quality horses, while his triumphant entry into Babylon in 331 BCE netted another sixteen thousand. As Karunanithy points out astutely, "[T]he very fact that Alexander regularly and often deliberately drove his cavalry horses so hard suggests that he had the reassurance of knowing that the army kept a fair-sized reserve of spare mounts as standby. The system of obtaining them was well organized, adaptable and effective enough for adequate numbers to be quickly gathered in when required." Based on the available sources, it appears that Alexander had plenty of horses spread across various military formations, including service and supply units.

Sophisticated and highly organized baggage trains using horses, donkeys, mules, and camels were a staple of the Macedonian military. King Philip abandoned the use of wheeled vehicles (and oxen), which were clunky, slow, and awkward, especially in the difficult and heavy mountainous terrain of Greece. While these modifications may be the banal aspects of war, they reduced the baggage train by 65 percent, allowing for streamlined, rapid movement. Under this new equine pack-heavy logistical system, the entire Macedonian camp could cover thirteen miles a day, while independent cavalry could easily cover forty miles. The horse unlocked the shackles of supply.

No other professional military of antiquity possessing equivalent punching power could move so far so fast. The armies of Philip and Alexander were the most mobile, versatile, and low-maintenance to ever take to the field. By overthrowing the tyranny of logistics, Philip's innovations set the precedent for Alexander and all future armies of Europe and the Middle East. The elevated performance of Alexander's supply lines, which included seaborne coastal drops, is evident in the breakneck speed and vast distances of his campaigns, spanning a total of twenty-two thousand miles across diverse environments.†

---

* The battle was fought on the Mediterranean coast near the Turkish city of Iskenderun, historically known as Alexandretta—founded by Alexander in 333 BCE after his victory over Darius III.

† By comparison, the equatorial circumference of the Earth is 24,902 miles.

THE EDUCATION OF ALEXANDER

Fighting alongside his father, Alexander gained invaluable training, confidence, and momentum. He quickly earned respect and repute as a fierce, admirable leader who inspired loyalty, courage, and devotion by fighting at the front of his ranks. The most important takeaway for Alexander was that cavalry charges against infantry could be decisive if delivered at the opportune time and location. Wars can be lost with a moment's hesitation. His personal bravery, or egotistical recklessness, certainly enhanced his genius as a cavalry commander, but it was also his innate prudence and precise instincts for when and where to strike that set him apart from his peers. *This* was his genius.

Alexander heeded Xenophon's advice that "It is always necessary for the commander to hit on the right thing at the right moment, to think of the present situation and to carry out what is expedient in view of it." His cavalry charges became the hallmark of his victories. "Alexander, in short," declares Robert Drews, "exploited the full potential of shock cavalry." The young prince possessed the appetite, intellect, and ability for war, and his sudden ascension to the throne of Macedon was close at hand.

Having united Greece under his rule—except for the recalcitrant but weak and largely irrelevant Spartans, whom Alexander ridiculed as "mice"—an anxious Philip smartly conjured up a common cause for all Greeks to rally behind by dredging up an old archenemy. It was time, he declared in 336 BCE, for a united Greece to march on Persia. Philip, however, would not steer the invasion. While presiding over his daughter's wedding, the king was assassinated by a disgruntled bodyguard.

Alexander, unexpectedly assuming the throne at twenty years of age, prepared to carry out his father's vision of conquest to unimagined heights. Having consolidated his domestic rule, in 334 BCE Alexander mustered his combined Macedonian and Greek force of no more than forty thousand soldiers, including five thousand cavalry, crossed the Hellespont, and marched on Persia. "With his small but excellently trained and organized army," wrote Major General Carl von Clausewitz, the brilliant Prussian strategist in his immortal 1832 treatise *On War*, "Alexander shattered the brittle states of Asia. Ruthlessly, without pause, he advanced through the vast expanse of Asia until he reached India."

From the ancient sources, historical reconstructions, and academic appraisals available, we can glean a relatively accurate representation of Alexander's cavalry actions and combined arms tactics during his legendary string of victories, including the major battles of Granicus (334 BCE), Issus (333 BCE), and Gaugamela (331 BCE) against Darius III and the Persians; Jaxartes River (329 BCE), confronting the unconventional Scythians; and, lastly, against the valiant warrior King Porus and his elephants at Hydaspes River in India (326 BCE).

Part of Alexander's genius rested with his innate ability to read the battlefield and react. He altered his approaches based on the formation, tactics, and movements of his adversary and the ebb and flow of battle. He did not apply the same successful methods that he had used against Darius in Persia against the Scythians in Central Asia or Porus in India. With crafty deployments and timely charges during all engagements, his warhorses always delivered the final blows of victory.

Outnumbered roughly two to one at both Granicus and Issus, Alexander positioned himself and his Companion Cavalry on the far right, as customary, with his elite infantry holding the end of the line. Alexander purposefully concentrated his best cavalry and infantry directly around him for protection, ease of command, and striking power. His loyal Thessalian cavalry and light infantry closed the left flank.

Alexander mosaic: Alexander the Great and Bucephalus (*left*) confront Persian emperor Darius III (*center*) at the Battle of Issus. Floor mosaic from the House of the Faun, Pompeii, Italy, circa 100 BCE. *(Prisma Archivo/Alamy Stock Photo)*

Alexander seized the initiative at Granicus by conducting a bold and timely advance against a vulnerable point of the Persian line, followed by coordinated and decisive cavalry assaults on both flanks. At Issus, his skilled use of cavalry to strike diagonally at the crucial spot at the critical moment was a stroke of genius.

Following his anointment as a god by the Egyptians, who viewed him as their liberator from Persian rule, Alexander drove his forces into the Persian heartland, instigating his overdue pursuit of Darius.* In 331 BCE on an open plain at Gaugamela, slightly east of the ancient Assyrian city of Nineveh (Mosul), their armies collided for the final showdown. Again, Persia enjoyed numerical superiority. As the opposing lines converged, Alexander beckoned for Bucephalus. During the marshaling stage, he had been riding another mount, as Bucephalus "was now past his prime." The old warhorse would carry his dear friend once more unto the breach.

Alexander immediately ordered his infantry to attack the center of the Persian position, while he and his Companion Cavalry occupied the far right of the line. He intended to draw the bulk of the Persian cavalry to his flanks, leaving the center of the Persian line fractured and exposed. It worked. Darius continued to commit resources to press the Macedonian flanks. Alexander had set the trap; now he waited patiently to engage the spring and strike. It was exactly as he had scripted, and he launched his main cavalry assault at the perfect place and time.

With Alexander riding Bucephalus at the sharp end, his reinforced cavalry formed a giant wedge and penetrated a gap forming in the center of the Persian line. "When the Macedonian cavalry, commanded by Alexander himself, pressed on vigorously, thrusting themselves against the Persians," reports Arrian, "all things together appeared full of terror to Darius, who had already long been in a state of fear, so that he was the first to turn and flee." With little motivation to continue fighting, disgruntled Persian soldiers simply drifted home. Others rebelled against Darius, who was assassinated shortly after his irreparable defeat.

---

* In preparation for traversing the unforgiving deserts, Alexander purposefully procured camels for his supply trains, having read about their benefits in books on military history and in the zoological writings of his tutor Aristotle.

At Gaugamela, Alexander's genius and unrivaled use of cavalry was on full display. "Gaugamela also influenced other military leaders who adopted the Greek tactics that had defeated far more numerous foes," notes military historian Michael Lanning. "These methods of military operation strongly influenced the leaders of the future Roman Empire and also provided inspiration and knowledge for the Napoleonic conquests more than two [millennia] later." Napoléon acknowledged that he "read over and over again the campaigns of Alexander. . . . This is the only way to become a great general and master the secrets of the art of war."

It had been just over three years since Alexander crossed the Hellespont into Asia. His battle record stood at a perfect 11–0. Stirred by his insatiable ego and Achilles complex, Alexander was hell-bent on conquering what Aristotle labeled "the ends of the world and the Great Outer Sea."

Following his consolidation of the former Persian Empire, he drove his forces through Central Asia. During a brief campaign in 330 BCE against the Amardi, a branch of Scythians occupying the mountainous south shore of the Caspian Sea, a group of raiders stole a herd of grazing Macedonian horses—including Bucephalus. An enraged Alexander vowed to ravage the lands and slaughter or enslave the entire population if his horse was not returned unharmed immediately. Bucephalus was back with his friend and master just as fast as the Amardi delegation could get the stubborn old warhorse to move. With Bucephalus in tow, Alexander headed straight into the heart of Central Asia toward the Fergana valley and its prized horses, where he settled a Greek diaspora at Alexandria Eschate.

His progress was impeded in 329 BCE by a faction of headstrong Scythians along the shores of the Jaxartes River, which straddles the borders of Uzbekistan, Tajikistan, Kyrgyzstan, and Kazakhstan. "This episode is of particular military interest," observes Robert Gaebel, "because it shows Alexander at his best, effectively employing various arms and technologies." After crossing the river under a barrage of catapults, Alexander stunned and scattered the Scythian horsemen with a quick cavalry charge. They quickly reformed, and adopted the traditional

Scythian method of riding rings around the trapped Macedonian cavalry. In doing so, they had taken Alexander's bait.

Having analyzed Scythian tactics, including those used against his dad and Darius I, Alexander had anticipated this response. The Scythians were unaware that the bulk of the Macedonian army remained concealed behind the riverbank. Alexander immediately led a decisive cavalry charge, disrupting the Scythian encirclement at numerous spots. Again, his timing was impeccable. Surprised and overwhelmed, the Scythians fled in terror, melting into the steppe.

Using his accrued education and unrivaled military acumen, Alexander had bested the mighty Scythians he had read about as a boy in the works of Aristophanes, Herodotus, Xenophon, and Aristotle, not to mention listening to his father's personal accounts. "People had believed the Scythians invincible," declared the Roman historian Quintus Curtius Rufus in his work from the first century CE, *Histories of Alexander the Great*. "But after this crushing defeat, they had to admit that no race was a match for Macedonian arms." When news of Alexander's march to India reverberated across the region, Scythian horsemen flocked to his banner, unable to resist the lure of riches and adventure. By this time, his forces had been fighting continuously without defeat (17-0) for nine years.

In the spring of 326 BCE, at the Hydaspes along the Indus River system, Alexander's sickly and spent army faced King Porus and his fearsome war elephants in what turned out to be its final large-scale battle. Maintaining secrecy through a series of brilliant deceptions, false movements, and feints, Alexander kept Porus preoccupied and blind to his true intentions. With the main Macedonian line standing firm, Alexander and a cavalry-heavy strike force waded ashore on the Indian side of the river undetected.

He immediately launched a cavalry detachment in a wide wheel to threaten the Indian position from the rear. As Porus scrambled his troops to counter this flanking movement, Alexander, with his impeccable battlefield instinct, led his Companion Cavalry in a beautifully choreographed charge that sent the Indian cavalry fleeing into their own line of elephants, causing confusion and disarray. Having seized the initiative with his expert horsemen, Alexander ordered his heavy infantry to advance and pin down the exhausted Indian force, while the relatively unimpeded cavalry maneuvered around the flanks to strike the final blow from the rear—flawless execution of the "hammer and anvil" movement.

For Alexander, the masterpiece victory was bittersweet. Following the battle, his faithful warhorse, Bucephalus, died of old age, fatigue, or fatal wounds (depending on the source). Alexander honored his devoted horse by founding a city on the Hydaspes River: Alexandria Bucephalus.* The beloved equine died at about age thirty, a little older, and perhaps wiser, than his master, who celebrated his own thirtieth birthday a few weeks later.

Although Alexander believed the ends of the world were within reach, it was not to be. The tomb of mighty Bucephalus, four thousand miles from home, roughly marks the eastern limit line of Alexander's enormous empire. A fitting tribute of immortality for his beloved and trusted friend and one of the most time-honored and famous horses in history.

Shortly after the death of Bucephalus, Alexander sensibly turned his tired and ailing army around and headed west. His homesick troops simply "longed to again see their parents, their wives and children, their homeland." Over the course of nine years, they had followed their leader across more than seventeen thousand blood- and sweat-soaked miles. The promised land at the ends of the Earth and the Great Outer Sea were still nowhere in sight, and they refused to aimlessly trudge any farther. Alexander was also aware that his supply lines were overstretched and fraying, his available pool of Macedonians and Greeks was evaporating, and his triumphs progressively tougher to attain.

His next target would have been the mighty Nanda Empire of northern India, which fielded a force of two hundred thousand infantry, twenty thousand cavalry, two thousand chariots, and three thousand war elephants (which spooked the Macedonian horses). Victory was not a foregone conclusion. Not even Alexander the Great, boasting a perfect

---

* The city, also known as Alexandria Bucephalia, among other, similar names, is believed to be near the modern community of Jalalpur Sharif in Punjab, Pakistan.

20–0 battle record, could circumvent these coalescing military complexities and impediments. In hindsight, given Alexander's untenable position, the temporary abandonment of his India campaign proved to be a perceptively cautious decision.

While Alexander's arrival in India did not stimulate war, which had long been serious business in the region, it did inspire notions of national unification and empire building. It also initiated an intense period of contact and creative interaction between Greek and Indian cultures. For example, it has been suggested that later portions of the Sanskrit masterpiece *Ramayana*, one of two principal Hindu epics, was heavily influenced by the *Iliad* and the *Odyssey*.

Four years after Alexander reversed course, Chandragupta Maurya founded an empire by conquering vast territories, including the Nanda and those satrapies (provinces) sustained by Alexander and his successors.* He united the Indian subcontinent and created the largest imperial domain in Indian history. This vibrant and prosperous kingdom, which lasted 130 years, paved the way for the modern consolidated Indian state. The Maurya Empire also nurtured the dissemination of Buddhism.

This philosophical faith based on the teachings of Siddhārtha Gautama—the Buddha—had remained confined to eastern Afghanistan, Pakistan, and northern India, until it was embraced by the Mauryan emperor Ashoka around 260 BCE. He sponsored spiritual missions across India and Central Asia, where Buddhism became firmly entrenched by the first century BCE. These delegations, along with horse-riding Indo-Europeans, were instrumental in extending this spiritual belief through the Tarim Basin into China by the first century CE. In addition to the philosophical traditions of Buddhism, the subsequent Maurya Empire cultivated academia and the arts.

During his overthrow of the Nanda, Chandragupta was counseled by his mentor and chief advisor, Chanakya. A brilliant polymath, he was a pioneer in the fields of economics, industrial production, state welfare and governance, taxation, international relations, military strategy, and other matters of statecraft. He bundled all of these into his landmark

---

* Chandragupta is said to have chatted cordially with Alexander not far from Bucephalia.

treatise the *Arthashastra,* written in Indo-European Sanskrit on palm leaves, which would be rediscovered in 1905.

His writings served as a blueprint for Chandragupta's confederation of India and the bureaucratic, commercial, and military administration of a centralized empire. Plutarch claims that Chandragupta "overran and subdued the whole of India with an army of 600,000 men," including 30,000 skilled cavalrymen and lingering Greek mercenaries.

The horse, cavalry, and equestrian training, influenced heavily by Greek literature and vocabulary, features prominently in Chanakya's writings. He states that the proximity of both the Persian and Macedonian Empires to India spurred the dissemination of horses and equestrian disciplines throughout the subcontinent long before the establishment of the Maurya Empire. It is evident that a string of outside influencers had successively shaped Indian horse culture: from the original mounted Indo-Europeans to the Scythians, Persians, and Greeks. "Kautilya [Chanakya] wrote just after Alexander the Great's death," explains Ann Hyland in *The Horse in the Ancient World.* "From the equestrian content, it must be taken as a distillation of what was current at, and for a considerable time prior to, Alexander's era. . . . So detailed are these that every aspect of equestrianism can be extrapolated."

Indian horses raised on royal stud farms stood fourteen to fifteen hands, with their temperament graded as "furious, mild, stupid, or slow." The superintendent of horses, echoing the posts established by the Assyrians and Macedonians, registered and assigned all horses to suitable duties. I assume the "stupid or slow" were not conscripted into cavalry service.

In 303 BCE a Macedonian ambassador to the Mauryan court was struck by the sophistication of the training methods and professionalism of Indian horse masters, who possessed "a strong hand as well as a thorough understanding of horses." The multifaceted roles of Indian cavalry follow the designs instituted by Alexander. Mirroring the earlier writings of Xenophon, the *Arthashastra* details the complex education of individual horses and the cutting-edge employment of cavalry. "Some manoeuvres that have been credited to the European Renaissance horse masters," adds Hyland, "had already had an ancient history in India."

Although Alexander turned his army west, he was by no means satisfied with his exploits, nor was he ready to burn out or fade away. Stopping in Babylon, he gave orders to prepare for an invasion of Arabia and North Africa, with his eye trained on Carthage and the western Mediterranean. Europe via the Rock of Gibraltar and Spain would have been in his sights. He simultaneously tasked reconnaissance missions to the shores of the Pontic-Caspian Steppe to secure a line of departure for the eventual rekindling of his Asian quest. "There is one thing I think I can assert myself," claimed Arrian, "that none of Alexander's plans were small and petty and that, no matter what he had already conquered, he would not have stopped there quietly." Alexander, however, would never reach his paradise cities at the unknown ends of the world, at least not in this lifetime tour.

The larger-than-life Alexander the Great, who had survived no fewer than eight serious battle wounds, died in Babylon—likely of malaria—in 323 BCE at the age of thirty-two. His body, which was visited by venerating admirers and disciples, including Cleopatra, Julius Caesar, Pompey, Caesar Augustus, and Caligula, rested in his namesake city of Alexandria, Egypt, until the fourth century, when it disappeared from the historical record.*

More than 2,300 years later, Alexander is one of those rare historical figures who still resonates today, and has captured the imagination, adoration, and respect of admirers across time. His "Great" epithet now rolls off the tongue as a proper surname, and with good reason: by revolutionizing and legitimizing cavalry, and conquering the almost ends of the earth, Alexander and Bucephalus transformed our world.

While Alexander's territorial gains were quickly erased by infighting and the absence of centralized authority, the enlightening legacy of his

---

* The sadistic, perverse, and unhinged Roman emperor Caligula, who ruled between 37 and 41 CE, planned to appoint his favorite horse, Incitatus ("at full gallop") to the Roman consulship. Cloaked in a purple silk blanket and a gem-studded collar, Incitatus was tasked with inviting dignitaries to join him at extravagant banquets hosted in his marble and ivory stable, where he dined on handpicked oats mixed with gold flakes. Anyone who dared to decline was tortured. Caligula also frequently pranced around in public dressed up as Alexander, even wearing the deceased's golden breastplate—which he'd poached while visiting Alexander's sarcophagus.

Hellenistic empire endures to this day. Following his death, Greek sociocultural influence peaked across Europe, North Africa, the Middle East, and western Asia.

Through academia and action, he mastered the art of war by fusing professional cavalry with infantry and other combat trades to create the modern combined arms military doctrine. This guiding principle of war is as potent now as when he perfected it. Alexander also entrenched the value of cavalry on the battlefield for the next 2,200 years! As Louis DiMarco reminds us, "The similarities between Alexander's lance-armed companion cavalry and the Indian lancers of Allenby's cavalry divisions were far greater than the differences." Alexander's influence, however, stretches far beyond strictly military innovations.

His roaming wars and restless wanderings pollinated cross-cultural academia and innovation. He exposed his sprawling, eclectic empire to the attributes of Greek civilization and significantly broadened trade between East and West. Gold and silver coinage bearing the likenesses of Philip and Alexander, for example, have been found from France to Central Asia. "Macedon was not merely a military upstart created by ephemeral military genius," stresses Robert Gaebel, "but rather a large, wealthy, well-managed state that was a major economic and political force." His embryonic fusion of Eurasia would be reinforced 1,500 years later by horses ferrying European traders like Marco Polo across the Silk Roads transecting the immense Mongol Empire of Chinggis Khan and his unrivaled mounted archers. Alexander's dream passed from one horse lover to another.

During his conquests and explorations, Alexander emulated his idol Cyrus the Great's promotion of cultural, technological, and religious tolerance, reciprocation, and exchange, and, like Cyrus, he nurtured the arts, engineering, and scientific intrigue. Alexander took along writers to catalog new discoveries and ideas, sending vast collections and observations back to his aging master Aristotle, who died a year after his most famous apprentice.

Alexander also heeded the advice of his father: that it was better to "be called a good man for a long time than master for a short one." In place of authoritarian rule, Alexander retained local administrative systems and constructed extensive infrastructure. He erected twenty-four major cities, including Alexandria, Iskenderun (Alexandretta), Kandahar, Herat, Bagram, Merv, and Khujand (many of which still exist today), while renovating more than fifty others in Greek tradition. The remains of a Greek site at Ai Khanoum in northern Afghanistan, for example, contains Hellenic fortifications, a gymnasium, a theater, a treasury, a palace for its Greek king, and a library containing the works of Greek and other literary masters.

Although his stays were brief and transitory as he drove his army east, Alexander instigated the spread of Hellenic ideas across his empire while establishing lasting Greek colonies to control it. "If we wish not just to pass through Asia but to hold it," he instructed his inner circle, "we must show clemency to these people; it is their loyalty which will make our empire stable and permanent." Hold it he did. The legacy of his achievements is permanently stitched into the fabric of our modern world.

Greek culture flourished across his empire long after his death. As Alex Rowson acknowledges in *The Young Alexander*:

> Few historical figures have exerted such a persistent influence from beyond the grave.... His many self-named city foundations (Alexandrias) facilitated trade and exchange between East and West. Homer began to be read by Persian schoolboys, the tragedies of Sophocles and Euripides were staged in new, exotic locations, [the Oracle of] Delphi's words of wisdom were displayed in far-off Baktria (Afghanistan); the mixing of Greek and eastern cultures spawned a multitude of ideas and art forms. It was the dawn of a vibrant new age ... set between the Classical era of Greek city-states and the establishment of the Roman Empire, a time of great innovation and productivity that rivals any in human history.

There are still thriving diasporas of Pontic Greeks in the Donetsk Province (including the city of Mariupol) of Ukraine, to the Pontus region of northern Turkey, to Georgia, Kazakhstan, and Uzbekistan. Alexander is still making headlines, all thanks to his taming Bucephalus, his creative military mind, and his mighty cavalry horses.

He assembled a stretching network of libraries and think tanks (which came to include the Great Library of Alexandria erected around 285 to 280 BCE), where he deposited copies of Greek texts in addition to reproductions of those he had gathered or translated into Greek from other cultures. This massive undertaking and academic enterprise ensured the broad dissemination of diverse ideas across his increasingly vast, multicultural empire. "Without him, it is unlikely that Greek culture and language would have spread so far or had the chance to take root on such a scale. . . . Syria, Asia Minor, and Egypt remained Hellenistic in language and culture for more than a thousand years, until at least the end of the Roman Empire," reminds Adrian Goldsworthy. "It is worth recalling that as well as the Gospels being written in Greek, Jesus is the Greek form of the Aramaic, originally Hebrew, name Joshua." Christians in the Middle East still speak Greek. The Ptolemaic (Macedonian Royal dynasty) rule of Egypt and its cultural influence, for example, lasted until the death of Cleopatra in 30 BCE. This synthesis of culture and academic advancement is perhaps Alexander's greatest achievement.

Exploding from the heart of his former empire, Greek literature, architecture, science, mathematics, philosophy, and military strategy were disseminated across a wider berth and flourished in an age of intellectual prosperity and progress. In the great madrassas and libraries constructed across the blossoming Arab expanse during the Islamic Renaissance, scholars pondered the principles of Socrates, Plato, Aristotle, Xenophon, Herodotus, Hippocrates, Thucydides, Aristophanes, Archimedes, and Hypatia of Alexandria, whose works occupied the brimming shelves of archives that Alexander helped build.* During the cross-cultural exchange of the Crusades, Islamic academics extended Europe a scholarly ladder by reintroducing Greek and Roman literature and culture, as well

---

* Hypatia was the first recorded prominent female mathematician and philosopher. An earlier mathematician, Pandrosion of Alexandria, of whom little is known, may also have been female.

as their own refinements and academic advancements. This is the true legacy of Alexander and his vaunted cavalry horses.

Through his insatiable drive to reach the ends of the Earth, much of his adolescent reading list, compiled by Aristotle, has not only survived but thrived. It was composed of names that now echo in Indo-European and other languages across the classrooms and hallways of academic institutions all over the world. Successive generations of scholars, generals, and dreamers have used their concepts and discoveries as a springboard to see further into the limitless beyond. Even the great genius Sir Isaac Newton gravitated toward the notion that if we "have seen further it is by standing on the shoulders of Giants." We all owe Alexander, Bucephalus, and his fellow warhorses a debt of gratitude. If greatness is the measure of influence and importance, then he is certainly one of the greatest of all time. The world still lives among the lingering equine shadows of Alexander's empire.

The Romans, who viewed Alexander as the greatest of the greats, were eager beneficiaries of his Hellenic traditions. The Greco-Roman culture that followed dominated Europe, North Africa, and portions of the Middle East for the next seven hundred years, profoundly shaping the eras that followed. Many legal and political systems, for example, are an adaptation of Athenian or Roman law and democracy.

The Roman Empire also first martyred and subsequently facilitated the migration of Christianity across the unstable boundaries of Europe. "In his bold sweep east," notes anthropologist Pita Kelekna in *The Horse in Human History*, "Alexander's cavalry had bravely laid the ground for the emergence of the Rome-Byzantium era and the subsequent spread of Christianity." This new remedial faith found its way across the sprawling Roman Empire on an unprecedented network of roads clattering with the thriving sounds and snorts of wagon wheels and horses.

Far outpacing the efforts of the Assyrians or the Persians, the Romans constructed an unrivaled network of military roads to shuttle a standing army of three hundred thousand to five hundred thousand soldiers across a maturing empire. A Roman legion could march upwards of thirty miles a day, while those on horseback could cover more than a hundred miles. At its territorial zenith in 117 CE, this transportation system snaked across 2 million square miles of holdings, home to roughly 75 million subjects (21 percent of the global population). The Romans constructed over 250,000 total miles of thoroughfare, including 53,625 miles of permanently paved, four-foot-thick trunk roads, most of which are still in service today. To put this feat into perspective, the current US Interstate Highway System totals 46,876 miles.

Based on the constraints of their massive, four-wheeled, oxen- or horse-drawn vehicles, where at all possible, Roman roads ran in perfectly straight lines. "Horses were essential for the many pack trains and, along with oxen, were hitched to wagons," explains Bill Cooke, former director of the International Museum of the Horse in Lexington, Kentucky. "Without a fifth wheel, which allows the front axle to rotate, it is extremely difficult to turn a four-wheeled vehicle, possibly explaining why the Romans almost invariably laid out their roads in a straight line." These roads not only allowed for the rapid movement of military forces, but also for extensive lines of communication and continuous trade of goods and services across the imperial realm and beyond.

Modeled on the Persian Angarium postal system, messengers and dispatch riders of the impressive Roman *cursus publicus* ("the public way") crisscrossed the empire delivering mail, packages, and people. Relay stables holding as many as forty horses were sited every 8 to 12 miles, while full-service stations/garrisons were spotted in intervals of 20 to 30 miles. Government dispatches marked "Urgent" could travel upwards of 240 miles a day.

The *cursus publicus* also shuttled official personnel, baggage and freight, as well as military equipment and supplies. Public "stagecoach" transit could cover 60 miles a day. A Christian pilgrim who traversed the 3,000-mile overland route from Bordeaux, France, to Jerusalem in 333 and 334 CE noted in his diary that he encountered 305 posts along the way—roughly one every ten miles. The "Pilgrim of Bordeaux" carefully distinguishes between the small quick-change horse stables, the more comprehensive stopover stations, and those housing military garrisons.

The adoption of Alexander's Hellenistic culture also stimulated Roman scholarship. Horses are vividly represented by some of the most celebrated academics and authors of the Roman era, including Varro,

Virgil, Strabo, and Pliny the Elder. More specifically, *On the Nature of Animals* and *Historical Miscellany*, composed by the writer known as Aelian in the early third century CE, detail particulars on horse breeds, propagation, and management, while Vegetius's late-fourth-century CE *Digesta Artis Mulomedicinae* provides a comprehensive guide to veterinary medicine. The Romans also left their permanent mark on horsemanship and cavalry through their introduction of horned, or pommeled, treed saddles and the first horseshoes, known as *hipposandals* in the adopted Greek (*soleae ferreae*, or "iron soles," in Latin), worn much like slide-on slippers or flip-flops.

Across Roman manuscripts and records, more than fifty distinct horse breeds are mentioned by name, including the top-tier Thessalian, Thracian, Nisean, and Turkoman, all standing fifteen to sixteen hands. Commercial horses were generally categorized according to three grades: noble, breeding, and common stock. The bustling horse markets also peddled donkeys, mules, and camels, all important sources of traction for agriculture, trade, transportation, and war across the extensive, but easily accessible, empire.

Horses, however, did not dominate the Roman world as much as one might rightly assume. "The Romans were not great equestrians," Juliet Clutton-Brock affirms bluntly. "Although they used horses for riding, warfare, and racing, the horse was not as much a part of Roman life as it appears to have been in classical Greece." As a result, the handful of Roman commanders who employed effective cavalry, including Julius Caesar, were the exception. "If the Hellenistic-style cavalry tactics were characterized by maneuver and mobility," points out military historian Jeremiah McCall, "Roman tactics certainly were not." As we will see, these equestrian deficiencies would come back to haunt Rome when a merciless horde of Hun horsemen descended from the steppe.

The near annihilation of seven Roman legions under general and statesman Marcus Licinius Crassus at the hands of ten thousand Scythian-styled Parthian mounted archers at Carrhae on the Turkish-Syrian border in 53 BCE illustrates the manifest shortcomings of Roman cavalry. Crassus was killed, along with twenty thousand of his men. Another ten thousand legionaries were captured and trudged east into the flesh markets of Alexandria Margiana (Merv), one of Alexander's refurbished cities and a major trading center in Turkmenistan along the Steppe Roads.

According to legend, grizzled veterans of this "Lost Roman Legion" supposedly drifted farther east and were later spotted fighting as mercenaries alongside Xiongnu horsemen in one of their many forays against the coalescing Chinese Han dynasty. While this apocryphal tale lacks all historical credibility, Rome and China did slowly piece together the puzzling fragmentary map of each other's existence. "The central period of Classical Antiquity, from the third century BC to the third century AD, was marked most notably by the development of the Roman and Chinese Empires," explains Christopher Beckwith. "They expanded to great size until they dominated the western and eastern extremes of the Eurasian continent. Both expanded deep into Central Asia." Inevitably, their worlds collided through swelling commerce, including the prized commodity of silk, seeping across the Steppe, or newly christened Silk, Roads.

While Rome was extending its imperial tentacles across the increasingly interconnected steppe, the warring Chinese states were fighting for their existence in the face of a formidable, mounted adversary. Although seemingly strange bedfellows, the fates of Rome and China were fused, bound to the nomadic Xiongnu (Huns) and other horse-based societies of the steppe. The rise of China directly influenced the fall of Rome.

The appearance of a relaxed but robust Xiongnu federation on the eastern steppe, coupled with a changing climate, set off a domino effect of displacement rippling westward toward Europe. Waves of nomadic mounted migrants found their way into the heart of Europe and marched on the eternal city of Rome.

The capricious quick-striking raids of the Scythian-styled Xiongnu also threatened to overrun the war-weary Chinese states. Walls and chariots could not defeat this common enemy or stem the crisis unfolding on their northern border. The very survival of China rested with two saviors: confederation and the cavalry horse.

CHAPTER 9

My Kingdom for a Horse
The Hitched Fates of the Chinese and Roman Empires

The embryonic Chinese nation was teetering on the precipice of disaster. Its very existence hung in the balance and was dependent on one thing: the Chinese desperately needed horses. The plundering raids and demands for tribute exacted by their unruly, nomadic Xiongnu neighbors to the north became more than the young kingdom could bear. "Horses are the foundations of military might, the greatest resource of the state," declared the brilliant Han politician and general Ma Yuan. Without horses, he concluded, "the state will fall." As a matter of national survival, the Chinese desperately needed horses.

Although only fifteen years old when he ascended to the throne in 141 BCE, Emperor Wu of the Han dynasty decided that enough was enough. Rumors of heavenly, blood-sweating horses from the uncharted regions beyond the forbidding western deserts and snow-crowned mountains offered deliverance from Xiongnu despoliation.* Though Wu could not have known it, these were the very same majestic horses Alexander had sought when he penetrated the Fergana valley and established the remote eastern outpost of Alexandria Eschate.

Salvation and the ability to shrug off the yoke of Xiongnu domination rested on securing these enigmatic "flying dragon-like horses." The eighth-century CE poet Li Bai captured the Chinese fascination with these mysterious beasts: "The Heavenly Horse sprang from a Tocharian

---

* The "blood" refers to a subcutaneous parasite common in Central Asia that causes slight hemorrhaging of the skin when horses are active. When mixed with sweat, it creates a discolored pinkish foam or bloody paste.

cave: / With tiger-striped back and dragon-wing bones. / Neighing to clouds in the blue, he shook his green mane. / An orchid-veined courser, he ran off with a flash / Up the Kunlun Mountains vanishing over the Western horizon." The teenaged emperor was obsessed with seizing these horses by any means necessary.

In 138 BCE Emperor Wu's trusted soldier, envoy, and adventurer, Zhang Qian, was tasked with leading a diplomatic delegation to establish political and economic relations with the great powers and peoples to the west. The overriding objective of this perilous and seemingly impossible mission, however, was to procure the prized horses of the Fergana valley. This was the first of numerous intrepid consular and military quests to secure horses. These enterprising western expeditions had a profound influence on the development of China, Central Asia, and Europe, while also defining the modern era.

On the backs of these heavenly horses, China metamorphosed from a collection of diminutive warring states to a world superpower. The

Flying horse of Gansu: This statue (circa 200 CE) representing the prized heavenly horses has become an iconic emblem of China. *(Wikimedia Creative Commons)*

horse shaped the modern world order, with China as a central player. "Its influence upon many aspects of Chinese history has been tremendous," wrote the acclaimed sinologist at the University of Chicago Herrlee Creel. "For some two thousand years, China's foreign relations, military policy, economic well-being, and indeed its very existence as an independent state were importantly conditioned by the horse." Although horses were introduced to China by Indo-Europeans through the Tarim Basin between 1500 and 1200 BCE, they were still rare beasts, acquired only through trade or raid with the nomadic Xiongnu to the north.

Like the Greeks, the Chinese were relatively slow to appreciate the capabilities of cavalry and balked at the thought of mounted warfare. These were the crude practices of northern steppe barbarians, broadly referred to as Hu (Xiongnu, or Hsiung-nu, or Huns), and not worthy of emulation by enlightened Chinese society. To the Chinese, the Xiongnu were as uncivilized as the Scythians were to the Greeks, or, later, as Indigenous peoples were to the Americans.

This ancient xenophobia aside, the Hu (and their name) have a complicated but important lineage for another ancestral bloodline of adroit horsemen who ravaged Europe and helped expedite the ruin of the decaying Roman Empire: Attila and his merciless Huns. Recent DNA studies, including those led by geneticist Christine Keyser, consistently reveal a direct bloodline of "Scytho-Siberians as ancestors of the Xiongnu and Huns as their descendants."*

A wealth of knowledge about the Xiongnu can be gleaned from the *Records of the Grand Historian* (*Shiji*) completed by the preeminent ancient Chinese historian Sima Qian in 94 BCE. It is considered the foundational work of Chinese culture, equivalent to *The Histories* by Herodotus or the historian Livy's monumental chronicle of Rome written between 27 BCE and 9 BCE. His extensive observations on the culture and customs of the Xiongnu mirror those of Herodotus when depicting their Scythian ancestors at the other end of the Eurasian Steppe.

The Xiongnu also fashioned the skulls of their slain enemies into

---

* *Xwn* (Hun), or *Hu*, is the root of their name in the Mongolian, *Hunnu*; the Indian Sanskrit, *Huna*; and the Chinese cognate, *Xiongnu*. They are known as the *Hsiung-nu* in an older and outmoded system of English translation.

wine goblets, used scalps and skin for horse garlands and human garments, and sported traditional tattoos. Like their Scythian kin, the primary weapon of the Xiongnu was the recurved bow, but with one ingenious modification: the lower limb was shorter than its upper counterpart. While this does not affect its performance, it affords the bow additional freedom of movement over a horse's neck when switching direction of fire—an essential alteration for the Xiongnu, who could shoot with both hands.

True to their Scythian roots, the Xiongnu were also formidable mounted warriors. "They lacked many elements of statehood, yet they forged the first nomadic empire—the third greatest land empire in history before the rise of the modern super-states (the first and second being the Mongol empire and the medieval Muslim empire)," highlights historian John Man in *Empire of Horses: The First Nomadic Civilization and the Making of China*. "They are the reason China reaches so far westward. They inspired one of the world's best-known monuments, the Great Wall."

The Chinese adoption of mounted warfare and the construction of the first great walls coincide with ever-increasing despoliation raids into Chinese territory by the Xiongnu. Facing the bleak prospect of annihilation, subjugation, and poverty through endless tributes of treasure and human capital, the rival Chinese states were forced to embrace the steppe-style, quick-strike warfare and mounted archery of their enemy.

Although access to horses was limited and unpredictable, in 307 BCE King Wuling of Zhao (a contemporary of Alexander the Great's) began to train select troops, dressed in Hu trousers and tight robes, in the skills of mounted archery. His experimental cavalry spurned all Chinese conventions and norms. Wuling was admonished by his advisors for adopting the dress and drill of uncivilized northern reavers. "People of faraway countries admire and learn from here, barbarians emulate the way things are done here. Now your majesty is giving up our high standards to follow the clothing style of outsiders, thereby changing the teachings of our ancestors and the ancient ways," rebuked a senior counselor. "This will upset your people and make scholars angry, as it deviates from the values of the Middle Kingdom [China]. Your majesty's subject wishes you to reconsider your decision."

Wuling appreciated fully that he would be "exposed to the hatred of the vulgar people" and that "later generations will surely criticize me for this, but what can I do? . . . My ancestor built a wall where our lands touch those of the nomads and named it the Gate of No Horizon. . . . Since benevolence, righteousness, and ritual will not subdue the barbarous Hu, we must go and defeat them." Despite mounting objection and ridicule, Wuling remained steadfast in his decision to raise Zhao horsemen based upon the Hu steppe model. "A talent for following the ways of yesterday," he informed his court, "is not sufficient to improve the world order of today."

With his conversion to cavalry and khakis, the fighting capabilities and fortunes of Wuling's forces improved immediately. Crossing the Gate of No Horizon, the visionary king of Zhao "dressed in barbarian garments, led his horsemen against the Hu." After a series of victories against the Xiongnu, Wuling constructed a larger wall on the northern side of the Yellow River at the Ordos Loop (a rectangular bend in northern China). According to a Chinese text from 244 BCE, "The Xiongnu still did not dare to come close to the cities on the border of Zhao." The prescient leadership of Wuling and his pants-wearing mounted archers paid off.

For the Chinese, cavalry was now directly tied to statecraft and sovereignty. "China's very survival relied on its equestrian prowess. From the fourth century BC forward, the empire's greatest threat came from its nomadic neighbors to the north and west," states Bill Cooke. "These northern tribes fielded some of the finest cavalry the world would ever see, while providing a constant thorn in the side of the Chinese." To combat this spiraling danger, the twin pillars of Chinese military policy—constructing walls and amassing cavalry horses—dominated the country's defense strategy for the next five hundred years.

While assembling walls was a Herculean task, acquiring horses proved even more difficult. The major Chinese cultural and political centers of Xi'an, Luoyang, Beijing, and Chengdu were 2,000 to 2,500 miles east of the major horse-trading hotbeds of Central Asia, separated by great distance, remote deserts, rising mountains, and endless untamed steppe occupied by fantastic beasts and ferocious barbarians. The gaze of the sparring Chinese states slowly drifted northwest as their covetous thoughts were increasingly consumed by horses. Before these kingdoms could project their power, however, their centuries-long conflict needed to be resolved.

These protracted rivalries were finally settled during the period of the Warring States (475–221 BCE), reminiscent of the Peloponnesian Wars in Greece. Shihuangdi, leader of the battle-hardened Qin, ruthlessly subjugated and annexed surrounding territory. According to Sima Qian, his armies "ate up their neighbors as a silk-worm devours a mulberry leaf." Between 234 and 221 BCE, one by one the other six contending states were conquered and absorbed by Qin into one unified body politic referred to as the Central Nation or the Middle Kingdom—names still in use in China today. Outsiders, however, associated this region with its dominant founding power of Qin ("Chin") or China.

Chinese Dynasties

Warring States period	475–221 BCE
Qin dynasty	221–206 BCE
Warring interregnum	206–202 BCE
Han dynasty	202 BCE–9 CE
Xin dynasty	9–23 CE
Han dynasty	23–220 CE

Although the cruel and despotic rule of the Qin dynasty was short lived, surviving its founder and first emperor, Shihuangdi, by only four years, it laid the foundations of centralized governance, imperial designs, and modern infrastructure—including canals, roads, palaces, walls, and the emperor's unrivaled terra-cotta tomb complex. Managed by a streamlined administrative system and subsidized by taxes from forty-two million Qin subjects representing 28 percent of the global population, a national conscription of labor fabricated the first massive construction projects in China.*

Like the Persians and Romans, Shihuangdi initiated the creation of a sophisticated 4,700-mile Chinese road system. Express lanes with full-service rest stops and official relay stations were designated for authorized couriers ferrying messages for a refined bureaucracy employing more than one hundred thousand government officials. Oxen, and to a lesser extent horses, donkeys, and mules, powered commerce, communication, and centralized administrative control.

The 460-mile "Straight Road" was constructed by a crew of 7,500 laborers between 214 and 210 BCE. It shot north from the first emperor's royal seat at Xi'an, through the heart of Inner Mongolia (Ordos), and toward the Yellow River and Xiongnu territory. This new demarcation was secured by the construction of the Great Wall. "I have travelled to the northern border and returned by the Straight Road," reported our grand historian, Sima Qian. "As I went along, I saw the outposts of the Great Wall."

It is worth remembering that this colossal building project, unparalleled in human history, was also an unrivaled response to seemingly invincible Xiongnu horsemen. "Their raids against the Chinese," write J. P. Mallory and Victor Mair in *The Tarim Mummies: Ancient China and the Mystery of the Earliest Peoples from the West*, "prompted the erection of the nucleus of that most famous of Chinese monuments."

The Great Wall snakes like a main circuit cable across the serene and diverse landscapes of deserts, grasslands, forests, and mountains of northern China. It is also an angry, ancient scar and a tangible reminder of the shuddering terror instilled in the Chinese people and their rulers by plundering mounted steppe archers.

In addition to consolidating the Chinese nation-state and the Great Wall, perhaps the first emperor's ultimate building legacy was not meant for this world but for the next. The army of the terra-cotta warriors, assembled by 700,000 prisoners and slaves from 246 BCE until his death

---

* The nomadic population of Mongolia during the domination of the Xiongnu and, later, during the imperial era of Chinggis Khan never exceeded one million to one and a half million.

Terra-cotta army of Shihuangdi: Bronze chariot and horses at the tomb of the first emperor, Xi'an, China, circa 210 BCE. *(Wikimedia Creative Commons)*

in 210 BCE, escorted Shihuangdi to the afterlife. This earthenware military parade contains the life-sized and individualized likenesses of an estimated 8,000 soldiers, 130 chariots pulled by 520 horses, 150 cavalry horses, and more than 40,000 operative bronze weapons. Shihuangdi's personal army of (or perhaps for) the dead has no comparison in world history.*

His draconian administration, unpopular military conscription, and exhausting forced labor, however, alienated the populace, igniting insurrection and civil war. This bitter Warring interregnum, from 206 to 202 BC, concluded with a convincing Han victory. The Han dynasty, which ruled from 202 BCE to 220 CE (with a brief interruption from 9 to 23 CE), lives on in the majority ethnic Han population (92 percent) of modern China.

The Han were far more successful in preserving their identity in the face of an onslaught of predatory horsemen than were the Neolithic farmers of Old Europe, who were overrun and erased. Horses made all

---

* This archeological treasure went undiscovered for more than two thousand years until 1974, when a group of peasants digging a well unearthed fragments of a clay warrior.

the difference. Emperor Wu's obsession to secure heavenly horses from the West would eventually safeguard Han dominance against the nomadic imperialism of the Xiongnu.

During the initial consolidation of Han power, however, war with the Xiongnu was a losing proposition. A charismatic and intelligent leader named Modu Shanyu initiated a Xiongnu renaissance by merging the mounted steppe power of the disparate clans into a cohesive and powerful nomadic confederacy or "shadow empire" to counter that of the nascent Chinese state.* The combined strength of the mounted Xiongnu was now an even greater menace to the security of the Middle Kingdom.

In nomadic societies like the Scythians, Xiongnu, Huns, Comanche, and Sioux, every adult male, and in some cases females, were trained in the cross-discipline arts of mounted hunting and warfare. As herders, they were not tied to a centralized bureaucracy, static cities, or farming fields. When detailing the customs of the Xiongnu, Sima Qian reported:

> The animals they raise consist mainly of horses, cows, and sheep, but include such rare beasts as camels, asses, mules, and wild horses. They move about in search of water and pasture and have no walled cities or fixed dwellings, nor do they engage in any kind of agriculture. . . . The little boys start out by learning to ride sheep and shoot birds and rats with a bow and arrow, and when they get a little older, they shoot foxes and hares, which are used for food. Thus, all the young men are able to use a bow and act as armed cavalry in time of war.

The ratio of warriors to the general population within these horse-based cultures reached one to four. By contrast, during the global twentieth-century crises of the World Wars, even with conscription, the maximum estimated ratio was only one to ten!

Within their crafty management of the borderlands, the Xiongnu made a habit of launching fierce attacks against the Chinese as a

---

* Honoring his brilliance, *Shanyu* became the official title bestowed upon all future Xiongnu rulers.

reminder of their military leverage. These lightning strikes were immediately followed by overtures of peace attached to inflated tributes, creating a cyclical pattern of raiding and trading, or "trick-or-treat" diplomacy. The Xiongnu price-gouged the Chinese at every opportunity, demanding the exorbitant price of one million bolts of silk for one hundred thousand horses.* This expenditure could not be sustained without bankrupting the Han dynasty. The Xiongnu also expanded their borders, carving out a broad land base (centered in north-central Mongolia) 1.3 times the size of Han China.

They stretched their tentacles to control the natural topographical passes of commerce through the narrow five-hundred-mile-long strategic bottleneck of the Gansu Corridor (also called the Hexi Corridor) connecting the plains of China to the caravan trails surrounding the Tarim Basin and its western approaches. This expansion allowed them entrance into the long-established markets peppered across the ancient Steppe Roads of Eurasia, including the strategic melting pot of Xinjiang. "It was a prelude to the growth of the lateral east-west networks that were soon to develop," notes Barry Cunliffe, "creating the intricate cross-continent connectivity popularly referred to as the Silk Road." The Xiongnu now controlled the vast stretches of the steppe from the Yellow Sea in the east, through the Gansu Corridor, to the lake districts in the eastern slices of Kazakhstan and Kyrgyzstan. This also meant command of the money-spinning silk and horse trade.†

The Xiongnu were now firmly entrenched between China and the prized horses of Central Asia they long coveted. China's survival depended on securing direct access to the heavenly horses of Fergana. As Chao Cuo, a Han official nicknamed "the Wisdom Bag" for his sound counsel, conceded in 169 BCE, "In climbing up and down mountains, and crossing ravines and mountain torrents, the horses of China cannot compare with those of the Xiongnu." Their warriors and horses, he

---

* One bolt of silk equated to a fifty-five-foot roll weighing 5.3 pounds.

† The earliest indications of silk in the Mediterranean compass fall somewhere in the fifth and fourth centuries BCE, corresponding with Scythian Pazyryk burials (and their silks) mentioned earlier. These silks came from wild or domesticated sources in India, quickly followed by those from China.

added, could withstand "rain and storm, exhaustion and fatigue, and hunger and thirst.... If they ever suffer a setback, they simply disappear without a trace like a cloud." Without a stable supply of steppe horses, the Chinese would never break free from the shackles of Xiongnu subjugation. Unification without sovereignty was not a viable path forward. Horses were the only means of freedom.*

To remedy this situation and confront the Xiongnu, successive Chinese emperors, beginning with Wu, sent exceedingly costly expeditions through the Gansu Corridor to Central Asia to acquire heavenly horses. According to Herrlee Creel: "By virtue of it [the horse], the nomads had become a deadly threat and were able at times to invade Chinese territory almost at will. The Chinese had to develop cavalry to counter the nomads, and even though they made great economic sacrifices to breed cavalry horses, they still had to secure additional mounts from outside their borders. Both to secure horses and to outflank the Xiongnu, they pushed far into Central Asia, opening a new chapter in China's political and military history and in its foreign relations."

When Wu became emperor in 141 BCE at the ripe old age of fifteen, he was counseled immediately on possible approaches to the Xiongnu predicament. "When the Han conclude a peace agreement with the Xiongnu, usually, after a few years, the Xiongnu violate the treaty," his cabinet informed him. "It would be better to reject their promises and send soldiers to attack them." There was one glaring hitch in this plan: a radical shift in policy from appeasement toward military action would require large numbers of horses that the Chinese still did not possess. Perhaps more vexing was the fact that any proximate or potential stocks were handled by the very enemy they sought to expel.

Following fierce debate, the Chinese decided to circumvent the Xiongnu by securing horses directly from the West. The emperor had learned from wayfaring traders that the regions beyond the spiking mountains and desert dunes of Xinjiang offered the salvation China sought so desperately: bountiful herds of heavenly horses. To solidify

---

* Within the Han dynasty, the minister responsible for the procurement of horses ranked eighth in the corridors of power.

policy and garner support, Emperor Wu conveniently proclaimed that a vision presaged that "divine horses are due to appear from the northwest." If the Xiongnu, or any other belligerent, stood in the way or sought war in the process, then so be it.

The Martial Emperor, who reigned for an astounding fifty-four years, thrust China into a series of foreign adventures, plunged headlong into total war against the Xiongnu, and deployed far-fluing western military campaigns to secure horses. "His answer to the Xiongnu menace was to set the boundaries of China wider," explains John Man, "and to do that, he decided to escalate the rumbling rivalry into a full-scale war. From this decision flowed many consequences—for the Xiongnu, for China"— and eventually for Rome. The flames of war and the ensuing embers of modern China began to flicker on the endless horizon of the steppe.

The strategic blueprint was simple. According to Sima Qian, the Chinese needed to "cut off the right arm" (western trading bases) of the Xiongnu in the Gansu Corridor and Tarim Basin "to create a split between the Xiongnu and the states to the west which had up to this time supported them." As a windfall to this military thrust, the Chinese would have unrestricted access to horses and other exotic goods by controlling the oasis trading hubs ringing the Tarim Basin, including Kashgar, Hotan, and Urumqi. Emperor Wu decided to act. The Chinese desperately needed horses.

His trusted advisor, Zhang Qian, and a deputation of ninety-nine others set out from the capital of Xi'an in 138 BCE on a daring mission to procure the legendary heavenly horses and deliverance from the Xiongnu. Pausing for a brief supply stop at the Great Wall fort of Longxi— the last Han outpost on the western frontier—the party advanced northwest through hostile Xiongnu territory toward the Tarim Basin. While leapfrogging from oasis to oasis across the Gobi Desert toward the Gansu Corridor, Zhang Qian was captured by the Xiongnu, who dispatched or enslaved his emaciated delegation.

After ten years in captivity, he escaped, determined to complete his mission and return to China with horses. Zhang Qian pushed his way west across the despondent deserts of Xinjiang, over its parched mountain passes, eventually landing in the verdant pastures of the Fergana valley. Zhang gathered valuable geographic, commercial, and cultural information about Central Asia, India, the Pontic-Caspian Steppe, Parthia, Mesopotamia, and even glints of Rome. In pursuit of horses, he accrued knowledge of at least thirty-six distinct cultures and the diverse trade routes along the Steppe/Silk Roads, previously unknown to the Chinese, forever changing the geographic and demographic configuration of the region and, by extension, the world.

While roaming the markets, Zhang was amazed and bewildered at the international selection of merchandise. In Bactria, for example, he was astonished to find Chinese bamboo products and woven fabrics. These goods had made the 2,200-mile journey from Sichuan Province via the southern trade routes through India, where, he was told, much to his incredulity, they used "elephants in warfare." After roughly a year exploring the trading hubs of Central Asia, Zhang Qian began his long journey home. Escaping the clutches of the Xiongnu for a second time, he finally returned to Xi'an in 125 BCE after a remarkable thirteen-year adventure covering more than 8,000 miles.

Delighting a captivated audience, Zhang Qian delivered the first thorough Chinese account of the West to Emperor Wu and his advisors. His colorful description of commercial thoroughfares and tangled bazaars, including the Fergana valley and its fabled blood-sweating horses, stimulated as much conversation as it did curiosity. It also triggered military action. Wu's expansive imperial vision rivaled that of Alexander. The young Chinese emperor, however, would launch his crusade to the ends of the world toward the West and not from it.

After years of low-intensity conflict on the northern borderlands, the Sino-Xiongnu War picked up tempo. In 121 BCE Wu dispatched his most trusted cavalry commander, Huo Qubing, to wrestle the strategic commercial artery of the Gansu Corridor away from the Xiongnu. At only nineteen years of age, Huo had already earned a glittering reputation as an exceptional horseman and archer. He also recognized the simple truth that now dominated strategic dialogue: "Whoever wishes to rule China must rule the Gansu Corridor."

After a sixty-day cavalry march and a series of fierce skirmishes claiming thirty thousand Xiongnu, Huo penetrated the corridor to the

My kingdom for a horse: Emperor Wu (*on horseback*) blessing Zhang Qian (*kneeling to the left*) and his western expedition to procure heavenly horses. Mural in Mogao Caves, Dunhuang, China, seventh century CE. *(Picture Art Collection/Alamy Stock Photo)*

fringes of the Tarim Basin. As customary, almost two million Chinese were relocated to these newly conquered multiethnic lands, where they quickly became the dominant majority. This colonizing Han population was slightly larger than that of the entire Xiongnu Confederacy. To cement these gains, the Great Wall was extended to enclose the Gansu Corridor. Chinese influence was slowly and methodically spreading west.

Following these decisive victories, in January 119 BCE the Chinese launched a massive two-pronged northern offensive consisting of over three hundred thousand troops sent deep into Xiongnu territory. Huo Qubing orchestrated another resounding victory. The Xiongnu were "shocked to the core" and cut to pieces. Xiongnu losses, estimated at seventy thousand, were so severe that entire clans ceased to exist. With the brilliant performance of the Chinese columns during the Battle of Mobei, the defeat and retreat of the Xiongnu were complete and severe.*

With their glory days behind them, the peripatetic Xiongnu Empire entered an irreparable decline. The forfeiture of the western regions

---

* Now something of a celebrity in China, General Huo died of illness in 117 BCE at age twenty-three.

destabilized their economic power and political cohesion. They were driven into northern Mongolia and the Transbaikal, away from the southern pasturelands. Their depleted horse and cattle herds never recovered. The fragile cohesion of the Xiongnu fragmented in the icy barren landscapes of exile. Over the next fifty years, however, the outcast Xiongnu continued to reject half-hearted Chinese overtures for peace.

These Chinese achievements came at enormous cost, however. It is estimated that during these campaigns the country lost an astounding 80 percent of its military horses. Direct access to the Central Asian markets was the only viable lifeline to replenish Chinese stocks. This crusade for equine reinforcements, however, was a high-priced enterprise.

With expenditures reaching 50 percent of available capital, the Han government introduced a series of supplemental taxes, adding to the hardships of an already overburdened and impoverished populace numbering sixty million (30 percent of the global population). During the ensuing decades, according to the *huji/hukou* registration (census for tax revenue and conscription), the Chinese population dropped precipitously due to starvation and disease. Undeterred and relentless in his pursuit of the heavenly horses, Emperor Wu continued to raise taxes to fund further excursions. It can safely be said that during this period, the value of a prospective horse outweighed that of a peasant. Once again the emperor turned to Zhang Qian.

―――

Shortly after the Battle of Mobei, Zhang set out on a second expedition and struck an alliance with the Wusun of the Ili River valley in southeast Kazakhstan in 115 BCE. The Wusun, along with the Yuezhi, were Indo-European peoples who had been displaced from the Tarim Basin by the Xiongnu during the imperial forays of Modu Shanyu. They drifted into Central Asia, triggering a knock-on effect of western migrations that was ultimately felt as far afield as India, Europe, and across the Roman Empire.*

―――

* The Yuezhi and Wusun settled in Bactria (the intersection of the Stans), where they dislodged and intermingled with Scythians and the loitering Greek diaspora of Alexandria

To solidify the partnership, Princess Xijun—a royal relative of Emperor Wu—was shipped five thousand miles away to wed the seventy-year-old Wusun leader in exchange for a thousand horses. She is credited with having composed the famous poem "Song of Sorrow": "My family has married me off to the ends of the earth, / To live far away in the alien land of the Asvin king. / A yurt is my dwelling, of felt are my walls; / For food I have meat, with koumiss to drink. / I'm always homesick and inside my heart aches; / I wish I were a yellow swan and could fly back home."

Recent DNA studies confirm that women from elite families, such as Xijun, wielded political power by serving as "princess emissaries" on the frontiers of empire. These women were buried in elaborate tombs containing horse-riding equipment, iron weapons, gold artifacts, glass beads, silk clothes, and other exotic goods.

While Zhang's remarkable expeditions were celebrated, they were not the only Chinese processions pushing west. Between 114 and 108 BCE, Emperor Wu sent ten trading caravans per year to the western outer rim. There they exchanged silk for horses, jade, coral, and other wares, accounting for an impressive 30 percent of total imperial trade. One mission reached the major trading center of Alexandria Margiana (Merv) on the eastern border of the Parthian Empire.

Centered in Persia, the Parthians were a group of Scythian-styled Indo-Iranian speakers who fiercely safeguarded their lucrative geographic middle-ground position on the trade routes between Rome and China. "When the Han envoy first visited the Kingdom of Anxi (Parthia) the king of Anxi dispatched a party of twenty thousand horsemen to meet them at the eastern border of the Kingdom," writes Siam Qian. "When the Han envoys set out to return to China, the king of Anxi sent envoys of his own to accompany them."

Thanks to the quests for horses, China emerged from an isolated collection of warring states into a cohesive nation and an integral part of the global village. Emperor Wu, however, still yearned for his heavenly

---

Eschate in the Fergana valley, entering the Hellenic literary world as *Tocharii* (Tocharians), *Asii*, or *Seres*.

horses and in 104 BCE dispatched yet another western task force, igniting what became known as the War of the Heavenly Horses.

When his envoys arrived in Fergana, the Scythian king executed the entire delegation. He explained to his people why this was a sensible response:

> The Han is far away from us and on several occasions has lost men in the salt-water wastes [Tarim Basin] between our country and China. Yet if the Han parties go farther north, they will be harassed by the Xiongnu, while if they try to go south, they will suffer from lack of water and fodder. Moreover, there are many places along the route where there are no cities whatsoever, and they are apt to run out of provisions. The Han embassies that have come to us are made up of only a few hundred men, and yet they are always short of food, and over half the men die on the journey. Under such circumstances, how could the Han possibly send a large army against us? What have we to worry about? Furthermore, the horses of Ershi [Andijon, Uzbekistan] are one of the most valuable treasures of the state!

The king of Fergana fatally miscalculated the Chinese obsession with his horses. According to Sima Qian, an infuriated Emperor Wu was "seething with rage too fearful to behold." He immediately ordered a punitive expedition of twenty-six thousand men under General Li Guangli, the brother of his favorite concubine. In his estimation, if he did not avenge this insult, he "would become a laughingstock among the outer states." But with no tenable supply line, the expedition was swallowed by the unforgiving desert sands of Xinjiang, forcing the campaign to be aborted. A battered and weather-beaten parade of three thousand tattered scarecrow survivors returned to the frontier garrison in the Gansu Corridor demoralized and defeated—just as the Scythian king had predicted.

Undeterred by this disaster, Wu ordered Li to launch a second invasion of Fergana two years later, in 102 BCE. This time the augmented Chinese force was properly outfitted, or so they thought, for the arduous 3,100-mile journey. The massive expeditionary force consisted of 60,000 infantry and 30,000 cavalry, in addition to a snaking convoy of service and support elements filled by 180,000 potential Chinese homesteaders. A stretching supply column of roughly 30,000 additional horses, 100,000 oxen, and 20,000 donkeys and camels brought up the rear.

Despite losing half his force to starvation and desertion en route, General Li still arrived with an army vastly superior to that of his Fergana adversaries. The Chinese initiated a forty-day siege of Ershi, which was lifted only after the king was assassinated by his own people. His severed head was presented to General Li as a peace offering attached to one simple and nonnegotiable stipulation: the inhabitants of Fergana agreed to deliver horses and concede their position as a vassal state only if Li withdrew his army and departed. If the Chinese refused, they vowed to fight to the death and kill every horse in their possession. Realizing this was not a bluff, Li smartly acquiesced. He installed a puppet leader, secured provisions, corralled his horses—thirty of the "superior animals" and three thousand of "middling or lower quality"—and headed home.

The harrowing forays of Zhang Qian and Li Guangli provided China with unfettered access to the Central Asian horse markets, including Fergana. "As a result of China's subjugation of Fergana," relates Bill Cooke in *Imperial China: The Art of the Horse in Chinese History*, "other states in the Tarim Basin voluntarily submitted to the Han court, and the influence of the Empire was expanded. Within a short period of time, Chinese garrisons were established throughout the western region. In essence, this resulted in the establishment of regular contact between East and West, and the opening of the famous Silk Roads. The other result, however, was an Empire left near bankrupt by the enormity of these two campaigns."

The western expeditions between 138 and 101 BCE, including the War of the Heavenly Horses, also exacted a heavy human toll. In total, only ten thousand men and roughly a thousand of the prized Fergana breed made it back to China. In stark terms of financial expenditure and human lives, these horses no doubt rank as the most expensive in history.

For the emperor, however, he had finally acquired his coveted horses. To celebrate, he composed a triumphant hymn: "The Heavenly Horses are coming / Coming from the Far West / They crossed the Flowing

Sands / For the barbarians are conquered / The Heavenly Horses are coming. . . . Across the pastureless wilds / A thousand leagues at a stretch, / Following the eastern road. / The Heavenly Horses are coming."

These horses were not only the incentive for Chinese expansion but also they entrenched their permanent presence on the northern and western frontier by dealing a final blow to the stubborn Xiongnu. An overwhelming coalition of Chinese and Wusun forces invaded their western territories in 72 BCE. Unsympathetic nomadic tribes took this opportunity to pillage the Xiongnu, who were now besieged on all fronts.

The Chinese military advance was reinforced by shrewd divide-and-conquer maneuvering. A Chinese ambassador to the Wusun acknowledged that "the practice of China has always been to get barbarians to fight each other . . . using foreigners to attack foreigners." Split into northern and southern camps, by 60 BCE, the Xiongnu were engaged in a brutal civil war, sparking the immutable erosion of their political cohesion and military power.

After a series of further destabilizing clashes with their northern kin, in 53 BCE the outmatched and vulnerable southern Xiongnu surrendered their sovereignty to the shelter of the Han dynasty. China extended its territorial reach by incorporating the areas of northern Gansu and Inner Mongolia. Eventually, wayward bands of Xiongnu drifted west, becoming the ancestors to Attila and his wanton Huns. "Long after the death of Wu and the abandonment of his aggressive policies, the Hsiung-nu accepted China's terms," explains anthropologist at Boston University Thomas Barfield in *The Perilous Frontier: Nomadic Empires and China*. "From that time onward, no nomadic power on the steppe ever seriously objected to the tributary framework."

With these expeditions through some of the most diverse and austere terrain on the planet in search of the heavenly blood-sweating horses, and simultaneous campaigns to the South China Sea, a united Chinese nation introduced itself to the world. China opened its western gates, initiating the process of globalization and sustained East-West relations through the establishment and extension of the modern Silk Roads.

These horse excursions are also the reason why the current borders of China stretch so far west to incorporate the multiethnic (including the

Muslim Uighurs) and contentious Xinjiang ("new frontierland") Province. Its pertinent strategic boundaries press against those of Russia, Mongolia, the Stans, and India. China now occupies the third largest landmass (after Russia and Canada), at 6.54 percent of the planet. The Middle Kingdom boasts the longest aggregate land borders (13,743 miles) of any country, shared with fourteen other nations. Half of these line its broad, geopolitically relevant, and influential western boundary created by a yearning quest for horses.

Most scholars date the origins of uninterrupted trade along the dendritic Silk Roads—from China, through Central Asia, Persia, and Mesopotamia, to the Mediterranean powers of Europe—to the second century BCE. Traffic, including assiduous disciples of Buddhism and Christianity, was voluminous and flourishing by the second century CE. This four-hundred-year stretch witnessed the chronological symmetry and mirroring rises of the Roman and Chinese Empires.

They eventually found themselves commercially connected by the Chinese desire for the superior horses of Central Asia, the strategic linchpin coupling these two expanding polarities of the steppe. "Trade was now flourishing along the northern and southern routes, and ambassadors were being sent to explore Central Asia and the West," says Barry Cunliffe. "The Silk Road, as it was much later to be called, was now fully operational, and the Tarim Basin, once an inward-looking backwater, had become a vibrant part of the trans-Eurasian trading network."*

While attempts were made to establish direct relations from 100 BCE onward, we know of no Chinese or Roman traveler or envoy that navigated the entire expanse of the Steppe Roads. In 97 CE cavalry commander Ban Chao, bestowed the title "Marquis Who Stabilizes Distant Places" for solidifying Chinese control of the Tarim Basin, sent

---

* Although "letters of passage" appeared in the Persian Empire, the first modern passports were issued in the second century BCE by the Han dynasty. They included age, height, descriptions of physical features, and other characteristics. The passports, required for everyone over the age of one, determined where the holder was allowed to travel, while individualized visas regulated trade.

his emissary Gan Ying on a quest to reach Rome. When he neared "the great sea" (Mesopotamia), he was intercepted by the Parthians.

Eager to maintain their monopoly on shuttling goods between East and West, they spuriously told him the journey to Rome would take a few more years and that most foreigners who ventured forward never returned. Gan was dissuaded from continuing his mission and returned to China. Nevertheless, by penetrating the farthest west of any Chinese explorer, he gathered considerable information, albeit secondhand, about Europe and the Roman Empire. What neither Rome nor China could have possibly imagined was that the ultimate, and contrasting, fates of the two empires were invariably intertwined.

By this time, the Roman Empire (Da Qin) was well known to China. Indirect commercial connections had been established centuries earlier. A royal Han diary from 166 CE, however, reports that "The King of the Da Qin, Antun [Marcus Aurelius], sent envoys who offered ivory, rhinoceros horns, and tortoise-shells from the border of Annam: this was the first time they communicated with us. Their tribute contained no precious stones whatever and makes us suspect that the messengers kept them back." These Roman traders (not ambassadors) arrived by sea and not by way of the arduous overland Steppe Roads.

Arabian Sea ports running across Parthia (Iran) and India connected the Roman and Chinese Empires. "The beautiful large ships . . . come bringing gold, splashing the white foam on the waters of the Periyar [River]," recounts a contemporary Indian report, "and then return laden with pepper. Here the music of the surging sea never ceases, and the great king presents to visitors rare products of sea and mountain." Another poetically describes the permanent luxury waterfront villas built by Roman merchants: "The sun shone over the open terraces, over the warehouses near the harbour and over the turrets with windows like eyes of deer. In different places . . . the onlooker's attention was caught by the sight of the abodes of [the Westerners], whose prosperity never waned."

The Greco-Roman trading manual *The Periplus of the Erythraean Sea*, from approximately 50 CE, offers a detailed inventory of goods filtering through these ports. Exported from the Roman world to China were cloth, metals, ivory, silverware, glassware, carpets, saddles, wine, coins, slave musicians, and "pretty girls" offered up as exotic concubines to the Chinese nobility. The Chinese also imported spinach and pistachios from Persia; raisins, almonds, melons, and hazelnuts from Fergana; sesame, peas, onions, and coriander from Afghanistan; and saffron from India. The conventional chair was also introduced from Europe and Persia. While a diverse menagerie of domestic and exotic animals was exchanged, horses always topped the wish list of Chinese emperors and aristocracy.

Similarly, the Xiongnu and Scythians possessed and bartered a wide array of foreign goods from Greece, Central Asia, Persia, India, Egypt, and the Roman Mediterranean. For their part, the Xiongnu bequeathed an eternal piece of furniture to the world. Their horse-inspired "nomadic barbarian chair" is now a universal household item recognizable as our modern-day portable folding lawn chair or collapsible camping chair.

The Roman Empire was also a voracious consumer of Chinese goods, specifically silk, pearls, and an assortment of spices (primarily black pepper). By the first century CE, even the imposing Colosseum in Rome was generously adorned with silk drapes. Other commodities being trafficked from the Far East into Europe were narcotic drugs (marijuana, hashish, and opium), turquoise, lapis lazuli, onyx, ivory, and cotton cloth. The horse-driven trade sweeping the Steppe/Silk Roads between the Chinese and Roman Empires was vibrantly dynamic.

Once the floodgates of consumerism were opened, entrepreneurs and merchants scrambled to meet the growing demand. Steadfast horses shuttled what was fast becoming a global capitalist economy across the greater Eurasian Steppe. Shells and carnelian beads from the sea-sprayed coastal haunts of India exchanged hands on the hot desert sands of Xinjiang in transit to China. The Roman emperor Claudius I deployed Indian war elephants against the Celtic Britons. Nomads north of the Great Wall admired Roman glass bowls while resting on Persian carpets. Roman soldiers on the battlements of Gaul or their comrades perched upon Hadrian's Wall staring across the frosted moors of Scotland spiced up their mundane rations with black pepper from China and India. Hindu, Christian, and Buddhist prayers and dirges could be heard across the broad compass of Eurasia. As Christopher Beckwith mentions, "Late in the period, the movement of ideas along the trade routes, particularly the Buddhist and Christian faiths, had a great effect on both center and periphery." Thanks to the horse, it was quickly becoming a small world after all.

The extent of trade and emerging consumerism can be gathered from grave goods uncovered in Afghanistan (Bactria) in 1978. The crypts (c. 50 BCE) contained the gilded bodies of five female and one male Scythian warriors and more than a hundred thousand international artifacts: Scythian-style gold objects, Persian turquoise ornaments, Indian ivory, Roman glass bowls and other items, Buddhist images and coins, depictions of Greek gods and Greek inscriptions, Xiongnu-style headdresses, Scythian-designed boots, pants, and riding gear, Chinese bronze mirrors, Siberian daggers, and coins from India, Persia, Mesopotamia, and across the Roman Empire as far west as France. A similar find from Bagram revealed Roman glass vessels; bronze tableware and polished stone containers from Egypt; ostrich egg wine decanters and ivory-and-bone-inlaid furniture and plaques from India; as well as hundreds of painted lacquerware from China.

This exquisite collection vanished during the vicious decade-long conflict from 1979 to 1989 between the invading Soviet Union and the mujahideen fighters of Afghanistan. It was believed to have been pilfered by Russian troops or destroyed later by the Taliban. However, the treasures resurfaced in 2003, having been safely stashed away in the underground caverns of the Central Bank of Kabul.

The two-front geographic and commercial convergence of the Roman and Chinese Empires had unforeseen ramifications. "The aggressive foreign policy successes of the Chinese and Roman Empires ultimately had disastrous consequences," explains Beckwith. "Their destabilization of Central Eurasia by their incessant attacks resulted in internecine war in the region." The Chinese overthrow of the Xiongnu and its annexation of Xinjiang led to a series of lengthy knock-on migrations by nomadic Xiongnu, Scythian, and Germanic tribes toward western Europe.

Eventually, and in relatively rapid succession, these horsemen attacked the heart of Roman power and severed its sinewed lifelines of survival. "The descendants of the Xiongnu or close relations," remark

Mallory and Mair, "would soon be terrorizing the European world as the Huns." While the Chinese bested and banished the Xiongnu, extending their borders in the process, the Romans could not withstand the onslaught of their Hun descendants.

During the Crisis of the Third Century, the Roman Empire began its slow decay in the face of destabilized centralized authority, a fiscal system overburdened with debt and inflation, and a population ravaged by disease, producing widespread manpower shortages for both agricultural labor and the once-mighty legions. Assassinations of emperors and politicians by rogue military commanders added fuel to widespread revolts and civil war. Cascading waves of relocated ethnicities within the empire and hordes of outside predatory horsemen pounced on an internally weakened and unraveling empire rocked by earthquakes and natural disasters. The Roman Empire was an imploding colossus that could not be salvaged.

Nomadic horsemen from the steppe, including Huns, played an instrumental role in putting the finishing strokes on the now-crumbling dreams of the Eternal City. "Rome suffered in battle with cavalry armies and with mounted archers precisely because it could not match, and often could not withstand, the threat of horsemen armed with bows," writes Louis DiMarco. "Because cavalry was not historically central to the Roman Army—or the horse to Roman culture and tradition—the Romans struggled to win their empire and protect it with a small, often poorly handled, cavalry force."

The first to strike Rome during this Era of Migrations were the Germanic Visigoths, led by King Alaric. In 408 his mounted "barbarian" archers swept south through Italy and laid siege to the city of roughly one million cowering occupants on three separate occasions. When a Roman representative asked what would remain for the beleaguered citizens, Alaric answered sardonically, "Their lives." He lied.

The Romans were ransacked, robbed, butchered, and sold into slavery. Satisfied with their slaughter and bounty, the Visigoths quit the city and headed south, leaving a trail of wreckage in their wake. News of the catastrophe reverberated across the Mediterranean world. "The speaker's voice failed, and sobs interrupted his speech, the city that had conquered the whole world had itself been conquered," wrote Saint Jerome from Jerusalem. "Who could believe it? Who could believe that Rome, built up through the ages by the conquest of the world, had fallen, that the mother of nations had become their tomb?"

With Alaric's death from malaria in 410, the Visigoths consolidated their loot, headed north, and carved out a kingdom in Gaul. They would eventually help defend their new homeland from its next menace: Attila and his marauding Huns, described by a Roman contemporary as "the seedbed of evil."

Genomic studies consistently reveal a hereditary continuity between the Xiongnu and the Huns. For example, a 2022 DNA analysis conducted by a team of Hungarian geneticists reinforces that the Huns "can be traced back to early Xiongnu ancestors. Thus our data are in accordance with the Xiongnu ancestry of European Huns, claimed by several historians." The chain of westward relocations in the second century BCE initiated by the Xiongnu displacement of the Wusun and Yuezhi and the subsequent Chinese scattering of the Xiongnu, mentioned earlier, created an ebb and flow of nomadic movements across the Eurasian Steppe.

Following the Xiongnu Civil War, a splinter group of Xiongnu drifted west into eastern Kazakhstan. The Roman historian Tacitus tells us that by the end of the first century CE, these Hunnoi had penetrated the Pontic-Caspian Steppe. Over the next few centuries, these ferocious mounted archers continued their methodical westward march, creating a domino effect of migrations.

The Huns displaced local Germanic and Iranian tribes such as the Vandals, Goths, Franks, and Alans, who were all experienced horseman—or, in the case of the Goths, immediately embraced an equine culture during their migrations toward western Europe. The Romans, lacking an equivalent equestrian tradition or counter cavalry, faced an entirely new and unnerving strategic situation.

By this time, however, the Huns were most certainly a multiethnic horde of various nomadic horsemen. As they rode and ransacked their way to Europe, the Huns absorbed nomadic stragglers of sundry horse cultures—including the Massagetae, Scythians, Alans, and Goths—adding numbers and mounted proficiency to their swelling ranks. The acclaimed Roman historian Ammianus informs us that the Alans were "somewhat like the Huns, but in their manner of life and their habits they are less savage."

The quick-striking Huns were unrivaled horsemen who terrified European populations with their fearsome tattooed arms, faces carved with patterned scars, and elongated skulls from having been bound between boards as infants. The Romans ranked the Huns (and Alans) as the most ferocious of all barbarians they encountered.

They initiated their prolonged invasion of eastern Europe toward the Hungarian Danube around 370. "A hitherto unknown race of men had appeared from some remote corner of the earth," recorded Ammianus, "uprooting and destroying everything in its path like a whirlwind descending from high mountains. The nation of the Huns surpasses all other Barbarians in wildness of life." The "smelly, bandy-legged, nasty, brutish, and revoltingly short" Huns were said to be forged into their saddles, remaining mounted to eat, sleep, defecate, and even fornicate. Their world was the horse.

As their raids intensified, the Huns extracted enormous Xiongnu-style tributes from Constantinople to spare the Eastern Roman Empire.* With 22,500 pounds of gold arriving from the timorous east between 435 and 450, a bold and ambitious new leader, Attila, projected his power westward over the Austrian Alps. "The emergence of Attila as the commander of a huge nomadic confederacy was totally dependent upon his ability to maintain a largely parasitic relationship with the Roman world," explains Barry Cunliffe. "This required him constantly to demonstrate his powers, both to the Romans and his own followers." Attila mastered the trick-or-treat diplomacy of the Xiongnu and took raiding and extortion to a new level, ensuring obedience through terror and retribution. It was only a matter of time before his skilled cavalry attacked Rome.

At the head of a massive army that entered Gaul in 451, he introduced himself to a Christian priest. "I am Attila," he announced, "the scourge of God!" His invasion was rebuffed on the bloody Catalaunian Fields by

---

* Constantinople, formerly the Greek colony of Byzantium, is now known as Istanbul.

a coalition of Romans, Visigoths, and Franks. He immediately turned southward and commenced a swift invasion of Italy, despoiling town and country along the way. Attila skillfully familiarized Europe with this foreign quick-strike mounted steppe warfare by ravaging or extorting its populations to settle for peace. The dire situation facing the Roman Empire, however, also forced his hand.

Attila aborted his marauding mission as famine and malaria stalked the peninsula. More immediate was the issue of suitable pasture for Hun horses. Grazing grounds, precious at the best of times, were now virtually nonexistent, as drought haunted Italy. "It may have been the pull of his flocks and herds that explains his mysterious departure from Italy in 452, when the peninsula lay undefended before him," remarks John Keegan. "A return to the grasslands in such circumstances would have made logistical sense." In this context, the needs of the horses forced a retreat and saved Rome from the ravages of Attila and his capacious Huns. In the immediate aftermath of Attila's inglorious death from complications triggered by acute alcoholism the following year, division and infighting followed, and the temperamental, tribal Huns abandoned their fragile unity and faded from history. Rome, however, was not spared.

As the Huns welcomed themselves to eastern Europe, the Vandals, a miscellaneous posse of pillaging Germanic tribes, cut a swath of despoliation clear across Europe to Spain. The modern English word *vandal*, meaning "deliberate destruction or defacement of property," perpetuates their legendary reputation. Under their warrior-king Gaiseric, in 429 more than twenty thousand Vandals were shuttled across the Strait of Gibraltar to North Africa. Launching into the Mediterranean Sea, Vandal activity was so viciously abundant that its Old English name was *Wendelsae* (Sea of Vandals).

Facing a dual threat from both the Huns and the Vandals, Rome recalled its legions from Britain. Sensing an opportunity, the Angles and the Saxons from Denmark and Germany teamed up and invaded Britain in the 440s. Anglo-Saxon culture quickly replaced that of the Celtic peoples and the vestiges of Roman occupation. This episode gave birth to the legends of King Arthur and his Knights of the Round Table.

These evolving oral stories and literary traditions, enriched with numerous character additions and plot twists throughout the centuries, contain diverse cultural elements (even Indo-European), codes of chivalry (from the French *cheval*, for horse), and spiritual lore. The prototype template of the Arthurian epic still permeates our modern mass media, cultural philosophies, and knightfall bedtime stories—from Indiana Jones and his last crusade, to the lightsaber-wielding Jedi knights of *Star Wars*. "The dramatic upheavals expressed in the Arthurian legends reflected the continentwide *völkerwanderung* that swept aside classic Rome and propelled Europe into the Middle Ages," observes Pita Kelekna. "The disappearance of centralized Roman rule from western Europe resulted in political fragmentation that fostered the development of feudalism in military response to territorial aggression." The Huns and the Vandals were two of the most destructive ensembles during the "people wanderings," or barbarian invasions, of Europe.

Two years after the death of Attila, Gaiseric and his Vandals landed in Italy and marched on Rome. After amassing a substantial fortune, including slaves, they took their leave. The Vandal sack of Rome was not quite as sadistic as legend would have us believe. Like the disintegration and scattering of the Huns following Attila's death, Vandal dominance in the Mediterranean region eroded after Gaiseric's passing in 477. By the time the Germanic Ostrogoths, another migrating and militant faction of horsemen, carved out a kingdom in Italy in the 490s, there had not been a Western Roman emperor for nearly two decades—and, as events unfolded, there never would be again.

The dream of Rome awoke in pieces partly because of the Chinese quest for horses on the opposite end of the steppe. The Sino-Xiongnu War coupled to the Chinese pursuit of horses had a profound impact on the fall of Rome. The westward nomadic migrations spurred by these events led directly to its downfall. While Rome collapsed under the combined trampling weight of the descendants of the Xiongnu and other displaced horsemen, the opposite was true for China.

Under Emperor Wu, the Han dynasty rapidly pushed its borders north, south, and, most importantly, west through the Gansu Corridor to the ethnic melting pot of Xinjiang in order to secure the heavenly blood-sweating horses. In the process, China gained control of the lucrative eastern hubs and approaches of the Silk Roads, announcing itself to the world as a major geopolitical player. Its relentless, single-minded pursuit of horses beginning in the second century BCE helped create the modern Chinese economic and military superpower of today.

The role of horses in safeguarding, stabilizing, and expanding the unified Middle Kingdom also had profound technological repercussions. During their imperial clashes with the Xiongnu, the Chinese originally sought to emulate their opponent by mustering light, highly mobile cavalry. It was only in the fourth century CE, after neutralizing the Xiongnu threat and obtaining unhindered access to external horse markets, that the Chinese adopted a culture of heavy cavalry. This transformation coincided with their game-changing invention of the stirrup.

Given the importance of the horse, it seems strange that it took so long for the stirrup to materialize; it was first represented in statuettes around 300 and in writing by 477. With the advent of stirrups, the human-horse centaur became an even more powerful killing machine within ever-swelling and gathering armies competing for commercial ascendancy.

This simple but revolutionary apparatus appears to have been devised in northern China before quickly disseminating across the region to Korea and Japan, signaling the arrival of their own cavalry cultures. The earliest record of *kiba minzoka* (cavalry) in Japan dates to the fourth century. Massive Scythian-styled kurgan burials with all the trappings of a thriving horse culture from the fourth and fifth centuries have been unearthed in both Korea and Japan, which were engaged in intermittent diplomatic cultural exchange, trade, and war.

By the fifth century, the Chinese refined their invention into a cast-iron stirrup. This modification transformed cavalry. Disseminated by steppe nomads, including the Avars (also descendants of the Xiongnu), the use of stirrups spread across Eurasia and the Middle East between the sixth and ninth centuries. In an illustration contained in *Commentary on the Apocalypse*, written by the Spanish monk Beatus of Liébana in 776, all Four Horsemen are exacting divine retribution utilizing stirrups. Interestingly, Arab riders were slow to appreciate their value, believing that stirrups somehow compromised the position and stability of the rider.

In fact, the stirrup allows the rider greater control of the horse while wielding larger, longer, and heavier weapons, as well as increasing velocity and accuracy with the bow—all without fear of becoming unseated during combat. In short, with the stirrup, the rider truly became one with the horse. The stable and secure riding platform it provided also allowed cavalrymen to wear heavier armor on progressively larger (and armored) horses.

Stirrups enabled riders to mount their steeds more quickly and far more safely. "It is now believed that the first true stirrup was developed not to increase the stability of the rider but to provide him with an easier and safer method for mounting his horse," explains Cooke. Previously, cavalrymen mounted their horses by using their spears to pole-vault or by stepping on a peg sticking out of the side of the spear, often maiming or killing themselves in the process. The mounting stirrup, which "did not come in pairs," Cooke continues, "and was attached to only one side of the saddle," reduced the mortality rates associated with hoisting oneself onto a horse.

The word for stirrup in numerous languages reflects its original purpose. In its birthplace of China, it is composed of the root words *ma* (horse) and *den* (to mount). The first European literary use of the stirrup comes from the sixth-century *Strategikon*, attributed to the successful Byzantine general and emperor Maurice. In this prominent manual on war, which influenced military practice for more than a millennium, he used the Greek term *skala* (step or ladder). The modern English *stirrup* stems from the Saxon *stigan* (to climb) and *rap* (rope). The literal translation of the German *Steigbügel* is "stephanger." Archaeological evidence suggests dual riding stirrups followed close behind their single archetype.

When the stirrup found its way to Europe during the late seventh century, this Chinese invention was a decisive factor in the emergence of European feudalism. Under the foot of a Frankish duke and his heavy cavalry, the stirrup would also turn back the Muslim invasion of western Europe, branding Charles "the Hammer" Martel as the savior of European Christendom. His history-defining victory at Tours in 732 set the stage for the domination of heavy cavalry. More than three hundred years

Commentary on the Apocalypse: The Four Horsemen in stirrups. From Beatus of Liébana, circa 776. *(Photo 12/Alamy Stock Photo)*

later, in 1066, the ingenious use of heavy cavalry by William of Normandy at the Battle of Hastings gave birth to modern England.

With the power of the horse, the foundational building blocks of the modern global order were being hauled into place: from China and India, across Central Asia and the Middle East, to France and Britain.

CHAPTER 10

Dark Horses
Feudal Knights and Contending Faiths

In the immediate aftermath of the collapse of the Roman Empire, Europe turned inward upon itself. The dictatorial feudalism of monarchies and lordships reigned supreme. A recoiling European population bunkered down as centralized government, trade, manufacturing, and academia slipped into the abyss and the knowledge of the ancients vanished from collective memory.

While Europe was shackled by religious and social upheaval, another spiritual and political order flowered in the Middle East. Inspired by the revelations of the Prophet Muhammad, the appearance of Islam in Mecca and Medina (Saudi Arabia) in the early seventh century spawned an inspired cultural and intellectual renaissance across the maturing Muslim expanse.

The cosmopolitan city of Baghdad was home to a million people and more than a hundred booksellers by 900. "It was the perfect symbol of the Islamic world's affluence, the heart of royal power, patronage, and prestige," states historian Peter Frankopan. "It marked a new center of gravity for the successors of Muhammad, the political and economic axis linking the Muslim lands in every direction." Correspondingly, the population of the dazzling Muslim capital of al-Andalus (the Iberian Peninsula) at Córdoba reached a half million.* As a center of academia and

---

* The Iberian Peninsula encompasses modern Spain, Portugal, Andorra, Gibraltar, and a small portion of southern France.

the arts, the splendid library housing more than four hundred thousand volumes served as the vibrant hub of intellectual life.

Conversely, the largest cities in Europe had been reduced to towns. The metropolis of Rome, home to more than a million souls by the second century, shrank to a mere thirty-five thousand, while Paris sheltered roughly twenty thousand. When William the Conqueror and his Norman cavalry invaded England in 1066, the population of London was fifteen thousand at most. Europe was on fragile footing.

Unlike the slow and labored spread of Christianity, Islam quickly won over the Arabian Peninsula and rapidly expanded its territorial base across the Middle East. Powered by fleet-footed Arabian horses, this upstart faith cascaded across North Africa and poured into the Byzantine and Persian worlds under the Umayyad caliphate.

In the process of creating the seventh largest empire in history between 661 and 750, the Umayyad dynasty hastened the spread of Islam. It stretched from the Indus River and the Fergana valley (including strikes into Chinese Xinjiang) in the east, across Turkey and the entire Middle East through North Africa, before occupying Spain, Portugal, and southern France in the west—knocking on the door of Paris. And this was only the beginning.

Arabic sources outline an ambitious Alexander-like campaign map for the army of Islam. From its jump-off in Spain, the projected line of conquest stretched east across Europe before subduing the Byzantine Empire and arriving back at Damascus, the capital of the Umayyad caliphate. Given the lightning speed of the Muslim advance thus far, they rightfully believed this grand ambition was achievable.

Exactly a century after the death of Muhammad, the only thing standing in their way was the Frankish duke Charles Martel and his feudal bands of heavy cavalry. The fate of Europe would be decided on the battlefield at Tours. Islam, however, was pushing its axis on two fronts: in Spain and France in the west and in Central Asia in the east.

The Muslim quest for Allah and affluence had guided them east to the gates of China. Conversely, the earlier Chinese quest for horses to confront the Xiongnu had compelled them west to the foothills of Central Asia. Imperial territorial boundaries and commercial rivalries over the bustling caravan markets along the Silk Roads, including Bukhara and Samarkand, were settled by one of the most pivotal—and overlooked—battles in history.* The Chinese and Islamic Empires came to blows at the Battle of Talas River in 751 on the border of Kazakhstan and Kyrgyzstan.

While both sides claimed victory, the first and only instance of direct combat between Arab and Chinese forces resulted in a strategic stalemate.† The Chinese withdrew to Kashgar and the Tarim Basin, never to venture this far west again. And no future Arab army ever penetrated east of the Fergana or launched any forays toward the increasingly fixed Chinese border. "The Battle of Taraz or Talas, like the battle of Poitiers [Tours] in 732 in the west," writes historian at the University of St. Andrews Hugh Kennedy in *The Great Arab Conquests: How the Spread of Islam Changed the World We Live In*, "were to mark the furthest limits of Arab expansion in their areas."‡

Farther west, the Umayyad Empire sought to press its claims into Europe. A Muslim force of twelve thousand mounted North African Berbers (Moors), supported by elite Arab units, sailed the Strait of Gibraltar and invaded Spain in 711, crushing its Visigoth occupants. Buoyed by an additional eight thousand skilled horsemen, this reinforced army systematically conquered the Iberian Peninsula by 718.

Although the Pyrenees Mountains formed a natural barrier, the Arabs quickly solidified control of the entire Mediterranean coast as far as Marseille in southern France. Roving bands of Moorish raiders plied

---

* The population of Samarkand is indicative of the surging trade and imperial expansion along the Steppe/Silk Roads. During the Xiongnu/Chinese expansion of the second century BCE, the city had twenty thousand inhabitants, increasing to fifty thousand by the sixth century. With the eastern expansion of the Arab world, by the eleventh century, Samarkand was home to a diverse collection of two hundred thousand people. At the height of the Mongol Empire in the thirteenth century, flourishing commercial traffic boosted the population to three hundred thousand.

† Prisoners brought back to Baghdad after Talas introduced cheap and easy Chinese papermaking methods to the Arab, and later European, worlds, profoundly influencing academic advancement. By the thirteenth century, manufactured paper was a staple across Europe. The Song dynasty of China introduced paper money in 1024 and within a century was issuing ten million paper notes per annum.

‡ The battle is known as both Poitiers and Tours. The latter is given preference to avoid confusion with the British victory over the French at the Battle of Poitiers in 1356 during the Hundred Years' War.

their trade north, penetrating deep into Frankish territory, and, by 725, were within seventy miles of Paris. Under the remarkable general Abdul Rahman al-Ghafiqi, the main Islamic thrust through the lush lands of Aquitaine brought them to Tours by 732.

The Franks belonged to a cultural swath of Germanic peoples that were heavily influenced by Indo-European and Scythian practices. During the fifth-century völkerwanderung of Roman decline, they migrated west into France, fashioning the Merovingian dynasty. Between 715 and 724, Charles Martel fought eight engagements to solidify control over the splintered Frankish kingdom and consolidate his rule as the duke and prince of the Franks. The Duchy of Aquitaine, ruled by Eudes, a rival Frankish duke, remained a thorn in his side and outside of his authority. Aquitaine, however, was also the initial target of the Moorish invasion. Although never a king in the true sense, Charles was in a stronger position both politically and militarily than any previous Frankish ruler to withstand the gathering storm in Spain. During his rise to power, he had assembled the footings of a formidable heavy cavalry.

The Franks cultivated a strong equestrian tradition long before the Muslim army sailed the Strait of Gibraltar. "He who has stayed at school till the age of twelve and never ridden a horse," scorns a Frankish proverb, "is fit only to be a priest." Frankish cavalry fought alongside both the Romans *and* Attila and his Huns on the Catalaunian Fields in 451. Frankish warriors were often buried with their mounts. One epitaph candidly (or frankly, as it were) reads: "Kill my mother, I don't care! Never will I give up to you the horse you demand. He was never made for the reins of a wretch like you!"

The proficiency and military sophistication of Frankish horsemen was acknowledged as early as the sixth century in the ten-book *Historia Francorum* by Gregory of Tours and in the Byzantine *Strategikon*. Building on this foundation, Charles reorganized Frankish society for war. "Charles's Franks went to war on horseback in unprecedented numbers," notes historian at New York University and twice winner of the Pulitzer Prize David Levering Lewis in *God's Crucible: Islam and the Making of Europe, 570–1215*. "Franks were as much cowboys as were Americans of the nineteenth century West." This equestrian transformation included the nascent system of feudalism and military estates—land tenure in return for military service, including heavy armored cavalry—based on a social hierarchy of monarchs, noble lords, vassals, and serfs.

While historians do not agree universally on a concrete definition of feudalism, in its basic agricultural-economic-military formula, the Crown granted the nobility lands in exchange for military service and the raising of levy armies. Portions of these lands (fiefs) were farmed by vassals, who were promised military, legal, and territorial protection. In return, these freeman tenants paid the lord through military service as heavy cavalry or knights and a share of the harvest collected through an excise called a tallage. This relationship was sealed with an oath of fealty. The horse, by way of heavy cavalry, was an instrumental agent of feudalism, which dominated all facets of medieval European society between the eighth and fifteenth centuries. The French term *chevalier* (horse rider) was anglicized to *cniht*, or *knight*.

Essentially, knights were the intimidating muscle for the landed gentry, terrorizing dependent farmers (serfs, or villeins) to produce profit for their masters. These peasants were not freemen and were bound to the lord of the manor or military estate. Knights were more the equivalent of the mobsters doing the bidding of Al Capone, Pablo Escobar, or Tony Soprano than they were of the mythologized Arthurian saviors of damsels in distress and the guardians of greater Christendom.

Driven by the desire for heavy cavalry, the relatively moneyless feudal system framed by Charles Martel ensured a continuous upward flow of wealth and military service—from the bottom rung of peasants, through vassals and lords, to the top-tier monarchy. "The Franks had always used cavalry," explains historian at the University of Manchester Paul Fouracre in *The Age of Charles Martel*. "As more wealth was gathered in the countryside in the course of the later seventh and eighth centuries, an increasing number of people could afford the expensive horses and more sophisticated equipment needed to become mounted warriors or enable others to do so."

Outfitting heavily armored cavalrymen was extremely expensive, a cost shouldered by the currency of horse-based feudalism. The outlay to furnish one heavy Frankish cavalryman was equivalent to twenty-five oxen, or the combined plow teams of ten peasant families. A comprehensive ninth-century register lists the individual pieces of a knight's wardrobe and weaponry on a scale of value against a cow. A helmet was worth six cows; a sword and scabbard, seven; spear and shield, two; greaves (leggings), six; and a byrnie, or hauberk (chain mail tunic), twelve cows. The standard fourteen-to-fifteen-hand horse traded for fifteen cows—comparable to an entry-level Ferrari ringing in at roughly $250,000. Stripped of its mythical chivalry, a knighthood was valued at forty-eight cows!

By the eleventh century, a long shift of chain mail interwoven with more than twenty-five thousand steel rings was worth as much as the annual income from a typical farming village of one hundred inhabitants or a full year at the University of Oxford, including tuition, lodging, food, and robes. The higher the "thread count" of the chain mail, the heftier the price tag. Equally expensive was a pattern-welded long sword that took more than two hundred hours of skilled labor to craft. By this time, the full dress of a knight weighed between forty-five and fifty-five pounds!

Given these exorbitant costs, cavalry quickly became associated with the aristocracy. "No animal is more noble than the horse, since it is by horses that princes, magnates, and knights are separated from lesser people," wrote Italian nobleman Giordano Ruffo in *De Medicina Equorum* (c. 1250), one of the earliest European horse-specific veterinary manuals, "and because a lord cannot fittingly be seen among private citizens except through the mediation of a horse." Fighting from the back of a noble steed was viewed as more "civilized" than slogging it out as infantry.

The gallant medieval cavalry charge came to epitomize the heroic virtue and historic romance of the horse. The charging knights of the European Middle Ages depended on one piece of humble equestrian technology that changed the world: the stirrup. "The medieval knight in armour did not become a fighting force with which to be reckoned until the introduction of stirrups," notes historian David Black of Curtin University in Australia. "Their adoption was revolutionary, for it enabled a heavily armored horseman to retain his balance in the saddle whilst using a weighty spear, sword or lance."

The stirrup created a secure fighting platform, allowing the knight to

Heavy cavalry and the cudgel of feudalism: A medieval knight and horse in full armor, Germany, circa 1540. *(Metropolitan Museum of Art/Internet Archive Open Access)*

use a couched lance: a jousting-style position with the lance tucked firmly under the armpit. The horse's forward velocity was transferred through the long spear to a target—as opposed to throwing or thrusting. What we envision to be the classic storybook medieval knight finally occupied its square on the chessboard of the clashing armies of Europe. "The adoption of iron stirrups in the eighth and ninth centuries," states Robert Drews, "brought mounted warfare into its golden age." The stirrup completed the melding of human and horse, solidifying the physical attributes of our Centaurian Pact. With this modification, the heavily armored medieval knight dominated the battlefield.

Stirrups were also partially responsible for the shift to feudalism built on the foundation of the heavy cavalry revolution realized at the Battle of Tours. "Historians readily acknowledge the importance of the horse," remarks historian Michael Prestwich of Durham University, "which links the emergence of heavy cavalry and feudal society to the introduction of the stirrup."

Frankish cavalry adopted stirrups in the late seventh century from Avar horsemen wandering the Hungarian Plain on their eastern border.* "While semifeudal relationships and institutions had long been scattered thickly over the civilized world, it was the Franks alone—presumably by led by Charles Martel's genius—who fully grasped the possibilities inherent in the stirrup and created a new type of warfare supported by a novel structure of society which we call feudalism," stresses historian Lynn White. "Few inventions have been so simple but few have had so catalytic an influence on history." Martel, described by his contemporaries as "uncommonly well-educated and effective in battle," reformed Frankish society to meet the demands of the day, which included a large Islamic army on his doorstep equipped with a substantial quick-striking cavalry component.

During the empire building of the Umayyad caliphate, cavalry played an increasingly vital role in successive military operations. Given the scarcity of horses across Arabia, cavalry was in short supply during the initial Islamic outgrowth of the mid-seventh century. Expansion into Central Asia and North Africa, however, yielded a stable supply of quality horses, and Arab armies quickly adopted a more refined cavalry tradition from their Persian, Turkic, Byzantine, and Berber opponents. "Cavalry played a minor role in the first Arab-Islamic conquests," observes military historian David Nicolle, "whereas the conquest of al-Andalus and the subsequent raids into France differed because of the presence of large numbers of horse-riding Berber warriors." The mounted Arab forces that trotted into France, motivated by pillage and plunder, were experienced and formidable.

---

* By this time, the Avars were an amalgam of numerous nomadic groups, including their Hun cousins and a hodgepodge of roving Turkic and Germanic peoples who intermingled in the folds of the Pontic-Caspian Steppe during the push-and-pull migrations of the fifth and sixth centuries.

The elite crack troops were foreign mercenaries or slaves. These ancient special forces were put through years of rigorous full-spectrum military training and indoctrination. Given its value, the horse was inadvertently caught up as a de facto currency in the Arab slave trade. The price of a premier Arabian horse was twenty human slaves. The Arab cavalry confronting the Franks, however, lacked a critical piece of kit: stirrups.

In the spring of 732, General al-Ghafiqi steered his forces over the Pyrenees into Francia. He smashed through the independent kingdom of Aquitaine, scoring a quick victory over the forces of Duke Eudes. According to the contemporary anonymous scribe of the *Chronicle of 754* (*Mozarabic Chronicle*): "The Muslims pierced through the mountains, trampled over rough and level ground, plundered far into the country of the Franks, and smote all with the sword, insomuch that when Eudo came to battle with them at the River Garonne, he fled. . . . God alone knows the number of slain." The stampeding Arab cavalry was described as "a brush fire fanned by the winds." Painfully aware of the strategic threat to the entire region, Eudes immediately summoned his adversary Charles Martel for aid. After all, according to the *Chronicle*, Charles was "an expert in things military."

When the two rival leaders finally met in Paris, Duke Martel needed little convincing about the magnitude of the peril posed to both Aquitaine and his own kingdom. He issued a general *ban*, or military summons, to raise the largest army possible. Roughly fifteen thousand to twenty thousand Franks, augmented by the fleeing army of Aquitaine in addition to mounted Germanic mercenaries, answered the call to head off the advancing Arab army, which was estimated to be twenty thousand to twenty-five thousand strong.

Following the old Roman Road, Muslim forces were targeting basilicas and monasteries to plunder their sequestered riches. After burning the church at Poitiers, the army of Islam pressed north toward its prized target: the lavish Abbey of Tours, the most prestigious and opulent in western Europe.

Charles intercepted Abdul Rahman al-Ghafiqi at a place now christened Moussais-la-Bataille, and both commanders quickly took up defensive positions. From the limited sources, it appears that the Arabs dismissed, or were genuinely ignorant of, Frankish martial abilities and were surprised to see such a strong show of force blocking their advance. Up to this point, aside from a few minor speed bumps, they had sliced through the Iberian Peninsula and Aquitaine with relative ease. It now appeared that the barbarian Franks, or *Farang*, as they were known to the Muslims, were about to offer the first resistance.*

Charles assumed a tactically advantageous position on high ground, protected by woodland on both flanks, nullifying the maneuverability of the Moorish cavalry. He also relied on the hills to conceal his true strength. Al-Ghafiqi tried to provoke Martel using the favored Arab tactic of the era: establishing a strong defensive position, absorbing an enemy attack, and then launching a fierce cavalry-loaded counterattack. A patient and wary Charles refused to take the bait and stood his (high) ground.

The Arab army finally broke the weeklong stalemate on October 25, 732, attacking the disciplined phalanx-styled Frankish infantry formations. According to the *Chronicle*, "The northern peoples remained immobile like a wall, holding together like a glacier in the cold regions, and in the blink of an eye, annihilated the Arabs with sword . . . killed the king Abd al-Rahman, when they found him, striking him on the chest." Charles immediately ordered a heavy cavalry attack behind the forward lines with "more heavily armored Christian horseman having a great impact." A timely Aquitaine cavalry charge turned the opposite flank of the Arab position.

The *Continuations of the Chronicle of Fredegar* (c. 721–768), penned by an anonymous author likely from the Frankish polity of Burgundy,

---

* The term *Farang* has a long legacy and has since entered the lexicon of numerous disparate cultures courtesy of both the global spread of Islam and European colonization. In Persia, Europe was known as *Frangistan,* and in Arabic, syphilis became known as *al-'aya al-afranji*. When the Portuguese established their colony at Macau in southeastern China in 1557, the Chinese referred to them as *folangji*. The British invaders and occupying diasporas of India were known as *firangi* in Hindi, as *firingi* in neighboring Bangladesh, and as *faranji* in the Maldives. Across Southeast Asia (including during the Vietnam War), Americans and other Westerners were called *firangi*. In Thailand, *man farang* is a potato and *no mai farang* is asparagus among a basket of other Western items with the imported label of *farang*. Malaysians dubbed Europeans as *ferenggi*. Even *Star Trek* creator Gene Roddenberry borrowed the term to depict the foreign, devious alien species *Ferengi*, who are obsessed with accumulating wealth.

details this phase of the battle: "Prince Charles boldly drew up his battle line against them [Arabs] and the warrior rushed in against them. With Christ's help he overturned their tents, and hastened to battle to grind them small in slaughter. The King Abdirama [al-Ghafiqi] having been killed, he destroyed them, driving forth the army he fought and he won."

Dawn found the Muslim camp deserted save for vast stores of plunder. This suggests that the Frankish cavalry struck just as quickly as the Arabs scattered down the old Roman Road commemorated by the Muslim historian Ibn Idhari as the "Path of the Martyrs." The Frankish stand at Tours turned the tide. "The defeat of the Muslims at Poitiers effectively marked the end of large-scale raiding in France," concludes Hugh Kennedy. "It became clear that they were not going to be able to conquer the country, or even continue raiding with any degree of success."

News of the battle spread quickly. The Venerable Bede's *Ecclesiastical History of the English People* referenced the Frankish victory as the final bulwark against the "dreadful plague of Saracens."* Bede was confident enough in the reports trickling in to his distant and remote monastery in northern England to remark that "the Saracens who had devastated Gaul were punished for their perfidy." The *Continuations of the Chronicle of Fredegar* conferred beatification upon Martel as the savior of Christendom. According to the *Chronicle of Saint-Denis*, compiled in the thirteenth century, Charles received his immortal nickname *Martellus* ("the Hammer") because "as a martel breaks and crushes iron, steel and all other metals, so did he break up and crush his enemies." David Nicolle brands the engagement as "the breaking of the tidal wave of Islamic expansion in Western Europe."

In his book *100 Decisive Battles: From Ancient Times to the Present*, military historian Paul Davis neatly sums up the impact and relevance of Charles Martel and his triumph at Tours:

> Had the Moslems been victorious in the battle near Tours, it is difficult to suppose what population in western Europe

---

* During the Middle Ages, *Saracen* was a European synonym for "Muslim," or more generally for people from the greater Middle East.

could have organized to resist them. . . . Whether Charles Martel saved Europe for Christianity is a matter of some debate. What is sure, however, is that his victory ensured that the Franks would dominate Gaul for more than a century. . . . His grandson was also called Charles, later termed "the Great," or Charlemagne. Under his rule, the Franks rose to their greatest power both politically and militarily. The nature of the European military changed after this battle. The concept of heavy cavalry was forming in the eighth century. The introduction of the stirrup made stability on horseback possible, and stability was vital for both carrying an armored rider and using heavy lances. The age of the armored knight, a fighting machine that was both the result and the foundation of feudalism, was being born. . . . Thus, the establishment of Frankish power in western Europe shaped that continent's society and destiny, and the battle of Tours confirmed that power.

It should come as no surprise that Tours is consistently ranked among the most influential battles in history.

As it was, the 780-year Islamic presence in western Europe, and the sprawling Umayyad and Abbasid Empires spanning the epochs from 661 to 1517, also had lasting and profound global impacts on technology, culture, academia, and horses. "One could say that the nations which today adhere to the Muslim faith from Malaysia to Morocco were founded in the hoofprints of the Arabian horse," remarks historian John Brereton. "No war horse in history has had such far-reaching influence on man's affairs and on horse breeding in general." So intertwined is the horse with Arab culture that the Arabic word *siyasa* ("horse handling") is also synonymous with "statecraft" and "politics."*

The Prophet Muhammad made numerous references to horses and their emotional, spiritual, and military importance. On one occasion after a hard ride, Muhammad dismounted and proceeded to wipe the

---

* Likewise, the English word *management* is borrowed from the French via the Italian *managgi*, which means "to handle or train a horse."

muzzle of his horse with his *thawb* robe. When questioned by his bewildered companions, he replied that Allah had admonished him for not taking better care of his equine friend.

Another narrative has Muhammad as the direct breeder of all modern horses through the Arabian line. Apparently, he ordered a herd of horses to be denied water for seven days, after which the horses raced to the carefully placed water trough. Before they could drink, however, Muhammad sounded a war horn. Five mares immediately disregarded the water and answered the call to arms. According to legend, all pureblooded Arabian horses, and thus most current horses, are the offspring of these five obedient and faithful mares. While modern science cannot confirm this story, it can validate the Arab conquests in Europe as the direct catalyst in disseminating the now-dominant Arabian horse.

During the sustained Arab occupation of al-Andalus, Arabian stallions outmatched and outbred males from other bloodlines, transferring their Y chromosomes to all horses on the planet today. This genetic tapering created a modern horse that is substantially stronger and faster than those ridden by Attila the Hun, Muhammad, Charles Martel, William the Conqueror, and Chinggis Khan, not to mention those first domesticated on the Pontic-Caspian Steppe.*

Numerous genomic analyses, including an international study published in the journal *Cell* in 2019, revealed that "the introduction of new domestic lineages to the south of mainland Europe between the C7th–C9th, a time strikingly coincident with the peak of Arab raids . . . indicates that breeders increasingly chose specific stallions for breeding from the Middle Ages onward, consistent with the dominance of an ~700 to 1,000-year-old Arabian haplogroup in most modern studs. . . . [T]he early Islamic conquests significantly impacted breeding and exchange. The legacy of these historical events has persisted until now."

According to equine geneticist and coauthor of the report Ludovic Orlando, "It was a moment in history that reshaped the landscape of horses in Europe. If you look at what we today call Arabian horses, you

---

* The current genetic diversity of horses is the narrowest it has ever been, resulting in numerous potential diseases.

know that they have a different shape—and we know how popular this anatomy has been throughout history, including in racing horses. Based on genomic evidence, we propose that this horse was so successful and influential because it brought a new anatomy and perhaps other favorable traits"—strength, stamina, and above all else, speed.

All modern Thoroughbreds, for example, trace their pedigree to just three Arabian stallions imported to England: Byerley Turk (c. 1686), Darley Arabian (c. 1704), and Godolphin Arabian (c. 1730). In fact, only two horse lineages, or breeds—the Arabian and the now-extinct Turkoman referenced earlier—are responsible for almost all existing horses, including Thoroughbreds, American quarter horses, and German and English draught horses. These two ancestral cousins share very similar features, while evolving specific traits to adapt to their differing environments.

To compensate for sand, the Arabian possessed larger hooves to distribute its weight point across a larger surface area, and a wider frame, providing more stability for its rider to throw or thrust a lance or a sword. The Turkoman developed smaller hooves for added stability over the hard, rocky ground of the steppe. Its narrow body and long back allowed mounted archers to twist and turn far more effectively to shoot in any direction. Each horse evolved to its environment, and riders adapted their combat styles to best take advantage of their animals' unique traits. Speed, however, was still their greatest attribute.

On the backs of these nimble, fleet-footed horses, Muslims vectored new produce and products across their vast empire, including Europe. From China they distributed silk, oranges, lemons, paper, the magnetic compass, and the sternpost rudder.* India supplied silks, sugarcane, and cotton. Flowing from Central Asia were bananas, spinach, melons, and asparagus, among other exotic goods. These horses also carried the academic fruits of the Islamic Renaissance across Eurasia.

Between the eighth and fourteenth centuries, overlapping the Umayyad and Abbasid dynasties, the Islamic Golden Age propelled the world

---

* Modern processed paper was invented in China around 200 BCE from hemp and mulberry bark.

forward with pioneering scientific and mathematical advancements. "Rulers consciously strove to create a multicultural empire," explains historian at New York University Robert Hoyland. "In the furtherance of this end, all kinds of sciences were investigated, and large numbers of scientific texts from other cultures were studied and translated into Arabic." Islam placed a high priority on interdisciplinary education, scientific investigation, and the accumulation of knowledge—which was widely disseminated by their exquisite equine messengers.

Arabian horses conveyed the concept of zero from India to Europe in the ninth century. From the name of the genius Arab polymath al-Khwārizmī comes the word *algorithm*, and from his book *Kitab al-Jabr* (c. 820), the term *algebra*. He is responsible for the first systematic solutions of linear and quadratic equations, the decimal positional number system, and accurate trigonometry tables of sine, cosine, and tangent. Al-Khwārizmī also wrote or revised numerous works on astronomical calendars; in addition, he amended the Alexandrian mathematician and astronomer Ptolemy's book *Geography* (c. 150) by updating the positions of cities using the geographic coordinate system of longitude and latitude.

The organics of chemistry, another Arabic loan word from *al-kīmiyā*, or *alchemy*, also took root during the Islamic Renaissance. This extraordinary enlightenment spanned every academic field, from the remarkable astronomical achievements of al-Zarqali and the polymath al-Biruni, to the medical advancements of the Persian physician Abū Bakr al-Rāzī (Rhazes).*

Inspired by the works of Aristotle and other Greek academics disseminated by Alexander and Bucephalus, the scientific, mathematical, and philosophical endeavors of the extraordinary Persian polymath Ibn Sina (Avicenna) would influence the renowned seventeenth-century English philosophers Henry More, Thomas Hobbes, and John Locke. Eloquent, and at times racy, female poets like Mahsati Ganjavi and Rabi'a Balkhi (the maternity hospital in Kabul bears her name) are still recited and celebrated. "During the height of their empire, from the mid-

---

* Al-Biruni recognized that the Earth revolves around the sun on an axis and was later bestowed a debt of gratitude from Polish astronomer Nicolaus Copernicus.

seventh to the mid-eleventh century, the Arabs surpassed the level of prosperity and technological sophistication that the Middle East and North Africa had known in Greco-Roman times," states historian Daniel Headrick. "Only China was more developed."

Arab medical knowledge, including veterinary care, also had no contemporary equal. The physician and "father of modern surgery" Abū al-Qāsim al-Zahrāwī wrote a thirty-volume encyclopedic medical textbook that was the standard for specialized schools and the go-to for health practitioners for centuries. Other Arab physicians, botanists, and chemists pioneered the use of anesthesia for surgical procedures and isolated a catalog of medicinal compounds.

Like Avicenna, the Arab veterinary sciences were also highly influenced by previous Greek writings, including the equine-heavy *Treatise of Veterinary Medicine* by Theomnestus (c. 320)—a lasting and immeasurable legacy of Alexander's magnificent imperialistic cavalry. Stimulated by Aristotle's *History of Animals*, the Islamic scholar al-Jahiz, author of 140 works, published his seven-volume zoological treatise *The Book of Animals* (c. 840), detailing a vast array of creatures, including horses. Given the importance of all animals to Arab society, veterinarians were common even in small settlements such as pre-Islamic Mecca. In 1187 (four years before it was captured by the crusader-king Richard the Lionheart) the coastal trading center at Acre, for instance, boasted 140 veterinarian stalls in the bustling multiethnic market.

Curious European scholars traveled to Muslim schools and libraries on Arabian horses and on ships using the Arab-developed triangular lateen three-mast sail, which allowed vessels to tack against the wind. They translated Arab works, thereby collecting the technological discoveries and academic developments of not only the Muslim world but also the accumulated knowledge from China, India, and Central Asia, as well as the lost works of Greeks and Romans. These scholarly treasures were saved and stored in archives and academies across the Arab expanse. "Horse-sped, scientific knowledge diffused rapidly over large segments of the educated elites, across different regions of the Islamic world and beyond," emphasizes Pita Kelekna. "In Spain and Sicily, Arab, Jewish, and Christian scholars quickly translated Arabic works into Latin for the benefit of Europe." The Islamic Renaissance riding across Eurasia on Arabian horses fertilized cross-cultural exchange and academic advancement.

On the other hand, Charles Martel, his grandson Charlemagne, and their stirrup-inspired heavy cavalry helped define the territorial boundaries between the Christian and Muslim worlds and ushered in the era of European feudalism. Heavy cavalry and feudalistic reforms dominated the European landscape until the mid-fifteenth century, when the longbow and gunpowder unseated the gallant knight.

The most renowned and shattering example of the historical influence of heavy cavalry is that of William the Conqueror. The pioneering British military historian Sir Charles Oman described his victory at Hastings in 1066 as "the last great example of an endeavor to use old infantry tactics of the Teutonic races against the now fully developed cavalry of feudalism." William's world-altering ascension to the throne in the wake of the death of King Edward the Confessor put a stark end to the Anglo-Saxon rule of England.

The king's council, or witenagemot ("wise men"), crowned Edward's brother-in-law Harold Godwinson king, making him the first English monarch to be invested in Westminster Abbey. This decree was challenged immediately by two objecting suitors. Edward's cousin William Duke of Normandy claimed he had been secretly named as heir, while the king of Norway, Harald Hardrada, also marshaled his forces to invade England. On September 25, three weeks before Hastings, he was defeated decisively by Harold at Stamford Bridge near York, setting off a chain of events that would mark the epic end to both the Saxon and Norse sagas. William and his Frankish-inspired heavy cavalry ushered in the modern era of England and a brave new world order.

The Normans took considerable effort to sail their horses to England, indicating just how imperative they were to the overall composition of the invasion force. Overcoming the logistical complexities of collecting, corralling, and conveying two thousand to three thousand cavalry horses across the English Channel over the course of six weeks is a noteworthy achievement.

The average Norman horse stood roughly 15 hands, weighed 1,300 to 1,500 pounds, and required 25 pounds of feed per day. The basic care of these horses while waiting in muster on the coast of northern France would have been immense: 9,000 cartloads of feed and 750,000 gallons of fresh water, producing 700,000 gallons of urine and 5 million pounds of manure; 5,000 cartloads to remove this excrement; 8,000 to 12,000 horseshoes, and 50,000 to 75,000 nails produced from 8 tons of forged iron; plus ten farriers working ten-hour days for seven days a week to affix the shoes.

Then imagine the chaotic chronological scene that unfolded: loading these frightened animals onto 350 horse-specific transports, sailing the choppy Channel, and finally disembarking and setting up suitable accommodations in England. The intricacies, manpower, and resources needed to ready and transport the Norman heavy cavalry were a daunting task. This also illuminates just how important these horses were to the success of the campaign.

For the most part, the English and Norman military systems were similar in design and function. By this time, feudalism, which was firmly planted in France, had also taken hold in England. Building on the legal reforms of his grandfather Alfred the Great, under the reign of King Athelstan (924–39) the law required that "every man shall provide two well-mounted men for every plough . . . if a noble who holds land neglects military service, he shall pay 120s [shillings] and forfeit his land." Similarly, the codified Welsh *Laws of Hywel Dda* (the reign coincided with that of Athelstan) makes clear that horseshoes and stirrups for the smaller 13.2-to-14-hand English and Welsh horses (lacking the mainland Arabian influence) had been in use long enough to be taken for granted.

Neither army at Hastings was overly large, each consisting of an estimated eight thousand to ten thousand troops. The two main differences in the composition of the English and Norman armies *on the day of battle* were that the Normans possessed two thousand archers (including crossbows) and two thousand to three thousand heavy cavalrymen, whereas the English possessed few, if any, of either.* "The Norman cavalry was in large measure the means by which William overcame the English," emphasizes Ann Hyland. "And the Norman cavalry's role on the whole given due credit." The complete absence of English cavalry at Hastings was the result of coalescing factors.

For starters, Hastings was fought on October 14, a mere twenty days after the Battle of Stamford Bridge. Although victorious, King Harold had suffered heavy equine casualties. This drastically reduced the number of cavalrymen available to confront William, whose sudden arrival on the coast stunned the English. Harold immediately marshaled his battle-fatigued army and hastily marched south to London, covering an estimated twenty-seven miles per day over the course of a week.

The number of horses that eventually reached Hastings was exceedingly small, and most were unfit for immediate combat. "There can be no doubt that the English army fought as an entirely infantry army," states military historian at Brunel University London Jim Bradbury in *The Battle of Hastings*. "No source says otherwise." According to the earliest account of the clash, the epic poem *Carmen de Hastingae Proelio* (*Song of the Battle of Hastings*) was thought to have been composed within a year of the battle, "the English scorn the solace of horses trusting in their strength they stand fast on foot . . . all the men dismounted and left their horses in the rear."

Moreover, the strong high-ground *defensive* position occupied by Harold did not lend itself to deploying cavalry. "It is of course easy to criticize a man who was acting under terrible stress after conducting a campaign at the other end of England," explains University of Oxford historian David Douglas in *William the Conqueror: The Norman Impact upon England*. "As it was, he was compelled with depleted resources to fight an early defensive battle against an enemy who could not afford delay." Caught off guard, the English were forced into a reactionary defensive posture.

Harold nevertheless used the terrain available to great advantage by presenting William with a steep frontal slope. The English bookends and rear were protected from flanking cavalry sweeps by tangled forest.

---

* Originating in China around 650 BCE, the crossbow had a long history in Asia and, subsequently, the Roman Empire. By the twelfth century, they were common across the armies of Europe.

It was an ideal position for a static "shield wall" defense. As William of Poitiers, warrior-chaplain to the Normans, recorded in his account *Gesta Guillelmi ducis Normannorum et regis Anglorum* (c. 1071), "Their extraordinarily tight formation meant that those killed hardly had room to fall."

The *Carmen* reports that the Normans sounded off a disciplined and methodical advance: "The duke and his men, in no way frightened by the difficulty of the place, began slowly to climb the steep slope." Having come into range, the forward archers and crossbowmen opened the battle by unleashing a barrage of effective volleys, "provoking the English and causing wounds and death with their missiles . . . against the crossbow bolts, shields are of no avail!" While the archers "transfixed bodies with their shafts," the crossbowmen "destroyed the shields as if by a hailstorm, shattered them by countless blows." The English, however, bravely stood their ground, trading "missile with missile [javelins], sword-stroke with sword-stroke" with the Norman infantry.

At this point, William committed his cavalry to the attack, and, as his chaplain-scribe recounts, "the mounted warriors came to the rescue and those who had been in the rear found themselves in front." The exquisite Bayeux Tapestry (c. 1080), measuring 230 feet long and 20 inches tall, depicts the Norman charge being met with a shower of javelins and clubs. It also clearly indicates Norman cavalrymen holding their lances with both an overhand grip for throwing or thrusting and an underhand grip, couched under the arm. These horsemen eventually turned the tide of battle.

Climbing the slope, the first cavalry charge was thrown back. William was reported dead, and general confusion and panic gripped the Norman front. All the advancements in cavalry—bits, reins, saddles, stirrups, iron shoes, and slightly larger horses—cannot overcome the inherent difficulty and heightened dangers of cavalry charging uphill against determined, unyielding rank-and-file infantry. It was always a risky business, even for Alexander and Bucephalus 1,400 years earlier.

In a famous scene depicted on the tapestry, within the general melee that followed the failed attack, William rode into view and lifted his helmet to show his face. He rallied and wheeled his cavalry for a quick counterattack, cutting down the disorderly English; "as the meek sheep fall before the ravening lion," vividly narrates the *Carmen*, "so the accursed

The birth of a nation: Norman heavy cavalry charge the English shield wall at the Battle of Hastings, 1066. From the Bayeux Tapestry, circa 1080. *(Robert Harding Images)*

rabble went down, fated to die." Charging forward, the Norman cavalry penetrated the English line in several places.

Usually this breach, as we have seen with Alexander, would cause the enemy line to collapse completely, yet somehow, against all odds, the English held on, mitigated the damage, and re-formed. Sensing the walls closing in, the Norman cavalry shrewdly disengaged, performing a feigned retreat to entice the English infantry to break formation and pursue on open, level ground. "They therefore withdrew, deliberately pretending to turn in flight. They remembered how, a little earlier, flight had led to the success they desired," wrote William of Poitiers. "As before, several thousands were bold enough to rush forward, as if on wings, to pursue those who they took to be fleeing, when the Normans suddenly turned their horses' heads, stopped them [the English] in their tracks, crushed them completely, and massacred them down to the last man."

This deception was certainly not a new tactic, having been employed by both Philip and Alexander and by the crafty Xiongnu leader Modu Shanyu. It was a staple in the Scythian war chest and had also been utilized by the Carthaginians, Huns, Franks, and previously by the Normans during the initial shock waves of the Crusades. "Nothing that happened at Hastings would have surprised the Athenian horsemen that

Xenophon wrote for or Alexander's Companions (although Alexander might have wondered what took the Normans so long)," notes Philip Sidnell in *Warhorse*. "Cavalry, of course, continued to develop and change in its technical details throughout the Middle Ages and beyond, but the fundamental relationship governing its use had been established centuries before on the bloodsoaked battlefields of the ancient world."

At Hastings, it was William's cavalry feint that provided the pivotal swing in battlefield momentum. "The Norman knights," stresses Douglas, "were responsible for the victory." All sources assent that William's heavy cavalry won the day (and the next one thousand years of English superpower status) at Hastings. As daylight faded and surrendered to the sunset flames of dusk, the English line finally buckled under the combined weight of suppressing archer volleys and overpowering cavalry strikes. "The Norman cavalry, again inspired by the example of William still leading in person like some latter-day Alexander," states Sidnell, "charged in and cut down Harold and those that remained clustered round his banner."

Hastings was the watershed moment where cavalry finally trampled and transcended the stubborn infantry. As Sidnell sees it, "The battle of Hastings in 1066 has often been seen as the coming of age of the feudal knight—not its apogee; that was still some distance in the future—but the point at which many see the elements of the 'revolutionary new way of doing battle' as essentially in place. . . . At Hastings, the Saxon shield wall was broken and trampled by Norman knights; surely proof positive that the full effects of the stirrup-instigated revolution had matured (a mere five hundred years after the Avars brought stirrups across the steppe)." Hastings also marked the last successful invasion of England.

The Saxon-Scandinavian cultural era, which had dominated England since the extraction of Roman occupation in the fifth century, was replaced with a thoroughly more European cultural package. A bespoke, more despotic version of feudalism revolving around the king as the primary power broker became the mainstay of English politics. Included in these draconian reforms was the revolutionary "Great Survey" of 1086 known as the Domesday Book. This census was an assertion of royal power to assess land holdings and wealth for the purposes of taxation and redistribution of estates following the Norman Conquest.

Under William I, England transformed from a loose confederation of nobles into a truly unified kingdom with a centralized conviction, outreach, and government. With the Norman Conquest, England became part of Europe rather than an isolated floating island—or as poet John Donne wrote in 1624, no longer "an island entire of itself [but] a place of the continent, a part of the main." With their victory at Hastings, William and his heavy horses ensured that England was hitched to Europe through language, institutions, religion, governance, culture, economics, trade, and war. "The Norman invasion of England completely altered the future of the British Isles, determining the heritage of its people and the nature of its political systems," writes Paul Davis. "The resulting society then proceeded to have major impacts on events in Europe and, ultimately, the world . . . with all the ramifications that has had on the course of world history." These include the twin democratic pillars of common law and the Magna Carta, unrivaled colonization with mind-boggling consequences, and an unsurpassed empire propagating British culture and society.*

At its zenith in 1913, the British Empire covered 13.7 million square miles (24 percent of the planet). In 1920 the total population living under some facet of British control was an astounding 458 million people (23 percent of the world population). In this sense, the modern world order—and, of course, William Shakespeare, James Bond, the Beatles, and Harry Potter—was birthed in 1066 by William and his heavy cavalry horses at Hastings.

---

* On June 15, 1215, at Runnymede, England, King John was forced to concede to the demands of the rebellious barons and sign the groundbreaking Magna Carta Libertatum—"The Great Charter of Liberties." The now-universal attachment of the catchall slogan "No one is above the law" to the Magna Carta, however, is a misconception. Nowhere among its sixty-three clauses does this phrase exist. The modern interpretation and construction of this expression can be loosely cobbled together from two articles: "*39. No free man shall be seized or imprisoned, or stripped of his rights or possessions, or outlawed or exiled, or deprived of his standing in any other way, nor will we proceed with force against him, or send others to do so, except by the lawful judgement of his equals or by the law of the land. 40. To no one will we sell, to no one deny or delay right or justice.*" These concepts, regardless of their meaning in 1215, ushered in the age of modern democracy, common law, and the footing for the universal unalienable rights of the individual to life, liberty, and the protection of property. Its influence echoes throughout the constitutions of modern democracies, including the Bill of Rights in the United States, the Canadian Charter of Rights and Freedoms, and internationally in the United Nations' 1948 Universal Declaration of Human Rights.

CHAPTER 11

Road Apples

The Medieval Agricultural Revolution and the Making of Modern Europe

In the aftermath of Hastings, William upended Saxon rule and granted loyal Norman barons and allied lords sprawling estates and large tracts of land across England. This feudal tenurial obligation of *servitium debitum* secured the fealty of more than five thousand knights. Imposing harsh taxation upon his newly sworn subjects, the new king also raised bulky contingents of mercenaries to consolidate his rule. In his newly annexed territory, however, humans outnumbered horses by a margin of one hundred to one. With England relying predominantly upon oxen as beasts of burden, the Normans continued to import stronger stallions from across the Channel, while crossbreeding with domestic stocks.

These measures gradually sired stouter horses that were more anatomically suited to the rigors of heavy husbandry. Stocky horses with a lower center of gravity and shorter drawing axis expend less traction energy than their taller counterparts. Guided by the human hand, burly draught and plow horses, including the shire, Clydesdale, Flemish, Belgian, and Percheron, were bred to be bulkier and stronger, befitting their role as beasts of burden rather than the swift and sleek Arabian warhorse.*

Alfred the Great's *Old English Orosius* of the late ninth century records the first reference to horse plowing in England. When Pope

---

* The *Acte for Bryde of Horses* (Horses Act) issued by King Henry VIII in 1540 standardized the measurement of a hand to four inches, while decreeing that all horses under fourteen hands should be eliminated.

Urban II issued his call to arms in 1095, exhorting Christians to unite and recapture the Holy Land from Muslim rule, igniting the First Crusade, he promised absolution not only to all men who marched to war but also to all *Christian* oxen and plow horses. This sanctified mention indicates that horses were being integrated onto European medieval farms in large numbers.

Like the original earth-shattering Agricultural Revolution ten thousand years earlier, this pastoral shift from oxen to horses was equally groundbreaking and, like its predecessor, permanently altered the course of human history. According to historian John Langdon in the meticulously researched *Horses, Oxen and Technological Innovation: The Use of Draught Animals in English Farming from 1066–1500*:

> It is possible that a certain difficulty occurs in imagining the introduction of the work-horse as a technological innovation, since, whatever its advantages, the simple substitution of one animal for another might appear to be technologically irrelevant. But in fact this "simple" transition marks the culmination of a series of intricate mechanical and biological changes. These involved not only developments in harnessing, shoeing and breeding, but also refinements in plough and vehicle design. It was, in essence, the substitution of one technological package for another, for which the physical replacement of the ox by the horse in front of the cart or plough was simply the most eye-catching step.

Throughout Europe horses became profit-spinning engines occupying a place of pride across medieval society. This horse-powered medieval Agricultural Revolution, however, was made possible only by cutting-edge Chinese equine technology that had worked its way across the steppe to Europe. The full-scale Chinese adoption of the horse in the second century BCE to counter the Xiongnu led to four of the most important and impactful equine equipment innovations in history: the effective harnessing system based on the breast strap; the horse collar; the whippletree; and the moldboard plow. "This society," stresses historian

Daniel Headrick, "produced a technological revolution in agriculture that set the stage for the modern world."

Like China's mastery of bronze and iron metallurgy and its development of the stirrup, these revolutionary inventions finally transferred the full potential of horsepower to agriculture, transport, and war, while recentering the path of human history. They also produced other unintended consequences. This horse-driven overhaul and increased productivity fueled population growth, urbanization (and spillover pathogens), the Commercial Revolution, intellectual sophistication, and, eventually, a Renaissance.

The original "throat and girth" harnessing system was devised for oxen. It was not anatomically suited to equids and failed to channel their full tractive power. The physiology of the two animals, as we have seen, is far from synonymous. The initial collar was not intended for the longer neck of a horse and put undue pressure on the windpipe. The compression impinged on its ability to breathe, inducing suffocation as it stepped forward to pull against the harness. The Romans, for example, adhered to a strict law that limited the weight of horse-drawn wagons to 717 pounds to prevent strangulation. Similarly, the empire's compilation of laws, the Codex Theodosianus, ratified in 438, listed the maximum legal weight hauled by a two-horse team at 1,100 pounds.

As a horse-specific substitute, in the fourth century BCE the Chinese adopted a hard yoke, or shaft, that ran across the chest. This was quickly replaced by a more economical breast strap, or "trace harness," designed to transfer the unrivaled strength and speed of the horse to a plow or vehicle.

The horse collar, devised in the first century BCE, further streamlined this breast-strap harness system, with traces (the straps on each side of the harness on which the animal pulls) connecting the horse directly to whatever was being towed. The padded and contoured collar rested on the withers, wrapping the chest, making it the simplest, yet most efficient, hauling apparatus designed specifically for horses. "How the Chinese were led to make one of the greatest breakthroughs in equestrian history is still a mystery," contends Bill Cooke of the International Museum of the Horse. "One theory holds that the invention of the breast-strap harnessing system for horses was derived from the human harness system used for boat haulage on China's canals and rivers. Men would have realized from their own experience that to be effective, pulling force must be exerted from the sternal and clavicular region in such a way as to permit free breathing." By allowing the horse to push with its shoulders, the horse collar increased drive power fivefold. The upgraded power output of the modern Chinese system was staggering.

These novel configurations fostered the development of sophisticated and advanced "shafted" horse-drawn vehicles—with four wheels, swiveling front axles, and brakes—that were far more economical than their predecessors. A single horse could now pull a Chinese chariot or carriage containing up to six passengers, as opposed to those in Greece or Rome that were fueled by four-horse teams accommodating only two people. Teams of four to eight horses easily pulled fourteen thousand to twenty thousand pounds.

With these revolutionary horse-specific amendments, the cost of transport dropped precipitously, heralding the dawn of the modern public transportation system and the mass movement of commercial goods and professional armies. They also facilitated the massive horse-powered Chinese construction projects mentioned earlier, including canals, Great Walls, palaces, roads, a postal system, and the first emperor's self-aggrandizing terra-cotta tomb.

Complementing these groundbreaking innovations, the Chinese added the whippletree mechanism in the third century CE, further increasing the horse's pastoral power and carriage capacity. Consisting of a loose horizontal bar (originally wooden) between the horse and its vehicle or plow, whippletrees are used to evenly distribute tension forces from the traces of draft animals to a hauled, drawn, or dragged load. Sometimes referred to as equalizers or shock absorbers, they also balance the pull from each side of the animal (minimizing alternate tugging), while preventing the weight point of the traces from collapsing into the sides of the animal.

A series of linked and interconnected whippletrees are utilized for teams, ensuring an equal share of the workload and a smooth turning radius, or pivot, preventing the vehicle or plow from upending. A swingletree (or singletree) is a term used to describe horse-specific systems.

We have all noticed these ingenious gadgets and dismissed them as another simple piece of equipment without thinking about their pragmatic function of increasing horsepower.

Correspondingly, the first moldboard plow also appeared in China in the third century. This design was more sophisticated than the scratch plow used by the Romans and other Western cultures. Its weight was transferred downward, digging deep furrows in the soil. This required increased traction power, which was generated by the horse and its new harnessing systems. This novel technique allowed the Chinese to exploit untapped fallow lands along riverbanks and recessed valleys with damp, thick soils previously too heavy to cultivate.

In combination, the trace harness, horse collar, whippletree, and moldboard plow quickly made Chinese agriculture the most productive on the planet. Between 300 and 400, the Chinese population increased by a whopping 32 percent. These four contributions to world culture had impacts far beyond their place of origin. When disseminated across the Eurasian Steppe, this game-changing package of equine innovations (along with the Chinese invention of paper) revolutionized agriculture and transport and, by extension, food production, distribution, demographics, economics, and academia, most notably in Europe, to where our story now shifts.

The breast-strap trace harness spread throughout Europe during the seventh and eighth centuries, with the horse collar arriving a few hundred years later. Together with the moldboard plow, which had been developed independently by Slavic peasants in the sixth century, maximum yields could now be extracted from the soggy, heavy loams of northern Europe.

The first appearance of nailed iron horseshoes in Siberia in the ninth century further elevated the position of the horse across European society. "This coincided with the start of selective breeding for a heavier horse," notes Ann Hyland. "Mounted warfare was the catalyst, for horses bred in wetter than steppe and/or desert conditions develop softer, shallower hooves. With the stresses imposed by the weight of a cavalryman and the pounding impact with the ground, hooves needed protection from splitting, cracking, laminae deterioration, and sole bruising." These tailored iron shoes, which quickly spread to China and Europe, were certainly a better fit than the previous slide-on Roman hipposandals.

This novel footwear allowed draft horses to operate in those heavy, wet soils by preventing their hooves from going soft and wearing down prematurely. Shoeing also prevented the same wear (while adding grip) on the hooves of transportation horses clip-clopping down now common hard-packed or cobblestone roads. Iron shoes significantly extended the working life of horses across their multivariate vocations, further increasing their numbers, employment, and value. The blacksmith instantly became an integral and vital part of any settlement. This multifaceted medieval transformation was accompanied by pervasive demographic, socioeconomic, and military repercussions across Europe.

With the adoption of harnessing systems, the horse collar, iron shoes, whippletrees, the moldboard plough, and three-crop rotation practices, horses propelled the medieval Agricultural Revolution forward. Horse-based husbandry steered the socioeconomic and demographic transformation of Europe.

A new class of cosmopolitan entrepreneurs, artisans, bankers, financiers, academics, tradesmen, and merchants emerged to venture into the commercial guilds and markets of medieval capitalism. "Thanks to the discovery of 'horse-power,' with its significant repercussions for warfare and agriculture, this was a time when the horse came into its own as a vital contributor to life at all levels of society," writes curator at the British Library Pamela Porter in *Medieval Warfare in Manuscripts*. "In one sense, it might even be said that the very foundations of medieval social structure rested on the horse. . . . The horse of the Middle Ages had become an indispensable member of society, making valuable contributions to numerous facets of everyday life." Riding the financial surplus of agrarian horsepower, Europe entered a profit-margined, cash-based economy. The traction capacity of medieval horses planted the seeds of our modern cultural and financial portfolios.

During the transitional period of the twelfth and thirteenth centuries, mixed teams of oxen and horses were customary until horses became more common and correspondingly cheaper. The same holds true

Medieval horsepower: English peasants with workhorse and harrow plow. Note the horse collar, harnessing, whippletree, and horseshoes. From the Luttrell Psalter, circa 1330. *(Heritage Image Partnership Ltd./Alamy Stock Photo)*

for the more expensive horses hauling wagons, carts, and carriages. As horses became more abundant, there was a steady increase in plowing speeds and a corresponding reduction in the size of cultivation teams. It was a win-win for farmers: amplified production with slashed overhead and maintenance costs. Horses maximized the profit ratio of traction power on medieval farms, estates, and demesnes. "Light two-horse plough teams of the fourteenth century," states economic historian at Queen's University Belfast Bruce Campbell, "were no less than 65 percent cheaper than conventional eight-ox teams."

The great Domesday survey of 1086, for example, counted horses as only 5 percent of draft animals across England and no more than 10 percent in any single county. By 1300, however, horsepower skyrocketed to 20 percent on estates and over 50 percent on peasant farms, with some regions reaching 75 percent. When horses replaced oxen, the energy output on European farms exploded.

Horses can work two to three more hours per day than an ox while moving 33 percent faster. Average working years also favored the horse: draft ox, 5.1; plow horse, 5.5; and cart horse, 7.0. In short, when hitched to a moldboard plow, one horse produced the traction power of two oxen, allowing farmers in the colder climes of northern Europe to exploit previously fallow lands.

This combination cut deep furrows into dense or clay-infused soils with one pass as opposed to multiple passes with the scratch plow (ard), or in the case of the heaviest loams, not at all. Repetitive plowing wasted enormous amounts of time, in addition to animal *and* human energy. The thick ruts scored by iron moldboard plows allowed excess water to drain off the crops to drown and kill strangling weeds, as did the trampling iron-shod hooves of industrious traction horses. Consequently, farmers did not need to share plow teams to till their lands in sequence.

Advancements in deep mining and forging techniques also ushered in the widespread use of iron scythes, hoes, shovels, and pitchforks, enhancing human efficiency. The wheelbarrow, invented in China in the second century, first appeared in Europe around 1200, adding to agricultural proficiency. The interwoven application of these innovations meant that more land was being turned over more quickly. The advent of the resourceful three-field crop rotation only quickened the pace.

This system of land management was a crucial component of the medieval Agricultural Revolution sweeping through Europe. In the past, farmers traditionally planted only half their fields each year, leaving the rest to replenish, alternating annually. Given that more land was now available to sow thanks to the horse and moldboard plow, they could now grow wheat, oats, barley, or rye for bread, soup, and brewing on a third of their land; peas and other legumes (amenable to storage) on a third; and leave the remaining third to rejuvenate, moonlighting as pasture for horses and cattle. "Compared to the old two-field rotation, this method increased the productivity of the soil by one-third to one-half with no additional labor," asserts Headrick. "It also reduced the risk of crop failure, provided more protein in the diets of humans, and produced oats for horses." Oats, containing more protein than wheat and barley, became the staple fuel for large European draft horses.

This rotational configuration produced a complementary rather than competitive environment between arable and grazing lands. It also reduced soil exhaustion, decreased land idleness, and diversified crops, leading to a healthier diet and a sustainable food source in leaner times. The nitrogen generated by the legumes and rich horse manure fertilized the soil within the cyclical interchange. "Substitution of legumes for bare fallows, more thorough application of fertilizers, better preparation of the seed-bed, improved weed control, and more scrupulous harvesting," summarizes Campbell, "were all possible with increased labour inputs"—courtesy of the horse.

Within the oxen-powered feudal system, most European peasant farmers were raising crops for subsistence and for the local tables of knights, vassals, lords, and the clergy. Attached to horses, food production became cheaper, a cost savings that was passed on to the consumer. Agricultural yields increased so much in certain regions that nutrition became an exportable commodity. By the late thirteenth century, for instance, foodstuffs accounted for 10 percent of English freight.

As a result of this affordable surplus, European populations and densities soared. According to historian Bernard Slicher van Bath in his seminal work *The Agrarian History of Western Europe*: "The tenth century saw the introduction of improved harness, which enabled horses to be used for ploughing. Thanks to three-course rotation, there was enough fodder to keep more horses. In all likelihood, it was through this augmentation of the sources of energy that the ensuing rise in population from the eleventh to the thirteenth century was made possible. Now it became necessary, and only now was it feasible, to start developing on a big scale." Populations, which had plummeted alongside the collapse of the Roman Empire, were rebounding due in part to the elevating impact of equine-based farming. Horses were progressively supplying the daily bread for surging urban, and increasingly specialized, populations.

By the twelfth century, modern towns and cities, although comprising no more than 10 percent to 12 percent of the total populace (increasing to 20 percent by 1300), began to spring up across Europe to accommodate horse-powered agricultural commercialism and its increasingly complex economic coattails. Paris, the most crowded city in Europe, reached 230,000 residents by 1300. The population of England surged from 2 million in 1086 to 5.5 million by 1300. London alone grew from 10,000 inhabitants to 80,000 during this same period. Across England, thirteen other cities topped 10,000 people.

In fact, the location of the towns and cities dotting the medieval European map would be almost indistinguishable from today. Urban centers sprang up along natural overland or waterway transportation routes, at the crossroads of communication, around monasteries, castles, and trading hubs, and adjacent to fertile agricultural hotbeds. Horses transformed the demographic, economic, academic, and physical landscapes of Europe.

Instead of rural farmers living in isolated hamlets, urban clusters developed around the needs of the horse to include blacksmiths, lumberjacks, carpenters, wheelwrights, masons, cobblers, tanners, veterinarians, stable masters, grooms, farriers, and other horse-related employment. These medieval characters spill off the pages of Geoffrey Chaucer's *The Canterbury Tales*, a collection of twenty-four stories written between 1387 and 1400.

As mentioned, in 1250 Giordano Ruffo penned his six-part *De Medicina Equorum*, one of the oldest comprehensive European horse-specific medical texts. The Company of Marshals, founded in London in 1356, changed its name to the Worshipful Company of Farriers to better reflect its high-demand craft.* The skilled trades were professionalized and controlled by tiered guilds with orderly Jedi-like ranks of apprentice, journeyman, and master.

Of course, these new city dwellers and suburban agriculturalists required the vital services of churches, taverns, mills, mines, banks, hotels, and, by the close of the eleventh century, universities, such as Bologna, Oxford, Salamanca, and the Sorbonne. Local languages began to replace Latin among the educated. Academic disciplines devoted to the analysis of humans, or the humanities, pulled up alongside the theological study of God.

When the first Bible rolled off Johannes Gutenberg's world-altering movable metal-type printing press in 1455, no more than a hundred thousand books could be found throughout all of Europe.† But within five years, more than six million volumes competed for space on crowded shelves. By 1500, 236 towns had print shops, and Europeans enjoyed access to an estimated twenty million books. Thanks in part to the power of the horse, Europe began to tinker with the ideas of humanism, which

---

* The military rank "marshal" is derived from the Old French *mareschal*, denoting "a person who tends to horses," particularly a craftsman who practiced *marescalcia*—the shoeing of equine hooves—known as a farrier.

† Bi Sheng, a Chinese commoner, invented movable type in 1045, and Koreans were employing metal type by the thirteenth century.

gave birth to magnificent and flourishing Renaissance cities such as Antwerp, Bruges, Toledo, Venice, and Florence.

The diversified economic specialization of the Commercial Revolution has been hailed, for better or for worse, as the "birth of modern capitalism." This rapid ascent of money was attended at all echelons of society by bankers, creditors, partnerships, accountants, masters of coin, and bookkeepers. The rate of European enterprise rose so quickly, however, that there was not enough coin currency in circulation to keep pace. Peppercorns, with a relatively standard weight and long shelf life, served as a convenient, albeit strange, substitute. The church's ban on usury (lending money at interest) was circumvented by the creation of contracts, which imposed stiff fines for late payments or defaults.*

The shift from Romanesque to Gothic architecture in the twelfth century is symbolic of the molten wealth circulating through the social ranks of merchants, guilds, nobles, kings, and, of course, the affluent aristocracy of the church. Medieval towns were dominated by towering cathedrals and bustling and noisy markets dripping with the waste of animals and swarming with a bourgeoning merchant class selling local agricultural surplus and commodities.

Massive annual or seasonal fairs supplemented daily or weekly markets. The busy Belgian port of Antwerp, for example, held four fairs per annum, each lasting three weeks. These were grand commercial events centered around wholesale markets where dealer negotiated directly with dealer. The entire community—from taverns and inns to blacksmiths and brothels—benefited from the massive influx of top-tier traders, small-time merchants, and a shady assortment of wanderers and swindlers peddling all types of wares, snake oils, and fortunes.

Many festivals were product specific. The annual wine fair held since the seventh century at the Saint-Denis monastery near Paris, for example, attracted vintners, sommeliers, vendors, and wine enthusiasts from across Europe. At Scania in southern Sweden, the star attraction was herring, while the cloth halls of Flanders showcased woolen fabrics.

---

* This competitive market and the population boom were also buoyed by the demise of slavery in England during the twelfth century.

Textiles spawned the origins of light industry, with horse- or water-powered mills driving presses and flails to extract oils, clean, and thicken organic cloth fibers.*

Networks of local stables and brokers acted as intermediaries for thriving horse fairs akin to our modern car shows. Like today, when American football legend Tom Brady's Bugatti Veyron costs more than my Chevrolet Equinox, the same could be said for medieval horse markets where both rich and poor bargained for animals within their budgetary range. An eyewitness from the marketplace adjacent to the Sorbonne in Paris reported seeing "over three thousand horses, and it is most remarkable that there should be so many, since markets are held twice a week." The horse fair in Whitson, Wales, across the River Severn from Bristol, was a three-day affair where the horses "were beautiful to see and profitable." Given their utilitarian value, horses themselves entered the commercial culture of the Middle Ages.

Horses became the economic backbone of medieval Europe. While they increasingly bore the brunt of agricultural labor, they also became a common sight pulling vehicles loaded with market goods, produce, livestock, or human passengers. By the end of the thirteenth century, horses accounted for 75 percent of English hauling. "On the basis of the weight of goods transported by animal, the pack-horse was patently inferior to the cart-horse," concludes John Langdon. "The maximum load for the former is thought to have been just over 400 lbs. On the other hand, a good cart-horse in the fourteenth century was seemingly capable of hauling over a ton on his own." Of course, this commercial traffic required an expanding network of manure-soiled roads.

Many of the old Roman roads, including 13,000 miles in France and 2,500 in Britain, were revamped, while new routes were constructed to transport the diverse and greedy commercial demands of horse-driven Europe. Given the volume of goods in transit and the uptick in capitalist ventures, most roads must have been reasonably passable throughout the year. "The Roman roads were the backbone of the system," explains British historian and archivist Geoffrey Martin, "and new roads had

---

* Windmills were also erected across Europe between 1180 and 1300.

come into existence, not through systematic construction, but simply through continued use during a period when transportation was becoming continually more important to most aspects of national life." The travels of the English kings John, Henry III, and Edward I, who ruled in succession from 1199 through 1307, are well documented. They encountered no difficulties covering great distances relatively quickly with their horses, wagons, and carriages during any season of the year.

As horse populations increased, urban and rural stables became standard infrastructure. While they were not quite as extravagant and sophisticated as the lavish "ideal stable," with its automatic cleaning and feeding devices envisioned by the animal-loving Leonardo da Vinci, they were pragmatically utilitarian. Most were rustic, wooden buildings with a raised hayloft and adjacent water cisterns or basins. Excrement was collected and mixed with other organic materials and stored in compost piles on the fringes of fields for use as fertilizer.

Manure itself became a commodity, especially the road apples produced by urban cart horses and stables. "Towns, with their concentration of animals and humans, were the single greatest supplementary source of organic fertilizer," writes Campbell in *English Seigniorial Agriculture, 1250–1450*. "Their demand for provisions provided farmers in the immediate countryside with a powerful incentive to intensify their output. Purchasing urban manure provided one means of achieving this." The supply, however, always exceeded the demand. Horse dung (and urine) was a compounding sanitation and health concern, and one of many cordial invitations extended to disease within the densely populated European urban landscape.

The 140 to 350 horses kept by the seven thousand residents of Luxembourg City during the fourteenth century, for example, produced between 165 and 405 gallons of urine and 1.8 to 4.5 tons of manure *per day* in an enclosed space covering no more than 119 acres. While some was collected and sold, most was left on the streets along with dead horses to fester and stink. "Also eight groats for removing a dead horse which lay under the castle," records one contemporary bill of service. "And some of the citizens requested to have it removed because it stank a lot." Luxembourg City was not the exception.

In most medieval cities, official cleanup occurred only before major events such as royal visits, fairs, and tournaments of knights.* As cities became more equestrian, and their human and horse populations boomed, "horse pollution" piled up and became, quite literally, an exponentially rising problem.

The horse and its medieval Agricultural Revolution transported the West to the threshold of the modern age. Europe, which had long been a splintered, impoverished, global backwater, finally caught up to the cosmopolitan and cultured polities of China, Southeast Asia, India, Japan, Persia, Anatolia, and Arabia. The transition to horse-powered agriculture and its technological counterparts supported and maintained an unprecedented period of academic, economic, demographic, and urban growth between 1100 and 1350. Eventually, this package, hitchhiking on the mercantilist tentacles of imperialism, was transferred to temperate zones in the Americas, southern Africa, Australia, and New Zealand, where the horse was unwittingly conscripted as a conquistador of colonization and in the subjugation and desolation of Indigenous peoples.

In 1492, when Christopher Columbus was eight thousand miles lost, and the last vestige of al-Andalus Islam was expelled from Spain during the Reconquista, Europeans controlled a mere 15 percent of the Earth's landmass. By 1800, they dominated more than 35 percent, and on the eve of the First World War in 1914, Europeans had colonized a staggering 84 percent of the world. As the requisite machine driving the arteries of agriculture, transportation, and war, the horse was not replaced until the rise of the machine at the dawn of the twentieth century.

Of course, if crops failed, this swelling medieval population was susceptible to mass starvation chaperoned by the Malthusian checks of disease and war.† Precipitated by disastrously wet and cold weather triggered

---

* King Richard the Lionheart legalized the "Tournament of Knights" in England in 1194 upon his return from the Third Crusade.

† In 1798 English cleric and scholar Thomas Malthus published his groundbreaking *An Essay on the Principle of Population*, defining his enduring ideas on political economy and demography. He argued that once an animal population has outpaced its resources, natural catastrophes or checks such as drought, famine, war, and disease will force a return to sustainable population levels and restore a healthy equilibrium.

by the five-year volcanic eruption of Mount Tarawera in New Zealand, the Great Famine between 1315 and 1317 marked the beginning of the end to this era of development and prosperity ushered in by horse-powered agriculture. Stalking its way across Europe, the famine culled an estimated 10 percent to 15 percent of the population.

Sheep and cattle herds were also cut down by disease. Roughly 80 percent of European cattle succumbed to the deadly rinderpest virus between 1315 and 1320.* "The floods of rain have rotted almost all the seed," lamented a chronicler in 1315. "In many places the hay lay so long under water that it could neither be mown nor gathered. Sheep generally died and other animals were killed in a sudden plague. . . . The dearth of grain was much increased. Such a scarcity has not been seen in our time in England, nor heard of for a hundred years . . . dogs and horses and other unclean things were eaten." The horror had only just begun.

This period of unprecedented growth and change came crashing down a few generations later amid the most lethal epidemic in human history. During the mid-fourteenth century, bubonic plague cut a murderous swath of wanton destruction through an already weakened and cowering population. The Black Death was delivered by the reaping whirlwind of Chinggis Khan and his Mongol hordes, on horseback.

---

* Thanks to modern vaccines and tireless global eradication campaigns, rinderpest was declared extinct in 2011.

PART III

GLOBAL TRAILS

CHAPTER 12

Shuttling the Silk Roads
Mongol Hordes and Eurasian Markets

You never get a second chance to make a first impression. Introductions leave indelible imprints and set the stage for future interactions. "I am Attila, the scourge of God" is a difficult greeting to outdo. But in March 1220 another even more fearsome and infamous warlord did just that.

In the bustling Silk Road market town of Bukhara, the Friday mosque was brimming with apprehensive and nervous patrons. The anxious crowd did not gather to answer the customary Muslim *adhan* call to prayer. After dismounting from his small steppe horse and demanding it be supplied with fodder, a foreign warrior dressed in striking clothing and lamellar armor entered the mosque and mounted the pulpit. He stood silently, surveying his frightened and stunned audience with intensely focused eyes. "I'm a flail of God," the stranger announced. "If you had not committed great sins, God would not have sent a punishment like me upon you." Not a bad opening line, and one that certainly rivals Attila's.

Following his own self-styled sermon to the inhabitants of Bukhara, Chinggis Khan and his "Devil's Horsemen" proceeded to systematically plunder the city. If plundering can be described as disciplined, then this was a disciplined display of pillage, rape, and mass murder. The residents were herded into the city squares, where more than thirty thousand were mercilessly slaughtered. The remainder were used as human shields in subduing the last remnants of resistance or marched into the bondage of slavery. Bukhara, the cultural and intelligentsia capital of the mighty

Khwarazmian Empire of Central Asia, was razed, with corpse-draped rubble smoldering for days.

When a wandering, shell-shocked refugee was asked by peoples of the neighboring Khorasan region what had transpired, he simply replied, "They came, they sapped, they burnt, they slew, they plundered, and they destroyed." A powerful first impression in its own right and a harbinger of what would come.

Within a year, Khorasan would feel the wrathful flail of the Mongols in one of the bloodiest campaigns in history. With a butcher's bill in the millions, according to one chronicler, no one was spared to even "piss against a wall." The Mongols, however, now controlled the precious trading hubs of Central Asia, including Kashgar, Samarkand, Bukhara, Merv, Herat, and Tashkent, connecting the frayed Eurasian bookends of the lucrative Silk Roads.

Almost eight hundred years after Atilla, when questioned about the greatest pleasures in life, Chinggis Khan decreed: "Man's greatest good fortune is to chase and defeat his enemy, seize his total possessions, leave his married women weeping and wailing, ride his gelding, use the bodies of his women as a nightshirt and support." Attila would have nodded in approval.

His harsh answer reveals the blunt motives behind the rapid Mongol conquest of the largest contiguous land empire in history. At its peak, it covered almost twelve million square miles, larger than Canada, the United States, Mexico, Central America, and the Caribbean islands—combined! Overlaid on a modern map, Mongol dominion enfolds thirty countries, encompassing almost half the global population. This vast empire, second only to the British in total territory, was procured with the power of the horse.

The lethal tandem of their cherished steppe horses and specialized composite bows was the very essence of Mongol culture. A Chinese chronicler declared, "The Mongols are by nature good at riding and shooting—they took possession of the world through the advantage of bow and horse." For the Mongols, who learned to ride and shoot as children, these survival skills were as instinctively natural as milking a horse, preparing the fermented beverage koumiss, or erecting a *ger*, or yurt.

Papal emissary John of Plano Carpini, who reached the Mongol capital of Karakorum in 1246, observed, "They are all, big and little, excellent archers, and their children begin as soon as they are two or three years old to ride and manage horses and to gallop on them, and they are given bows to suit their stature and are taught to shoot; they are extremely agile and also intrepid. Young girls and women ride and gallop on horseback with agility like men. We even saw them carrying bows and arrows." For the Mongols, the horse-and-bow synergy was merely an extension of being.

Mongol archers could deliver their arrows on target at a full gallop with a pull between 100 and 160 pounds-force. At a gallop, a horse's head extends farther forward, creating clearer lines of sight and unobstructed fields of fire. Unlike European archers, the Mongols used a thumb ring to maximize draw and torque on the bowstring while reducing drag and friction on a quicker and more accurate release. Their Scythian-styled bows were highly effective within 350 yards (roughly 100 yards farther than the heavier longbow) but could reach upwards of 500 yards. The Mongol military machine was consequently the purest horse archer army ever committed to war.

When the Mongols commenced their invasion of China in 1209, a Song dynasty official bemoaned:

> The reason why our enemies to the north and west are able to withstand China is precisely because they have many horses and their men are adept at riding; this is their strength. China has few horses, and its men are not accustomed to riding; this is China's weakness. . . . The court constantly tries, with our weakness, to oppose our enemies' strength, so that we lose every battle. . . . Those who propose remedies for this situation merely wish to increase our armed forces in order to overwhelm the enemy. They do not realize that, without horses, we can never create an effective military force.

It seems the Chinese had forgotten the hard-earned, lifesaving equine lessons of the Sino-Xiongnu War over a millennium earlier.

Flail of God: Mongol and Khwarazmian warriors clash at the Battle of Parwan, Afghanistan, 1221. From *Jāmi' al-tawārīkh* (*A Compendium of Chronicles*) by Persian scholar and historian Rashid al-Din Hamadani, circa 1310. *(Heritage Image Partnership Ltd./Alamy Stock Photo)*

Confronting a similar perilous situation on their northern frontier, this time there would be no Emperor Wu, Zhang Qian, or blood-sweating heavenly horses riding to the rescue. The Mongol invasion was swift and overwhelming.

The rise of the Mongol Empire to superpower status, however, sprang from the humble, poverty-stricken beginnings of a young boy named Temujin who was terrified of dogs and tested his mother's patience with his importunate crying. The life and times of the boy who grew into his title of Chinggis Khan (Universal Ruler), which would strike fear into the hearts of his enemies, were recorded shortly after his death in *The Secret History of the Mongols*, the oldest surviving work written in the Mongolian language.

Adopted by the Mongols to manage the mind-boggling volume of economic and military activity crisscrossing their vast and seemingly ever-expanding empire, this writing system (adapted from the Uighur alphabet) also required importing staggering amounts of paper. In August 1221, for example, Mongol scribes at Karakorum sent an urgent demand to their Korean subjects for one hundred thousand sheets of their high-quality paper. The *Secret History* is a wealth of primary information for its extraordinarily immediate and intimate portrait of Mongolia's most famous son, who was an unlikely candidate to lead the charge for world domination.

The austere and windswept grasslands of Mongolia were occupied by warring clans and duplicitous factions. Alliances within the five main coalitions were as capricious and fleeting as the whims of the blustery steppe breezes. Temujin was born around 1162 into this inhospitable landscape and rivalrous, clan-based society that revolved around vengeance, raiding, pillage, and, of course, horses. After enemy warriors fatally poisoned his father, his clan lost prestige within the political arenas of Mongol tribal power. Surviving in exiled destitution, Temujin and his family were reduced to hunting small marmots and scavenging the carrion of dead animals.

Vying to reestablish his family's honor, Temujin, now fifteen years old, was seized during an attack by his father's former allies. He escaped from enslavement and vowed revenge on an increasingly long list of enemies. "Chinggis's attitude was determined by three key experiences," states anthropologist Thomas Barfield: "the desertion of his family at the death of his father, the desertion of those kinsmen who had elected him khan, and disputes with relatives after he became supreme ruler." Although Temujin was reluctant to share authority, he appreciated his immediate circumstances and realized that securing ultimate power and prestige, as counseled by his mother, rested on strong and resilient alliances.

Following successful campaigns to unite the warring Mongol factions, Temujin secured his position as supreme leader and his mandate as Great Khan. By 1206, Chinggis controlled a stretching steppe domain housing a tartan collection of roughly one million to one and a half million nomadic people and fifteen million to twenty million herd animals, including horses. With his consolidation of power, he produced the most potent, devoutly loyal, and disciplined mounted combat strike force the world has ever seen.

Senior military and political positions were based on merit and acumen rather than the traditional conventions of clan membership and family nepotism. "Under Genghis Khan," writes Jack Weatherford in his acclaimed bestseller, *Genghis Khan and the Making of the Modern World*, "cowherds, shepherds, and camel boys advanced to become generals and rode at the front of armies of a thousand or ten thousand warriors." This commonsense approach safeguarded a loyal cadre of skilled and competent officers, while also allowing for organic battlefield promotion. "These men are more obedient to their masters," declared Friar Carpini, "than any other men in the world."

Tribal affiliations were purposefully disregarded in the formation of collective "Mongol" units. This savvy and cohesive integration undermined tribal intrigues and ensured solidarity across all tiers of the military, which was organized on base-ten units with a streamlined chain of command.* The Mongol army reached its peak in the mid-thirteenth century at 150,000 cavalrymen supported by 1.5 million horses.

Breaking from the contemporary narrative painting the Mongols as ruthless, uncivilized barbarians, in truth they were extremely malleable, open minded, and pragmatic. The Mongols were willing to evolve beyond their conventional customs and mores to incorporate new and eclectic cultural and military elements into their collective. As they advanced across Eurasia, the *tamma* (specially blended units with military, political, economic, and propaganda branches) studied and disseminated the knowledge and technology of conquered cultures to meet shifting strategic demands such as fortified cities, hefty rivers, and forested terrain. They assimilated a pliable international arsenal to tackle these obstacles, including siege armaments, counterweight catapults, experimental mortars and cannons, bridge layers, and even a navy.

---

* Like Russian dolls placed one inside the next, each of the three corps was made up of divisions of ten thousand mounted troops divided into ten regiments of one thousand each, further separated into ten squadrons of one hundred that were composed of ten patrols of ten cavalrymen. The Mongols may have borrowed this efficient base-ten order of battle from the Xiongnu, who used a similar military structure.

Premeditated fear was also a potent Mongol weapon, and they purposefully allowed their reputation to precede them. When people "beheld the surrounding countryside choked with horsemen and the air black as night with the dust of cavalry," wrote Persian scholar Ata-Malik Juvaini in his detailed eyewitness account, *The History of the World-Conqueror*, "fright and panic overcame them, and fear and dread prevailed. Whoever yields and submits to them is safe and free from the terror and disgrace of their severity." The Mongols encouraged the gossip-spinning rumor mill and ministered to its promulgation with relentless propaganda. Emissaries were then sent ahead of the main army to corroborate the unprecedented horrors and accept surrender.

The Mongols' greatest weapon, however, was their beloved horses. Utilizing overwhelming speed and surprise (the army could travel eighty miles a day), they encircled their hapless enemies with breathtaking momentum and ferocity. The Mongols differed from all other nomadic steppe armies in that they had a penchant for seeking out decisive engagements rather than the protracted cat-and-mouse campaigns employed by the Scythians and their Xiongnu descendants.

When applying their customary maneuvering, waves of Mongol archers launched frontal assaults: firing at an effective distance before retreating while letting loose a second volley of Parthian shots. The follow-on wave simultaneously advanced through the withdrawing units in a rhythmic, cyclical pattern of attack and sustained fire. They also excelled at luring their enemies into vulnerable positions before springing the classic pincer, or double flanking, movement (although wider than those of Alexander the Great) anchored to the center corps.

Matthew Paris, a Benedictine monk and historian at Saint Albans, England, was busy collecting and chronicling all the news and chatter trickling in from the East in his great compendium *Chronica Majora* (c. 1259). His firsthand account includes the oldest known mention of the Mongols in western Europe, referring to them as "an immense horde of that detestable race of Satan." Paris wrote that the Mongols warred and advanced "with the force of lightning into the territories of the Christians, laying waste the country, committing great slaughter, and striking inexpressible terror and alarm into every one."

The Mongols were the true architects and inspiration for what was later called blitzkrieg, or "lightning war." Prior to the outbreak of the Second World War, single-minded German, Soviet, and Japanese strategists all scoured the recently unearthed *Secret History*, determined to apply Mongol military tactics to modern mechanized warfare. Instead of Panther and Tiger panzers, however, the Mongols deployed the natural, highly evolved automation of stout and sturdy steppe horses standing thirteen to fourteen hands.

Mirroring numerous other equestrian armies, horses were overseen solely by the offices of Supervisor of Geldings and Supervisor of Horse Herds. Contrary to Europe, where stallions were preferred, the Mongols, like Attila and his Huns, exclusively rode geldings into combat. Stallions were castrated at the age of four, producing full-grown yet more mature and docile mounts for warfare. "Their horses are trained so perfectly," acknowledged the Venetian explorer Marco Polo, "that they will double hither and thither, just like a dog, in a way that is quite astonishing."

Individual warriors possessed four to six operational geldings at any given time. Stealing a horse was punishable by decapitation or breaking the thief's back. Discounting reserve horse herds and those employed for breeding and other noncombat duties, a battle-ready Mongol regiment of a thousand cavalrymen possessed between four thousand and six thousand horses.

All subjects of the Mongol Empire were governed by Chinggis Khan's *Yassa*, or codified law. Many regulations were dedicated to the management and care of horses. Warriors were prohibited from inserting mouth bits when on the move. This reduced injury while also allowing horses to opportunely drink or graze. Previously scouted grasslands and watering grounds were also figured into Mongol march tables and route maps. Since Mongol horses subsisted entirely on grazing, there was no need for a long, expensive, and time-consuming fodder-shuttling supply chain.

Europeans trekking east with the Mongols were advised to trade out their Western horses for steppe mounts, who were accustomed to foraging in all climates. As papal diplomat Carpini recalled, "They told us that if we led the horses we had into Tartary [Mongolia], they would not know how to dig grass from beneath the snow when it was deep, as Tartar horses do, and it would not be possible to find anything else for them to eat because the Tartars have neither straw nor hay nor fodder, and they would all die. So, we talked among ourselves and told the two boys who looked after them [the horses] to send them away."*

With imperial law and standardized taxation rates regulated by the Yassa, the Mongol Empire was the largest free-trade zone in world history. To drive commerce, fuel competitive pricing, and accrue wealth, taxes and tariffs were deliberately lowered across the empire, never exceeding 3 percent to 5 percent (compared with 10 percent to 30 percent at prominent ports outside its authority). Select professions, including scholars, lawyers, doctors, spiritual leaders of all faiths, and their respective institutions, were all tax-exempt.† "The Mongols shrewdly combined state power—over treaties, currency issuance, taxation, supervision of roads—with liberal exchange policies, in order to encourage commerce," notes historian at Paris Nanterre University Marie Favereau in *The Horde: How the Mongols Changed the World*, adding, "The Mongols built dense economic connections from the Mediterranean to the Caspian Sea to China." These carefully crafted and extremely egalitarian religious and economic policies often get buried under the narrative rubble of Mongol cruelty and callousness.

It is folly to view the Mongol Empire as a haphazard territory conquered and administered by barbarian hordes with no technology, recordkeeping, or cultural complexity. This depiction was sold to the Western world in the eighteenth century by the writers and philosophers of the Age of Enlightenment. While the Mongols did not leave any enduring edifices like the Pyramids or the Great Wall, or achieve academic breakthroughs like the Greeks or Romans, they were far from the

---

* The name *Tartar* or *Tatar* for the Mongols is a derivation of *Tartarus*—the infernal abyss of divine punishment and torment for the wicked and evil in Greek mythology.

† Chinggis Khan recognized the negative repercussions that religious volatility and violence would have on taxation and trade. Under the Yassa, all persons under Mongol rule had complete freedom of faith.

vicious, backward heathens portrayed by the French writer Voltaire and his truth-seeking contemporaries.*

As we have seen with previous imperial powers, including Persia, Rome, and China, a sophisticated horse-charged postal system was the glue that held together the military, administrative, and economic layers of empire. The Mongols ordered the construction of roads lined with shady willow trees to ease the passage of troops, traders, dignitaries, transients, and, with the recent adoption of a written language, official messengers. Similar in design to those of previous empires, more than ten thousand relay posts housing more than two hundred thousand horses were erected at intervals between twenty-five and fifty miles.

These hospitality stations operated as rest stops, military garrisons, post offices, and stables for state-sponsored couriers known as arrow riders. By changing horses at regular intervals along this international Mongol Pony Express known as the *Yam*, these messengers could cover upwards of 250 miles in a day. Harness bells announced their arrival so that the next horse or rider could be primed and readied, to save precious time. These elite Mongol dispatch riders could cover the 4,300-mile distance from Karakorum to the Hungarian Plain in about a month!

John of Plano Carpini, recording his journey to Karakorum in 1246, wrote:

> We went as fast as the horses could trot, for the horses were in no way spared since we had fresh ones several times a day, those which we left being sent back, as I have stated previously; and in this fashion we rode rapidly along without interruption. . . . [T]hey have such a number of horses and mares that I do not believe there are so many in all the rest of the world. . . . The horses the Tartars ride on one day they do not mount again for the next three or four days, consequently they do not mind if they tire them out seeing as they have such a great number of animals.

---

* Voltaire described Chinggis Khan as a "destructive tyrant . . . a wild Scythian soldier bred to arms / And practiced in the trade of blood."

In 1253, another friar, William of Rubruck, trekked to Karakorum on behalf of King Louis IX of France. He estimated that he traveled seventy miles a day through Central Asia, "and sometimes more, depending on our supply of horses. On occasions we changed horses two or three times in one day." His round trip covered more than eleven thousand miles.

Transportation and trade were enhanced further by massive bridge and canal construction projects. This included extending and revamping those in China, which was the first target of Chinggis and his confederated steppe cavalry. This initial southern campaign was, in part, the result of a mini–Ice Age. This mercury-dipping climate change detrimentally reduced the grasslands that sustained Mongol horses and their nomadic way of life. For the Mongols, it became expand or expire.

After uniting the Mongol clans, Chinggis and his adroit mounted archers launched a flurry of rapid assaults into fertile China to secure living space—and then some. In the process, he accomplished what Alexander and Bucephalus could not: binding together the two halves of the known world. While both warriors dreamed of bridging the "ends of the Earth" from Asia to Europe, it was Temujin, the discarded Mongol boy from the steppe, and his intrepid horses that ultimately fulfilled Alexander's youthful ambition. For the first time, the East formally met the West, albeit under harsh and hostile circumstances.

By 1221, in the wake of Chinggis's disquieting sermon and merciless encore at Bukhara, the Mongol Empire stretched from the Pacific coasts of Korea and China, through the Eurasian Steppe, to the Tigris River. At this time, the Mongol army was divided into two prongs. Chinggis led the main army back east through Afghanistan and northern India toward Mongolia. A secondary force of roughly thirty thousand horsemen punched north through the Caucasus into the Pontic-Caspian Steppe.

During this sadistic shock-and-awe campaign, the Mongols routed the local Kyivans, Bulgars, and Rus. "For our sins, there came unknown tribes, and some people called them Tartars," wrote a baffled Russian scribe. "Only God knows who these people are or from whence they came." Local populations were ravaged, murdered, or sold into slavery, and opposing soldiers were afforded no quarter. "Yet still did the mountains tremble and our eyes strain across the steppe to find the storm," recorded an observer from the Muscovy principality of Russia. "Then, like the onset of nightfall came the endless shadow of the Horde to blot our lands from view. A numberless multitude swept across the hill crests and like waves of a black ocean did they sweep down upon us. Their arrows fell like clouds of biting flies from the darkened sky. The death screams of our warriors were o'erwhelmed by the drumming of infinite hooves so that only the endless thunder was heard at the last. Our enemy then struck wide its wings and eclipsed the sun from Russia's plain forever." When the dust settled and the Mongol hoofbeats drummed in the distance, upwards of 80 percent of the local populations had been killed or enslaved. The Mongols probed Poland and Hungary to gather intelligence before quickly retreating east in the summer of 1223 to join the Mongolia-bound column under Chinggis Khan.

The infamous warrior died in August 1227 at the age of sixty-five and, in keeping with cultural norms, was buried with forty horses without pageantry or marker. Legend has it that the small funeral party killed anyone they met en route to conceal his final resting place and diverted a river over the grave—or alternatively, branded it into historical oblivion with eight hundred stampeding horses. Then these attendants were all executed by another band, who in turn were slain by yet another, until all knowledge of his resting place vanished from memory. Like Alexander, the body of the Great Khan has been lost to legend and lore.

Under Chinggis's son and successor, Ögedei, the Mongols carried out a ruthless campaign through Eastern Europe between 1236 and 1242, reaching Budapest and the Danube River on Christmas Day 1241. "They razed cities to the ground, burnt woods, pulled down castles, tore up the vine trees, destroyed gardens, and massacred the citizens and husbandmen," documented Matthew Paris in prose eerily similar to our refugee from Bukhara; "if by chance they did spare any who begged their lives, they compelled them, as slaves of the lowest condition, to fight in front of them against their own kindred." From Budapest, they continued their drive across Austria before ransacking their way across the Balkans. Suddenly, in March 1242, the Mongols mysteriously abandoned Europe as quickly as they had come, never to return. Thanks to modern science, we now know why.

During the summer and fall of 1241, the bulk of the Mongol forces was resting on the Hungarian Plain. The following year was unusually moist, with surging levels of precipitation turning the grasslands of Eastern Europe into a quagmire patrolled by malarious mosquitoes. "Documentary sources and tree-ring chronologies reveal warm and dry summers from 1238 to 1241, followed by cold and wet conditions in early 1242," explain climatologist Ulf Büntgen and historian Nicola Di Cosmo. "Marshy terrain across the Hungarian plain most likely reduced pastureland and decreased mobility, as well as the military effectiveness of the Mongol cavalry, while despoliation and depopulation ostensibly contributed to widespread famine. These circumstances arguably contributed to the determination of the Mongols to abandon Hungary." Not even the vaunted Mongols could circumvent these coalescing challenges. In this instance, their lethal tandem of horses and bows was their undoing.

This change in climate, compounded by the death of Ögedei in December 1241, produced an untenable strategic situation, rendering the Mongol military impotent. The high water table robbed them of essential grazing grounds for their horses, who fed solely on grass. A substitute fodder supply was simply not feasible. The heightened humidity also caused Mongol bows to falter. The stubborn glue refused to coagulate in the damp air, and the diminished tautness of their bowstrings nullified the advantage of increased velocity, distance, and accuracy. The Mongol hordes, "ferocious though they were, ultimately failed to translate their light cavalry power from the semi-temperate and desert regions where it flourished in to the high-rainfall zone of western Europe," affirms military historian John Keegan. "They had to admit defeat." It must also be added that compared with other potential targets, Europe was relatively bereft of wealth and was not an overly enticing destination for hedonistic despoliation.

With Europe in the rearview mirror, the Mongols focused their lusting eyes on more lucrative regions lying just beyond the fringes of their control. They sacked Baghdad in 1258, killing more than two hundred thousand—including the caliph, who was rolled up in a Persian carpet and trampled to death by horses, putting a harsh end to the Abbasid dynasty. Two years later, they launched their first campaign into the Holy Land, adding another contender to the moribund Crusades.

The Mongols' entrance into this flagging contest occurred during a respite between the disease-ridden Seventh and Eighth Crusades (1248–54 and 1270, respectively). Indicative of the confusion engulfing these final muddled aftershocks, coalitions among Muslim, Christian, and Mongol factions were fleeting and allegiances fickle.

While enjoying some limited successes, including brief stopovers in the Syrian cities of Aleppo and Damascus, the Mongols were rebuffed repeatedly by disease and powerful defensive coalitions. In September 1260, at the Battle of Ain Jalut on the apocalyptic plains of Megiddo, they were defeated decisively by the Turkic Mamluks of Egypt. The Euphrates River marked the effective extent of Mongol political and economic hegemony. Like the ecological limitations in Hungary, the deserts and arid terrain of Arabia did not agree with the Mongol way of war. The parched and thirsty landscape could not maintain the massive herds that powered the Mongol military machine.

Rebuffed in both Europe and the Levant, Kublai Khan, a grandson of Chinggis, sought to counter these setbacks by conquering the last independent vestiges of Asia east of the Himalayan Mountains. After crushing uprisings in Xinjiang and Tibet, he suffered the humiliation of two miserably rushed and poorly planned amphibious assaults on Japan. The massive Mongol invasion fleets and disembarked troops were thwarted and driven back by flawed ship design, violent kamikaze ("divine wind") typhoons, and legendary mounted samurai archers.*

Much like the original knights in Europe, the mounted samurai ("those who serve") were members of a powerful military caste attached to wealthy lords and landowners in feudal Japan. During the twelfth century, a brutal civil war raged among samurai factions and Buddhist warrior-monks. In 1185 the victorious samurai, Minamoto no Yoritomo, assumed the shogun title of military dictator. The balance of power shifted from the emperor to the samurai and their horses, fostering the feudal age in Japan. The supremacy of the samurai lasted until the Meiji Restoration of 1868 and the dissolution of the Japanese feudal system.

---

* The Mongols' use of gunpowder explosives in Japan marks the first such instance outside of China. They launched bombs from both catapults and cannons, as well as hand-thrown grenades.

The stinging defeat shattered Kublai's dreams of conquest and in the process tarnished the Mongols' towering reputation of invincibility. After nursing his bruised ego and scrapping a third invasion of Japan, in 1285 Kublai set loose the bulk of his might on southern China and Southeast Asia. His targets included the powerful Khmer civilization, or Angkor Empire. From its origins around 800, Angkor culture quickly spread across Cambodia, Laos, and Thailand, reaching its zenith at the dawn of the thirteenth century. The awe-inspiring and majestic ruins of Cambodia's Angkor Wat and Bayon temples stand as enduring monuments to Khmer sophistication and splendor.

While the horse made its way into Southeast Asia from China by the end of the third century CE, it remained a novel and exotic creature. The use of horses increased after the Mongol invasions but, given climactic and grazing limitations, never gained any meaningful commercial or cultural traction. The same can be said for the Indonesian archipelago, where the horse appears to have arrived around the ninth century CE.

During these southern campaigns, Kublai disregarded the time-honored tactic of extracting his forces to the north during the rainy season, always buzzing with voracious mosquitoes. As a result, his marching columns totaling roughly ninety thousand men were ravaged by malaria. When he abandoned his designs on the region in 1288, a sickly force of only twenty thousand survivors staggered northbound for Mongolia. "Thus, between 1242 and 1293, the Mongol expansion reached its maximum," says Weatherford, "and four battles marked the outer borders of the Mongol world: Poland, Egypt, Java, and Japan." Over the next century, political infighting, military losses, overextension, and disease enervated and corroded the once-indomitable Mongol Empire. Aside from a few loitering enclaves in the Caucasus and Crimea, by 1400, it had fragmented and collapsed into geopolitical irrelevancy.

This demise, however, gave rise to numerous nation-states on the periphery of Mongol horse-powered expansion. Facing a communal life-or-death reality and coalescing for shared defense, groups of people on these metaethnic frontiers began to identify with their broader "inside" commonalities, contrasted directly against the "outside" foreign invaders. Through consolidation to withstand the Mongol onslaught or as a consequence of subjugation, Korea, Japan, Thailand, Laos, Vietnam, Cambodia, and European Russia began to fan the embers of national identity and statehood, while China and India further entrenched their current borders.

For example, during the fourteenth and fifteenth centuries, Russia responded to the Mongol occupation of Eastern Europe, known as the Golden Horde, by transforming itself from a scattering of self-governing city-states and competing principalities into a united military front and expansive imperial power centered on Moscow through the gathering of the Russian lands. This creation of numerous modern nation-states, including Russia (and Mongolia itself), is an often overlooked yet momentous and enduring legacy of the Mongols and their horses.

The dynamic vitality of the Mongol Empire and the heritage of its steppe horsemen also still reside in today's global DNA. Geneticists believe that 8 percent of people living in its former Eurasian heartland are direct descendants of Chinggis Khan. To put this another way, roughly forty-five million to fifty million people currently on the planet are his progeny. If we collected all the descendants of Chinggis Khan into one country, it would be the thirty-fifth most populous nation in the world today, ahead of countries such as Canada, Poland, Afghanistan, Ukraine, and Australia.

Mongol bloodlines carried more than just genetics, however. While the Mongols did not overpower Europe, a pathogen originating in China did. This is perhaps the greatest impact of the Mongol Empire, hardened into existence by their horses. "The most important effect that the Mongol conquests had on the transformation of Europe," states Peter Frankopan, "did not just come from trade or warfare, culture, or currency. It was not just ferocious warriors, goods, precious metals, ideas, and fashions that flowed through the arteries connecting the world. In fact, something else entirely that entered the bloodstream had an even more radical impact: disease. . . . The Mongols had not destroyed the world, but it seemed quite possible that the Black Death would."

In 1331 an epidemic ruptured in northern China, killing more than

90 percent of the population there before bursting its borders of containment. The plague ricocheted across Pax Mongolica like a poison arrow gaining strength as it ripped across the Eurasian Steppe, passing through bustling ports and the rapid, international, horse-powered Mongol postal system. The Persian geographer and historian Ibn al-Wardi witnessed that the pestilence "sat like a king on a throne and swayed with power, killing daily one thousand or more and decimating the population," before he himself died of plague in 1349. Across Asia the death rate was roughly 55 percent, with the Middle East reaching 40 percent.

The disease marched forward on the backs and in the packs of Mongol warriors and traders peddling all sorts of goods, including squirrel and marmot furs. The plague, or *Yersinia pestis*, is a bacterium transmitted by fleas that hitch a ride on numerous ground rodents, including, in this case, rats. The legendary Silk Road was now a lethal highway to hell, with transmission hurried by the economic and military engines of horses, donkeys, mules, camels, and ships.

During the cordon of the Black Sea port city of Kaffa in 1346, the Mongols purportedly catapulted corpses infected with the bubonic plague over the city walls to contaminate the inhabitants and break the siege. More importantly, Kaffa was a bustling trading center frequented by Italian, and other European, cargo ships. Vessels carrying rats with infected fleas, or the plague-ridden sailors themselves, soon made port at Sicily in October 1347. Subsequent ships of death docked at Genoa and Venice in Italy and Marseilles, France, in January 1348. "Our hopes for the future," wrote the Italian poet-scholar Petrarch, "have been buried alongside our friends."

The plague peaked between 1347 and 1351 in Europe, where it registered a death toll of around 50 percent. By 1350, it had even island-hopped across the North Atlantic from the Faeroe Islands, through Iceland, to the barren windswept island of Greenland, likely hastening the final extinction of its forsaken and orphaned Norse outposts.* Death dominated Europe.

---

* Horses were present at the Norse colonies of Greenland (c. 985–1450) but did not make the trip to the colony of Vinland (L'Anse aux Meadows) in Newfoundland, Canada

In plain numbers, the Black Death collected the souls of 40 million Europeans, with global fatalities estimated conservatively to be around 150 million and perhaps as high as 200 million. Keep in mind that at this time, *global* excluded the Americas, Australia, New Zealand, and sub-Saharan Africa, as European colonizers and their cascading Rolodex of deadly diseases had not yet called upon these pathogen-free Indigenous peoples—for now. The numbers are simply mind-boggling and difficult to comprehend. It took two hundred years for the global population to recover to its previous level. The Black Death, in a league of its own, is the most shattering single Malthusian check in human history.*

To put this unrivaled carnage into context, the roving 1918–19 influenza epidemic, with truly global transmission, killed "only" 75 million to 100 million—still five times more than the First World War that helped it go viral. As of this writing, the coronavirus pandemic that began in late 2019 has taken the lives of some 7 million people, including roughly 2.2 million in Europe. Again, the Grim Reaper of the Black Death harvested *150 million to 200 million* human beings at a time when the human population was approximately 475 million—a mere 6 percent of what it is today.

For European survivors, however, the reverberations and repercussions of this catastrophic loss of life were remarkably positive. "Despite the horror it caused, the plague turned out to be the catalyst for social and economic change that was so profound that far from marking the death of Europe, it served as its making," acknowledges Frankopan. "The transformation provided an important pillar in the rise—and the triumph—of the west." For every action, there is an equal and opposite reaction, and the rise of Europe in the wake of such ruin was just as potent as its fall.

Large tracts of vacant and now unoccupied land transferred to the living, equating to increased prosperity. More land for fewer people also

---

(c. 1000–1005), founded by the Norse explorer Leif Erikson five hundred years before Columbus "discovered" America.

* Since the emergence of antibiotics in the 1880s and Alexander Fleming's penicillin breakthrough in 1928, the plague has mostly disappeared, killing roughly 120 unlucky people per year.

meant less demand for the staple crop of wheat, leading to a diversification of agricultural produce and tabling a more robust, complete diet. As food was more plentiful, affordable, and nutritionally balanced, healthy populations thrived and life expectancy shot up. Protein consumption also increased, as previously cultivated marginal lands returned to their former natural states of livestock pastures or reforested wildlife sanctuaries.

The trend of rising percentages of both plow horses and cart horses continued, as they became more numerous and replaced oxen in greater proportion. "It does appear that the period after the Black Death continued to experience an increase in the level of work-horses employed," explains Robert Langdon, "and that this increase was probably significant." Even during these dark days of the fourteenth century, reports indicate an increasing flow of horses moving west from Central Asian markets, including Fergana. The lucrative horse exchange, associated infrastructure, and equine-related trades were all economically reinforced to meet the rising demand.

Competition for jobs was reduced, which translated into higher negotiating leverage and wages for both skilled tradesmen and unskilled laborers. In England, for example, the year 1349 witnessed an unprecedented collapse in the cost of food and goods by an average of 45 percent, while wages increased by 24 percent. Birth rates continued to rise, and couples shed their disposable income to educate their children. Increased wealth, combined with less scholastic competition, allowed for a slow but steady growth of university enrollment and the general advancement of academia across Europe, which eventually led to the Renaissance and a global projection of imperial power. The impact of the Mongols and their hardy horses was far more than military conquest. Together they generated ripple effects that were felt outside of Eurasia.

The Mongols irrevocably changed the immunologic, sociocultural, demographic, and commercial configuration of the world. They were willing to allow traders, scholars, missionaries, and other travelers to freely navigate their empire, permanently opening China and the rest of Asia to Europeans, Arabs, Persians, Indians, and others. Thriving diasporas of Christian and Muslim converts took their place alongside the major Eastern faiths of Buddhism, Confucianism, and Hinduism. The invasions of Mongol horsemen, which lasted roughly three hundred years, created an immeasurably smaller global village by fusing previously distinct geographical worlds.

An imported metalworker from France forged the magnificent Silver Tree of Karakorum, a serpent-spouted drinking fountain at the Mongol capital. Intended to illustrate the extent and diversity of the Mongol Empire, this immodest and extravagant centerpiece dispensed four drinks: rice beer from China, grape wine from Persia, Slavic honey mead, and, of course, Mongolian koumiss. With a beverage in hand, people across the planet now holler the Mongol *"Hurray!"* as an exclamation of excitement or approval while cheering on their favorite sports team or racehorse. The Mongol Empire was an intersected and interconnected superhighway of foreign exchange of a vast array of goods, services, and skills.

Spices, silk, and exotic imports beyond imagination became mainstays on the shelves and stalls of European markets. In England, the chivalrous knights of the Most Noble Order of the Garter, founded by King Edward III in 1348, wore insignias, facings, and sleeves made from blue "Tatar cloth." Contenders at the 1331 Cheapside Tournament (of knights), wore "Tatar silks" and masks bearing the likenesses of wild Mongol warriors. Lightweight and portable paper money and playing cards found their way from China into the pockets of princes and peasants across Eurasia.

The Mongol diffusion of Chinese gunpowder also modernized war. Contrary to the popular fallacy that the Chinese used gunpowder only for fireworks, by the tenth century, it was also fused to rockets, grenades, bombs, mines, and flamethrowers. The Muslim world acquired gunpowder from the Mongols around 1250, with cannons following close behind. Gunpowder continued to burn its path west, first recorded in Europe by English scholar Roger Bacon in 1267 and in connection to weapons in 1326. Bacon likely gained his formula and knowledge of gunpowder from conversations (and demonstrations) with his friend William of Rubruck, who, you might remember, traversed the Silk Roads, including a residence at Karakorum, between 1253 and 1255.

Shuttling the Silk Roads: Ferried by camels and horses, Marco Polo and a group of traders explore the bustling market towns connecting East and West. From the *Catalan Atlas* by cartographer Abraham Cresques, 1375. *(Wikimedia Creative Commons)*

When William arrived at the Mongol capital, he was greeted in his native tongue by a woman from his homeland who had been captured during a Mongol raid fourteen years earlier.* Contemporary literature and archives reveal an incredibly safe and permeable Eurasian expanse for trekkers, moguls, academics, and merchants during what Jack Weatherford labels the Mongol Great Awakening. The unrivaled Mongol cavalry carved out the largest empire of horses the world has ever seen while fostering unprecedented commercial exchange and cultural hybridization across Eurasia. The travel stories of William, John of Plano Carpini, Marco Polo, and others fed the trading frenzy and the coin purses of Europe.

The celebrated narrative of Marco Polo was a product of mere coincidence. To break the boredom and monotony while incarcerated in

---

*William also met several Germans; a French silversmith; the nephew of an English bishop; and a woman, from France's Lorraine region, who cooked him Easter dinner.

Genoa between 1298 and 1299, he regaled a fellow inmate with stories of his adventures on the Silk Roads and of his stint as a Mongol foreign emissary in service of Kublai Khan spanning the years 1271 to 1295.* The enraptured convict, a romance writer by the name of Rustichello da Pisa, eventually published these epic tales in 1300 as the *Book of the Marvels of the World,* now commonly read as *The Travels of Marco Polo.*

A prized possession of Christopher Columbus was his well-worn and heavily annotated copy of Polo's book, which he carried on his voyages as a sort of inspirational Lonely Planet travel guide for when he reached "the lands of the Great Khan." A note scribbled in the margin claimed, "He has a very large palace, entirely roofed with fine gold." Columbus set sail with a satchel of royal letters of introduction and a stack of fill-in-the-blank trade agreements for tendering to enthusiastic Asian rulers. "In truth, the Columbian exchange," stresses Favereau, "should be seen in part as a legacy of the Mongol exchange." Polo's descriptions of the East and its limitless bounty inspired Columbus's attempt to reach the riches of Asia by sailing west. The overland Silk Road routes had been interrupted by the aggressive and combative expansion of the Ottoman Empire during the fourteenth and fifteenth centuries.

After six years of Columbus's pestering the monarchies of Europe for financing, the royal Spanish couple of King Ferdinand and Queen Isabella finally relented, providing him with a token investment—one-thirtieth of what they shelled out on their daughter's wedding. Columbus was forced to borrow the balance (roughly 25 percent) by shaking down wealthy Italian merchants.

The Spanish Crown's lack of faith is understandable, for by all rational estimates and valuation, this was a reckless and financially ruinous enterprise. Undeterred, in August 1492 Columbus set sail into the great unknown with three relatively small ships and ninety men, determined to reopen access to eastern markets. "In a sense," explains historian Barbara Rosenwein, "the Mongols initiated the search for exotic goods and

---

* Polo was imprisoned for commanding a Venetian ship during an outbreak of the intermittent Venetian-Genoese Wars spanning 1256 to 1381.

missionary opportunities that culminated in the European 'discovery' of a new world, the Americas."

Like Attila and Chinggis, Columbus and ceaseless colonizing waves of European explorers, conquistadors, settlers, and horses would also realize from their interactions with Indigenous peoples the raw, unfiltered power of first impressions. These initial contacts left deep scars and enduring imprints. The horse unwittingly became an integral factor of the Columbian Exchange and an overwhelming force in the ultimate triumph of European occupation. "Horses rather than cannon were the important cargoes of the conquistadors in their campaigns of conquest," emphasizes Keegan. "In a contest of hundreds against thousands, it was their horses that gave the invaders the decisive advantage."

By this time, the opportune domestication of a globally endangered animal on the Pontic-Caspian Steppe around 3500 BCE had permanently transformed our planet. The horse had cheated the evolutionary prospect of extinction by the propitious intervention of the human hand and ridden its way into the undisputed position as our most important invention and historic partner.

As a result, after a nine-thousand-year absence from the Americas, this magnificent creature was now on the cusp of returning to its original home and native land. The horse, not *Homo sapiens*, was the first animal to truly circumnavigate the globe. This reintroduced first impression would be nothing short of transformative.

CHAPTER 13

The Return of the Native
The Horse and the Columbian Exchange

Given the unknowns surrounding his first voyage, no horses—or any livestock, for that matter—were crammed aboard the three unremarkable ships steered by Columbus. The manifest for his massive second expedition in 1493, however, inventoried a wide selection of seeds, stem cuttings, plants, and animals, including horses. "We brought pigs, chickens, dogs, and cats," noted his friend and shipmate Michele de Cuneo. "Cattle, horses, sheep and goats do as with us."

According to the royal cedula (permit), in preparation for the second voyage, it was commanded "to go in these vessels there will be sent twenty lancers with horses . . . and five of them shall take two horses each, and these two horses which they take shall be mares." The hardy survivors of this intrepid herd pressed the first hoofprints into their ancestral lands in more than nine thousand years.

The armada of seventeen ships, 1,200 all-male passengers, and twenty-five horses pushed off from Cadiz, Spain, in September 1493. For these frightened and hysterical horses, the two-month forced voyage across the middle passages of the Atlantic on woefully ill-equipped and rickety ships must have been torturous. These early vessels were small, low waisted, and not suitably designed for transporting horses, let alone much else.

With little to no exercise, horses were suspended from the beams of the hold in hoisted hammocks or belly slings to avoid being thrown from the roll of the waves. Each suffocating breath tasted of the sludgy excrement-laden bilge water that showered the helpless animals with every pitch of the ship. On the smaller boats, panic-stricken horses were hobbled and strapped down to the main deck, with no reprieve from the harsh elements.

Upon leaving Spain, the first pit stop for these sea horses was the Canary Islands, off the coast of northwest Africa. The roughly nine-hundred-mile body of water separating Spain from this archipelago was named Golfo de Yeguas (Gulf of Mares) by the crews because so many of them were entombed in its watery grave. After weighing anchor at the Canaries, the horses confronted the most perilous and harrowing leg of their crossing.

Roughly mid-journey, the fleet entered the doldrums of the Atlantic, stretching approximately 30 to 35 degrees north and south of the equator. Caused by high pressure at the divergence of the trade winds and westerlies, these dead zones are characterized by blistering sun, stagnant winds, and scarce precipitation. Both humans and horses went mad, adrift for weeks, starving for an airstream. Many horses simply died of hunger and dehydration. To conserve rations and fresh water, others were eaten or hurled overboard. To lighten the load and "catch a wind," sickly or hapless horses were skidded down the gangway, spooked and shrieking as they entered the murky, shark-swirling waters. Average equine losses in transit are estimated at 25 percent to 35 percent, with some shipments reaching 50 percent. These windless belts are forever known as the horse latitudes.

During these early years, before wharves and harbors were built, the most common means of unloading the exhausted and emaciated horses upon arrival was to push them over the rail, or "jump and sink or swim." The lucky ones were gently lowered into the water with winches and waist slings. Finally, after this entire traumatic and terrifying ordeal, the handful of stubborn, skeletal survivors struggled ashore to greet their new life in the new world that Columbus created.

The first equine immigrants splashed onto the beach at Hispaniola in November 1493 and were immediately assigned to combat duty. According to Columbus, "only 200 Christians, 20 horse, and about 20 ferocious attack dogs" decisively defeated an Indigenous Taino force of 100,000. Columbus understood immediately the immense value of horses. He promptly dispatched to his benefactors in Spain a request that "each time there is sent here any type of boat there should be included some brood mares." Under horrific and appalling conditions, increasing numbers of horses were shipped to the Americas.

Like the colonists themselves, migrant horses quickly created a self-reproducing, country-born population supplemented by continuing consignments. In 1501, for example, Hispaniola was home to about thirty horses. Two years later, there were at least seventy. Offspring from this original stock were distributed to Puerto Rico, Cuba, Jamaica, Trinidad, and Panama, which all established their own breeding stables. By the close of the sixteenth century, the Jesuit missionary José de Acosta wrote that horses "multiplied in the Indies and became most excellent, in some places being even as good as the best in Spain, good not only for fast messenger work but also for war and the parade." For the horse, it was a bleak and somber welcome back to the Americas.

These horses, however, were not the untamed and evolving creatures that had once roamed these ancient lands. They disembarked fully loaded with all the enhanced features of domestication and selective breeding honed by five thousand years of biological engineering. Consequently, the horses reintroduced to the Americas were a modern prepackaged vehicle primed for instant use. "If the fifty-five-million-year history of horses in North America were condensed into a day, horses would be a native species right up until they became extinct at 11:59:43 p.m.," states David Philipps in *Wild Horse Country*. "They returned in the last second of the day." What a difference this fraction of time and brief disappearance made. After a nine-thousand-year absence and a split-second lifetime making history across greater Eurasia, the horse, upon its grand entrance, immediately became an active agent of the Columbian Exchange.

Coining this term with his seminal 1972 work *The Columbian Exchange: Biological and Cultural Consequences of 1492*, historian Alfred W. Crosby proposed that, whether by accident or by design, global ecosystems were irrevocably rearranged. Columbus and the ensuing surge of explorers, colonists, and enslaved people that sailed in his shadows initiated the largest cultural and biological interchange in human history. Globalization was the new reality.

Despite Columbus's miscalculations and blunders, in the wake of his flagship, the *Santa Maria,* humanity became wholly integrated for the first time. "Columbus changed the world not because he was right," chides Pulitzer Prize–winning author Tony Horwitz in *A Voyage Long and Strange: Rediscovering the New World,* "but because he was so stubbornly wrong. Convinced the globe was small, he began the process of making it so, by bringing a new world into orbit of the old." Although Columbus was eight thousand miles lost and staunchly adamant that he had reached the East Indies, his first stumbling step onto the island of Hispaniola in December 1492 was indeed a giant leap for both human and horse kind.

Columbus introduced the complete barnyard collection of zoonotic animals to their new world, where they thrived and flourished. Cattle, horses, donkeys, sheep, goats, pigs, and chickens quickly dominated pens and pastures across the Americas. Corn, potatoes, tomatoes, cacao, peanuts, and tobacco were uprooted from the Americas and shipped to fertile horse-plowed fields across the planet, while sugarcane, coffee, wheat, rice, apples, and various leafy greens found a sanguine home in the Americas. As a corollary to colonization, Indo-European languages quickly came to dominate all settler-state societies.* While his résumé is the stuff of Neverland nightmares, Columbus launched a novel and irreparable era in our ongoing human odyssey. For better or worse, his impact remains unshakable.

At the onset of the Columbian Exchange, Sir Thomas More's 1516 political satire *Utopia* exposed the validations underpinning intrusive European imperialism and the pervasive themes that dominated European-Indigenous relations:

> If the natives wish to live with Utopians, they are taken in. Since they join the colony willingly, they quickly adopt the

---

* An Indo-European language, or a creolized version, is the primary tongue of all the thirty-five sovereign nations and twenty-five overseas territories of the Americas, Australia, and New Zealand.

same institutions and customs. This is advantageous for both peoples. For by their policies and practices the Utopians make the land yield an abundance for all, which before seemed too small and barren for the natives alone. If the natives will not conform to their laws, they drive them out of the area they claim for themselves, waging war if they meet resistance. Indeed they account it a very just cause of war if a people possess land that they leave idle and uncultivated and refuse the use and occupancy of it to others who according to the law of nature ought to be supported from it.

Land, or the economic capital derived from it, became the common currency across colonial outposts. Clamoring for mercantilist imperial wealth, the avaricious powers of Spain, England, France, and, to a lesser extent, the Netherlands and Portugal, carried the cross of colonization, a genocidal cocktail of disease, and the poisonous epoch of African chattel slavery to the Americas and beyond.*

By now, most readers are familiar with the three foremost disparities between Indigenous and European cultures within the larger Columbian Exchange: anthropologist Jared Diamond's well-worn, but never worn-out, trifecta of guns, germs, and steel. To this destructive trio, I will tack on horses. None of what transpired during the European colonization of the Americas, weighs Crosby, "would have been possible without the horse." Germs, however, were the first to leave their lethal impression on nonimmune Indigenous populations.

Columbus's arrival breached the lengthy quarantine of the Americas. The Western Hemisphere had been maintaining a relatively strict policy of social distancing for millennia. Although the Norse preceded Colum-

---

* Mercantilism was an economic system practiced by European imperialist nations between the sixteenth and eighteenth centuries. Designed to maximize profits for the Mother Country, the natural resources of overseas colonies were exploited by way of African slave labor. These raw materials were transported to the home country and turned into manufactured goods, which were then traded for more African slaves or sold at inflated prices back to colonial populations. A greater collection of colonies meant not only an increased and more diverse inventory of natural resources but also a compounding market/population for homespun manufactured goods.

bus by five hundred years with their fleeting outpost on the island of Newfoundland, Canada, from approximately 1000 to 1005, their temporary stay does not appear to have made much of a pathogenic impact— or, at the very least, it remained limited or localized. Columbus, on the other hand, opened the doors to the Americas permanently. Unrelenting waves of conquistadors, settlers, slaves, and their smuggled pathogens clattered and surged upon the pitching shores of this "New World." As we discovered recently with the COVID-19 pandemic, during both past and present, travel and commerce are the most efficient carriers of communicable disease.

Multifarious and industrious Indigenous trade routes spanned and connected the entire breadth of the Americas. Peoples of the boundless interior plains of the United States and Canada adorned their clothes with seashell beads, although they had never tasted the stinging, salty breeze of an ocean beach. Coastal peoples who frolicked on those shores wore bison hides, yet they had never laid eyes on the stately creatures. Elders smoked ceremonial tobacco, only envisioning what the plant might look like. Great Lakes copper was fashioned into jewelry by South American craftsmen. Colonizing diseases were also exchanged along these extensive commercial corridors, ravaging Indigenous peoples long before they ever set eyes on a European.

These foreign pathogens shuddered through the virgin continents, laying waste to Indigenous populations until they teetered on the precipice of extermination. "Wherever the European has trod, death seems to pursue the aboriginal," observed Charles Darwin in 1846. "We may look to the wide extent of the Americas, Polynesia, the Cape of Good Hope and Australia, and we find the same result." By 1535, for example, only forty-three years after Columbus's own first impression on the Indigenous Taino people of Hispaniola, they neared extinction. This staggering loss of life was the contemporary equivalent to wiping out the entire population of the British Isles and then some. A sickening catalog of Indigenous cultures, including the Inca and the Aztec of Mesoamerica, the Cueva of Panama, the Caribbean Arawak, and the Beothuk of Newfoundland, would all suffer the same fate.

The potent deadliness of imported diseases on the Indigenous peoples of the Americas approached an extinction-level event. Of the estimated one hundred million inhabitants on the eve of the Columbian Exchange, by 1750, only five million remained. During the worst per capita catastrophe in human history, roughly 95 percent of the Indigenous residents of the Americas—representing 20 percent of the global population—had been erased from the planet in a mere 250 years. Adding to this misery and carnage, the shattered survivors faced an unrelenting merry-go-round of enslavement, massacres, wars, and forced relocations.

Correspondingly, even with the tragic collapse of Indigenous communities, the global population burst by a whopping 28 percent: from 425 million in 1500 to 545 million in 1600. European populations, rebounding from the Black Death, sought escape and wealth in the recently discovered lands of opportunity in the Americas. This influx of Europeans and their cargoes of African slaves marks the largest human relocation in history. While Columbus opened a new world for European economic and cultural expansion, it came at an excessive and terrible price: the near extermination of Indigenous peoples and the establishment of the transatlantic African slave trade.

As Indigenous peoples sank in the mire of disease, starvation, and brutality, an alternative pool of laborers was needed to fuel lucrative sugar, coffee, and tobacco plantations. Starting out as a trickle, as mercantilist extraction colonies proved increasingly profitable, the flow of African slaves became a steadily rising torrent of human trafficking. Between twelve million and fifteen million human beings were eventually delivered from Africa across the middle passages of the Atlantic and arrived *alive* into the shackled clutches of slavery in the Americas.*

Given its value, the horse was chained involuntarily to the African slave trade. "It is neither for slaves nor for tamed horses to reason about freedom," espoused the great eighteenth-century philosopher of liberty Jean-Jacques Rousseau. "They only know their broken state." The Portuguese hocked horses for African slaves to power their booming sugar and coffee plantations in colonial Brazil. These two commodities

---

* It is estimated that a total of twenty-five million to thirty million people were kidnapped from Africa, with nearly half dying en route to the coastal "barracoons" (slave forts) or during the transatlantic crossing.

produced profits of 400 percent to 500 percent on direct investment and accounted for 70 percent of Brazil's mercantilist economy. As the foremost destination for African slaves, Brazil accounted for 40 percent (roughly six million people) of the total transatlantic trade. During the sixteenth and seventeenth centuries, one horse fetched anywhere from six to twenty slaves.

Horses remained a relatively rare, and prized, possession in sub-Saharan Africa, as they did not survive particularly long in the heavily diseased and hothouse tropical landscape. They reinforced the supremacy of African elites as both a status symbol and through their use in relatively small but extremely profitable cavalry units. According to historian Robin Law in *The Horse in West African History*, "The connection between the horse and slave trades lay in their relation to war. Horses were valued primarily for their use in warfare and were perhaps especially useful in the pursuit and capture of fleeing enemies—that is, in securing slaves. Slaves, conversely, were most readily obtained through capture in warfare. The exchange of horses for slaves therefore tended to become, it is often suggested, a 'circular process': horses were purchased with slaves, and could then be used in military operations which yielded further slaves, and financed further purchases of horses." As a result, the horse was drafted into the horrific and dark human chapter of African chattel slavery.

Caught up in the whirlwinds of the Columbian Exchange, the first African slaves landed at Hispaniola in 1502 as an accessory to Columbus's fourth and final voyage alongside the Spanish priest Bartolomé de Las Casas. In his intense, raw, and scathing eyewitness narrative, *A Brief Account of the Destruction of the Indies*, Las Casas bears witness to the "seeping plagues" and condemns the atrocities committed by his fellow Spaniards against Indigenous peoples as "the climax of injustice and violence and tyranny." Columbus personally oversaw brutal acts of barbarity and sexual transgressions against Indigenous populations.

Adding to the perverse misery effected by disease, enslavement, and sexual predation, the Spanish military, notably its horsemen, wreaked havoc on the Indigenous inhabitants of the Americas. Horses do not always bring out the best in humans. The Spanish behaved like Indo-Europeans,

Huns, or Mongols on horseback, reminiscent of their own respective mounted terror campaigns of colonization in Old Europe, the Roman Empire, and greater Eurasia. The colonization of the Americas and other imperial outposts followed this all-too-familiar storyline. Horses, as usual, made all the difference.

Repeatedly, history books remind us that Indigenous weapons of stone, bone, and wood shattered against steel swords, axes, and armor wielded by Europeans. The recent Gunpowder Revolution had also furnished Europeans with explosive, albeit still primitive, projectile weaponry. Second to germs, however, horses proved to be a greater asset and military advantage than guns or steel. "The effective use of well-trained war horses provided Europeans with quick mobility and massive power," remarks historian Noble David Cook. "The horse was exceptionally effective in the conquest of regions without large mammals, such as the Caribbean islands and Mesoamerica. Savage attack dogs were employed as well, causing fear and bloody devastation." According to Las Casas, one Spanish cavalryman could annihilate two thousand Indigenous humans in an hour.

During the initial push of Spanish imperialism in the Caribbean, Mexico, and Central/South America (outside the habitat of the remaining large mammals, including bison, moose, bears, elk, and caribou), Indigenous peoples had never seen an animal this massive, and certainly not one mounted and controlled in synchronization with humans. The psychological impact, or "centaur shock," on Indigenous communities upon first witnessing humans riding horses is well documented in primary accounts across the multinational chronology of colonial first contacts.

European cavalrymen often decorated their horses with bells, armor, and other accessories to enhance the terrifying spectacle of this first impression. The Caribbean Arawak (and others) believed that horses ate humans, a fable that the Spaniards were all too eager to promote. While traversing New Mexico and Arizona in 1582–83, for example, the conquistador Antonio de Espejo reveled in the panicked reaction of the Pueblo when he told them this outrageous piece of propaganda. The Purépecha, a contemporary empire northwest of the Aztec, believed that horses could speak, having observed the conquistadors talking to their steeds. And the Inca of the Andes took the shine of their shoes to mean that horses had sacred and supernatural silver feet.

In some instances, the sudden appearance of one horse was enough to subdue a large village. While conquering Cuba in 1511, Pánfilo de Narváez, who had taken part in the conquest of Jamaica two years earlier, single-handedly scattered and slaughtered thousands of attacking Ciboney-Taino by jumping on his horse and ringing bells while cutting them down. He turned to Las Casas, who witnessed the ordeal, asking calmly, "What do you think about what our Spaniards have done?" The horrified priest responded, "I send both you and them to the Devil!" Moreover, Indigenous sentries or messengers could not outrun Spanish cavalry to warn their unsuspecting villages of impending assaults. The horse allowed the invaders to transfer soldiers, information, and supplies faster than their Indigenous targets could communicate, respond, and retaliate.

Returning to Diamond's paradigm plus one, if we line up the usual suspects of Christopher Columbus, Hernán Cortés, and Francisco Pizarro among other notorious, or nefarious, conquistadors, they all track a similar plot progression: an opening sequence of unbridled greed, followed by genocidal scenes of germs stalking Indigenous peoples, and, finally, a rising climax-conclusion of conquest abetted by guns, steel, and horses. From the Americas and Australia to New Zealand and Africa, imperialism followed a reliably consistent script. The global stages of European colonization were all directed by guns, germs, steel—and horses. The infamous duo of Cortés and Pizarro can serve as informative case studies.

―――

Hernán Cortés secured his mounts for his invasion of Aztec Mexico from Cuba, where he had settled in 1513. Jamaica supplied his copycat second cousin Francisco Pizarro with his fifty horses destined to sack the mighty Incan Empire. Many of the sixteenth-century adventurers, including both Cortés and Pizarro, came from the Extremadura region of Spain, north of Seville, celebrated as the Cradle of the Conquistadors.*

---

* Francisco de Orellana, a close friend and relative of Pizarro's (and a member of his expedition against the Inca), was also from the Extremadura region. He completed the first known European navigation of the entire length of the Amazon River between 1541 and

"In the earliest stages of the Columbian Exchange," points out Peter Mitchell in his comprehensive account *Horse Nations*, "horses were instrumental in Spain's overthrow of the Mexica (Aztec) and Inka Empires."

According to legend, in the early fourteenth century the Mexica received a divine vision of a celestial eagle perched upon a prickly pear cactus and devouring a snake, a depiction now emblazoned on the Mexican flag. The revelation included directions to the location where they were instructed to build a great Aztec civilization. The magnificent city of Tenochtitlán soon emerged from the muddy waters of Lake Texcoco. Rivaling the most illustrious cities of the world, with a population of 250,000, it was five times larger than London, and, in Europe, only Paris approached its size. It stood at the center of a loosely configured empire of alliances containing five million to six million people.

An intersecting and carefully plotted network of roads, swivel bridges, canals, and embankments spanned the twenty districts of Tenochtitlán. These transportation routes interchanged with three main thoroughfares ending in causeways (wide enough for ten horses) connecting the floating island city to the mainland. The streets were lined with towering, ornately carved, and gilded pyramids, temples, sculptures, palaces, ball courts, and brimming markets. These avenues were also brimming with beautiful botanical gardens instead of squalid refuse and ordure, thanks to a talented crew of a thousand green-thumbed rubbish collectors.

Aqueducts, dams, gates, and hydraulic systems controlled water levels and delivered a constant flow of fresh water from distant springs and snow-crowned mountains. More than thirty thousand acres of reclaimed "floating" fields, or *chinampas,* ringing the city produced seven cycles of crops per year—primarily the "three sisters" of corn, beans, and squash, as well as chilis and avocadoes. Each acre produced enough food to feed six people, a yield unmatched anywhere on the planet. Tenochtitlán was a marvel of modern engineering and human imagination.

1542. His party was attacked by a group of Indigenous Pira-Tapuya, led by women, in June 1542, reminding Orellana of the female Scythian Amazon warriors. He described this section of the waterway in Brazil as "the river of the Amazons."

To the south, the Inca, originating as pastoral peoples in the Andes highlands and intermontane valleys of Peru and Bolivia, created an equally impressive and imposing civilization radiating from their jeweled capital at Cuzco. Unlike the Aztec, who ruled through tributes, alliances, and regional autonomy, the Inca governed through an authoritarian monarchy guided by a centralized military bureaucracy resembling the classic European model of empire. By the fifteenth century, after conquering their neighbors, the Inca carved out a vast interconnected territory. The largest in the history of the Americas, it stretched across 770,000 square miles and encompassed upwards of ten million people.

Two parallel near-straight highways, one along the Pacific coast and the other punching through the Andes, ran the 2,700 miles from Chile in the south to Ecuador in the north. These strictly controlled Incan interstates were transected by a grid of smaller east-west roads. Along the main routes, ravines and gorges were spanned by anchored cable suspension bridges, and lowland rivers were traversed by sturdy pontoon crossings. Maintaining the shortest and most expedient route through straight-line construction, six-foot-wide passes were cut through solid rock, and stairs climbed steep cliff faces. As the roads neared cities, composition shifted from packed earth and pressed gravel to perfectly fitted cobblestone blocks, which continued into urban centers.

Licensed and authorized traders, travelers, soldiers, officials, and llamas (the only pack animal, and one of the few regionally domesticated animals in the Americas) trekked across more than ten thousand miles of Incan roads. Like the transportation infrastructure of Eurasian empires, the Inca erected state-sponsored rest stops (*tampus*) containing provisions, beds, and military stockpiles every twelve miles. Smaller water-stop shelters were erected every two to three miles for specially trained government and military dispatch runners called *chasquis*. "A message could travel 1,500 miles in a week, a speed comparable to that of the horseback-riding Persian or Roman messengers," notes historian Daniel Headrick. "Thanks to their roads and runners, the Incas were able to hold together an empire as large as any in the Eastern Hemisphere, despite the lack of large animals or navigable rivers."

Elaborate and glittering Incan cities adorned with precious metals, including Cuzco and Machu Picchu, showcased an impressive array of temples, statues, and palaces constructed with precisely cut individual stone blocks, each weighing a hundred tons or more. Fitted without mortar, these building blocks are aligned as tightly and seamlessly as achievable without the aid of modern machinery or History Channel aliens. To feed and provide water to a growing population, the Inca developed hydraulic irrigation and aqueducts, and built canals and cisterns.

Expert terraced agriculture enabled the mass production of dozens of different and nutritionally varied crops, from corn, potatoes, peppers, peanuts, and tomatoes, to quinoa, beans, cucumbers, avocadoes, and cacao (as well as cotton), many of which would not have flourished in the region otherwise. This highly advanced and complex civilization, along with the Aztec of Mexico, would soon be welcomed into the harsh reality of the spiraling Columbian Exchange by way of Spanish invasion radiating from its initial Caribbean colonies.

On his quest for glory, adoration, and unimaginable wealth, Hernán Cortés sailed from Cuba to Mexico in March 1519 accompanied by sixteen horses and 508 conquistadors armed with crossbows, swords, cannons, and long, heavy matchlock guns called arquebuses. These horses were the first to stamp hoofprints into the North American mainland in almost ten thousand years.

This Aztec expedition was chronicled by Bernal Díaz del Castillo in his vivid and monumental history *The True History of the Conquest of New Spain*. He documented the owners, names, colors, and competency of every horse in vibrant detail: "Captain Cortés had a dark chestnut stallion which died when we reached San Juan de Ulua. . . . Francisco de Montejo and Alonzo de Ávila had a parched sorrel, useless for war. . . . Gonzalo Dominguez, an excellent horseman, had a dark brown horse, good and a grand runner. . . . Baena, a settler of Trinidad, had a dark roan horse with white patches but he turned out worthless." Castillo, a soldier and accomplished horseman, did not receive a mount.

Cortés and his troops realized quickly the psychological and military impact of the horse after initial encounters with smaller groups of Indigenous peoples on the fringes of Aztec control. When the ambassadors of the Aztec emperor Montezuma II arrived, Cortés staged a spectacular demonstration for his guests. He launched a full mock cavalry charge, complete with gleaming swords and cannon fire. To enhance the frightening spectacle, the horses were outfitted with bells and were commanded to rear up and whinny when they reached the utterly terrified Aztec delegation. Having yet to see the magnificent creatures for himself, Montezuma was informed that they were "as high as rooftops [and] deer which bore the visitors on their backs." Díaz recounts, "The Indians thought the rider and the horse were the same body, as they had never seen a horse."

At Tenochtitlán, Cortés and his entourage were received cordially by Montezuma, who lavished them with golden gifts. Much like the Aztec's reaction to horses, the Spaniards were stupefied and spellbound by the beauty, sophistication, and splendor of the capital. An astounded Díaz exclaimed, "When we saw so many cities and villages built in the water and other great towns on dry land, we were amazed and said that it was like the enchantments . . . on account of the great towers and cues and buildings rising from the water, and all built of masonry. And some of our soldiers even asked whether the things that we saw were not a dream? . . . I do not know how to describe it, seeing things as we did that had never been heard of or seen before, not even dreamed about." A dazzled and enraptured Cortés stated, "In Spain, there is nothing to compare with it." Of course, this mutual wonderment did not last.

Cortés hurriedly returned to the coast to head off another competing Spanish expedition led by Pánfilo de Narváez. After a brief skirmish, he absorbed the defeated force before marching to conquer Tenochtitlán. Cortés rode at the head of his column, remarking that the horses were both "companions and salvation." By now, he fully appreciated the power of his horses and cannons. "Do you know, gentlemen, it appears to me that these Indians have a great fear of our horses," Cortés regaled his men. "They really think they are the ones who make war upon them, and the same with the cannon." During his final push and siege of the city, Cortés had a force of nine hundred Spaniards, more than seventy-five thousand Indigenous allies (a historically neglected, yet critical component), and eighty-six invaluable horses.

After a methodical three-month campaign of cordon, destruction, and attrition, Spanish cavalry finally entered the city, which had been thoroughly ravaged and gutted by smallpox, malaria, and famine. In the Spaniards' view, the epidemic was their God's will of eradicating all those "whom we would have to fight and deal with as enemies, and miraculously Our Lord killed them and removed them from before us."

This perspective was not shared by one of the few Aztec survivors, who pointed the finger squarely at the Spanish. Prior to their arrival, he lamented, "There was then no sickness; they had no aching bones; they had then no high fever; they had then no smallpox; they had then no burning chest; they had then no abdominal pain; they had then no consumption; they had then no headache. At that time the course of humanity was orderly. The foreigners made it otherwise when they arrived here."

A Catholic priest, Francisco López de Gómara, recorded that during the final annihilation of the Aztec at Tenochtitlán, "Our men utilized the advantage given them by the man on horseback and hurled themselves on the natives, killing and wounding many." To this, Díaz added nonchalantly that the cavalry "soon reached the natives and speared them as they desired." On August 13, 1521, the feast day of Saint Hippolytus, the patron saint of horses, the once-prodigious Aztecan civilization fell to Cortés. The imposing, stylish city of Tenochtitlán was reduced to a pile of smoldering rubble. Fourteen years later, it would be rebuilt and rebranded as Mexico City, the capital of a mighty Spanish Empire.

Cortés, however, refused to take credit for the successful carnage. "Next to God," he reasoned famously, "we owe our victory to our horses." Díaz was saddened and dismayed by the slaughter, but not for the reasons you might think. "It was the greatest grief to think upon the horses," he mourned, and not the millions of dead human beings massacred by his comrades or corroded by smallpox. The horses, he added, "were our only hope for survival."

In honor of these horses, Cortés built his own *estancias* (ranches) and breeding stables at three sites. By 1533, only a dozen years after the fall of Tenochtitlán, the fertile plains north of Mexico City, which marked the southern limits of the Great Plains stretching north into Canada, were home to vast public grazing grounds, interspersed with more than sixty stud farms teeming with over ten thousand horses. Inspired by the

Spanish horsemen breach the city: The fall of the Aztecan capital of Tenochtitlán in 1521. From the *Lienzo de Tlaxcala* (*History of Tlaxcala*), circa 1552. *(Smith Archive/Alamy Stock Photo)*

plunder of Cortés and Pizarro, another conquistador, Francisco Vásquez de Coronado, was easily able to outfit his 1540 excursion across the American Southwest in search of the legendary golden cities of Cíbola from Mexican horse stocks.

Hernán Cortés did not vanquish six million Aztec, just as Francisco Pizarro did not conquer ten million Inca. Europeans began their global push for world domination with the benefit of their pathogens—to which Indigenous peoples had no acquired immunity. Time and again, European triumphs, including those of Columbus, Cortés, and Pizarro, rode the coattails of infection, not the other way around. Their horse-shuttled militaries simply mopped up after the stealthy invasion of disease.

Between 1520 and 1600, the Indigenous peoples of Mexico endured no fewer than fourteen major epidemics, while those of Peru suffered through seventeen. The main culprits, but by no means an exhaustive list, were smallpox, measles, influenza, tuberculosis, and malaria. It is a wonder that any Indigenous people survived the onslaught of European disease and human destroyers. By 1600, only 1.5 million, or 7.5 percent, of Mexico's precontact Indigenous population of 20 million could be counted.

When Pizarro penetrated the coast of Peru in 1531, the devastation wrought by smallpox, introduced five years earlier, and an ensuing civil war allowed his meager force to annihilate an Incan civilization that had numbered in the millions only a decade before. In November 1532, Pizarro, bolstered by reinforcements led by another Spanish explorer, Hernando de Soto, marched on the Incan capital of Cuzco.

Prior to the decisive Battle of Cajamarca, Pizarro made a show of brandishing his power by riding in circles around senior Incan politicians and commanders before charging at Emperor Atahuallpa—halting and rearing only a few feet away. With a total strength of a mere 106 infantry and 62 cavalrymen, Pizzaro confronted 80,000 Incan warriors. Nevertheless, the Spaniards prevailed at Cajamarca against an army five hundred times their size, slaughtering thousands of Inca without suffering a single combat death.

Inca Garcilaso de la Vega, the illegitimate son of a conquistador and a royal Incan mother, published his masterpiece of ethnography, anthropology, and history, *The Royal Commentaries of the Inca*, in 1609. He observed: "Nothing convinced them to view the Spaniards as gods and submit to them . . . so much as seeing them fight upon such ferocious animals—as horses seemed to them—and seeing them shoot harquebuses and kill enemies two hundred or three hundred paces away." The resounding victory at Cajamarca was followed by a string of four others in which just 80, 30, 110, and 40 Spanish cavalrymen, respectively, killed thousands of Inca during each confrontation. "The weapons and tactics the Spaniards used in Peru were the same as those they had used in Mexico," notes Headrick. "Thanks to their horses, their swords and armor, their boats, their firearms, and the germs they carried, the Spaniards defeated the two most populous, highly organized, and militarily successful states in the New World."

The value placed on horses was evident when it came time to divvy up the spoils of war. Pizarro gifted each horseman with 9,000 gold pesos and 300 marcs of silver, an amount double that of his infantry. The cavalrymen also received a handsome bonus of 3,300 gold pesos (equivalent to sixty swords) for (or *from,* as the case may be) the expense and priceless service of each of their horses.

When Pizarro was greedily transporting his horde of Incan treasure, he faced a troubling predicament. He had worn out the shoes of his packhorses and had no iron replacements.* If he could not solve this impasse immediately, his ill-gotten booty would most certainly be stolen. The solution was literally staring him in the face: his horses continued to move their precious cargo shod in improvised silver shoes, just as the Inca had suspected.

Although not generally wearing silver slippers, horses propelled forward the insatiable European mercantilist appetite of the Columbian Exchange. "When we consider the advantages that Spaniards derived from horses, steel weapons, and armor against foot soldiers without metal, it should no longer surprise us that Spaniards consistently won battles against enormous odds," summarizes Diamond. "Time and again, accounts of Pizarro's subsequent battles with the Incas, Cortés's conquest of the Aztecs, and other early European campaigns against Native Americans describe encounters in which a few dozen European horsemen routed thousands of Indians with great slaughter. . . . The tremendous advantage that the Spanish gained from their horses leaps out of the eyewitness accounts." Diamond's assertion, while accurate, begs the question: If horses (and guns) were unknown to the Indigenous populations of the Americas, and were also the paramount European military advantage in ensuring the destruction or subjugation of these same peoples, how did they come to fall into enemy hands?

Across history, societies placed embargoes on selling or trading valuable military assets to potential adversaries. The restrictions on dealing in horses and guns were no different from the modern prohibitions on the exchange of nuclear technology and materials under the 1970 Treaty

---

* In the half century after Columbus, the total yield of gold and silver from New Spain (hauled to the awaiting galleons by horses and donkeys) was ten times more than that produced by the rest of the world.

on the Non-Proliferation of Nuclear Weapons (NPT), among other weapons sanctions.

Medieval Spain, England, France, and Russia, for example, all outlawed horse exports. During the sixteenth-century reign of the English Tudors, all shipments of horses to Scotland were banned. The Spanish Netherlands enacted a similar law toward England during the sixteenth and seventeenth centuries. However, the embargoes were rarely effective. If there is money to be made, individual smugglers, organized crime syndicates, parasitic states, or national governments surreptitiously circumvent or openly flout the laws.

In 1502 Ferdinand and Isabella of Spain enacted Law 31 forbidding the sale of arms and horses to Indigenous peoples. In 1568 their great-grandson King Philip II outlawed "Indians riding horses" under the penalty of death except with permission from colonial governors. Neither statute worked.

Throughout the sixteenth century, domestic and feral horse herds across the Spanish Empire swelled in numbers. "The horse found a home in Peru, Chile, and Paraguay; in the pampas of Río de la Plata, he found a paradise," describes Alfred Crosby in *The Columbian Exchange*. "What happened when the horse reached what is today Argentina and Uruguay is best described as a biological explosion: horses running free on the grassy vastness propagated in a manner similar to smallpox virus in the salubrious environment of Indian bodies."

By the 1580s, according to Spanish friar Antonio Vázquez de Espinosa, the Pampas was teeming with horse herds "in such numbers that they cover the face of the earth and when they cross the road it is necessary for travelers to wait and let them pass, for a whole day or more." The plains surrounding Buenos Aires, Argentina, where horses were first introduced in 1535 by the syphilitic conquistador Pedro de Mendoza during his ill-fated and abandoned settlement, were overrun with "escaped mares and horses in such numbers that when they go anywhere, they look like woods from a distance."

Spaniards embraced the two great swaths of fertile grasslands of South America: the 145,000-square-mile llanos in Venezuela and Columbia and the 460,000-square-miles of the Pampas, incorporating all of Uruguay, a large portion of Argentina, and a sliver of southern Brazil. The original cowboys in the Americas were Spanish ranchers driving their cattle across these open savannahs. Cattle quickly became the largest immigrant mammal population in the Americas.

Many of the initial gauchos were Spanish Jews who had signed on to these early expeditions, including that of Cortés, to escape the Spanish Inquisition. As an accessory to the larger Reconquista targeting the last stubborn enclaves of Islamic al-Andalus, this judicial institution was established in 1478 by Ferdinand and Isabella to uphold Catholic orthodoxy and combat heresy. "Jews also played a decisive role in the translation of Spanish equine knowledge into the technological culture of Indigenous North America," points out Ulrich Raulff in *Farewell to the Horse: A Cultural History*. They were not "just the first ranchers of the New World," he adds, but "also the first cowboys in America." They plied their trade (acquired from the Moors) across the expanding Spanish American Empire as far north as New Mexico.

The language and terminology of these Spanish vaqueros also came to dominate the profession: cowboy, bronco, mustang, buckaroo, rodeo, roping, cinch, desperado, cavalier, caballero, dude, ranch, stampede, corral, chaps, lasso, and lariat, among other terms, eventually migrated north, along with the horse, finding their way onto the even larger grasslands of the Great Plains of North America.

With the Pampas awash with horses, Indigenous peoples, most notably the Mapuche of Chile, quickly adapted and learned to harness the power of the horse. They also acquired firearms from indiscriminate traders and gunrunners, creating a lethal combination. Originally hunter-gatherers and part-time farmers, the Mapuche (formerly called Araucanians) were one of the few Indigenous groups west of the Andes to resist draconian Incan rule and retain their autonomy. Stealing horses in the 1550s from isolated Spanish settlements, they quickly mastered mounted warfare.

Utilizing more than four thousand horses, the Mapuche pushed east into the Argentinian Pampas, where they met or merged with other Indigenous equestrian cultures such as the Puelche, Pehuenche, and Tehuelche. As with the colonization of Eurasia by mounted Indo-Europeans, this diffusion and propagation of the Mapuche horse complex (including language and other cultural attributes) between 1600 and 1850 is referred to as the Araucanization of Patagonia.* In short order, the Mapuche became the premier cavalry force in the region.

Mapuche horsemen were soon more plentiful and skilled than their Spanish counterparts, as Félix de Azara, a Spanish brigadier general, observed in the late 1700s:

> The occupation of their lives is war. . . . When they assemble, either to attack their enemies or to invade the Christians, with whom they are at war, they collect large troops of horses and mares, and then, uttering the wild shriek of war, they start at a gallop. As soon as the horses they ride are tired they vault upon the bare backs of fresh ones, keeping their best until they positively see their enemies. The whole country affords pasture to their horses. . . . How different this style of warfare is from the march of our army of our brave, but limping, foot-sore men, crawling in the rain through muddy lanes, bending under their packs!

For 250 years, these adept horsemen ensured Mapuche independence through protracted, low-intensity conflicts, known collectively as the Arauco War, against the Spanish. However, the Mapuche gradually lost ground, devastated and destabilized by epidemics of smallpox (which was first documented in the Pampas in 1558), measles, and influenza, and an increasing disparity in demographics. During the 1870s and 1880s, after another round of epidemics and wars, their land base was eroded and ultimately annexed by Chile and Argentina.† This narrative of rapid horse adoption by Indigenous peoples, resistance to colonial expansion, and eventual defeat and subjugation would be mirrored, albeit in quicker sequence, on the Great Plains of North America.

---

* The innovative Mapuche developed a weighted rope snare, or lasso loop-pole (much like that of a dog or alligator catcher), to hunt llama-like guanaco and flightless birds from horseback. They also designed specialized fifteen-to-eighteen-foot lances and long spiked cudgels for mounted combat, as well as distinctive mobile shelters made from guanaco skins.

† Roughly 1.9 million Mapuche currently make up around 9 percent of the total Chilean population.

For certain Indigenous nations in the United States, notably the Apache, Comanche, Shoshone, Lakota, Crow, and Nez Perce, for a legendary but brief period, the horse was the great equalizer in their conflicts with the US Cavalry and encroaching homesteaders. As with the Mapuche, the horse imposed its will upon these cultures. By swapping their farming roots for an equestrian regime, these Indigenous nations effectively became Scythians of the plains and cowboy cavalry themselves.

They proficiently adopted a mounted, nomadic lifestyle based upon herding horses and hunting bison while attempting to safeguard their traditional territories and stem the tide of trespassing settlers and soldiers in the face of insidious disease. "The sole Native Americans able to resist European conquest for many centuries were those tribes that reduced military disparity by acquiring and mastering both horses and guns," explains Jared Diamond. "Thanks to their mastery of horses and rifles, the Plains Indians of North America, the Araucanian Indians of southern Chile, and the Pampas Indians of Argentina fought off the invading whites longer than any other Native Americans, succumbing only to massive army operations by white governments in the 1870s and 1880s."

Although a comparatively recent and previously unimaginable addition to Indigenous societies, the horse is now synonymous with those cultures. As a pivotal agent of the Columbian Exchange, the horse found its way back to the Americas and quickly dominated its pioneering multicultural landscapes. The colonial birth and seminal years of the United States were rocked in the bosom of a saddle on the back of a horse.

Imagine for a moment, if you can, the heroes, villains, and half-truth idols of American Hollywood History—such as Paul Revere, George Washington, Ulysses S. Grant, George Armstrong Custer, Sitting Bull, Crazy Horse, Billy the Kid, Wyatt Earp, William "Buffalo Bill" Cody, Calamity Jane, Wild Bill Hickok, Jesse James, and Butch Cassidy and the Sundance Kid—*without* a horse.

Although it is a relative newcomer to the star-spangled backdrop, the horse is now seen to be as American as apple pie, Uncle Sam, the bald eagle, McDonald's, Mickey Mouse, Budweiser, and baseball.

CHAPTER 14

Big Dogs of the Great Plains
Horses, Bison, and the Downfall of Indigenous Peoples

In 1872, while Othniel Marsh was scouring western fossil beds during the Bone Wars between rival paleontologists, former California governor Leland Stanford made a wager with some friends. Although he had no visual proof, he was convinced that all four hooves of a horse were off the ground at the same time while galloping. His pals believed otherwise.

Like so many others, Stanford had made his fortune during the California gold rush of 1849—not on the precious metal itself, but as a merchant, wholesaler, and railroad executive. His two passions, aside from money and politics, were wine and horses. He owned two vineyards as well as breeding stables for racehorses at Palo Alto Stock Farm, now Stanford University, still known colloquially as "the Farm." With a generous endowment equivalent to $1.3 billion in today's money, Stanford and his wife established the school in 1891 to honor their teenage son, who had died of typhoid.

Given his avid love of horses, Stanford was curious if all four hooves were elevated at any given moment. Unlike the ancient Egyptians, who had sought an answer to this same question, new technology—and some ingenuity on the part of Eadweard Muybridge, an eccentric and unconventional English photographer—finally caught up to the horse. After a series of blurry failures, in 1878, Muybridge delivered Stanford his answer and settled the bet.

It was the embryonic age of photography, and each glass plate had to be wet coated immediately prior to snapping a picture, significantly impacting shutter speeds. This made it impossible to take successive, rapid-fire shots. That is, impossible with only *one* camera. On June 10, 1878, Muybridge set up a row of twelve single-lens cameras at twenty-seven-inch intervals (later, twenty-four cameras at twelve-inch intervals) on a track at Palo Alto against a backdrop of white sheets, white walls, and white dirt to contrast with the galloping dark horse named Sallie Gardner. Muybridge laid across the track a series of electromagnetic circuits and trip wires attached to the camera shutters to time the consecutive exposures (accurate to two-thousandths of a second) to record a virtually instantaneous chronological sequence.

The findings were published as *Sallie Gardner at a Gallop* and later as the three-second clip *The Horse in Motion*. Stanford won the $25,000 bet and affluent bragging rights, although it is estimated that he spent nearly $50,000 to bankroll his investigation. All four feet of a horse are off the ground during each stride, collected under its barrel as it switches from "pulling" with the front legs to "pushing" with the back.

Two years later, Muybridge printed the serialized photographs on a round glass plate. By spinning this disc on what he called his

Betting the farm: Sallie Gardner running over the Palo Alto track as photographed by Eadweard Muybridge, 1878. Printed as *The Horse in Motion* cabinet card. *(Library of Congress)*

"zoopraxiscope," which projected the images in a repeating loop on a screen, Muybridge, Stanford, and Sallie Gardner gave birth to the motion picture. "Thomas Edison took some of the ideas for his 'kinetoscope' from Muybridge, and he also took advantage of the recent invention of celluloid photographic film to replace the glass plates," explains J. Edward Chamberlin in *Horse: How the Horse Has Shaped Civilization*. "The two of them even talked about adding a phonograph to the moving picture to provide sound.... So movies began with horses."

Muybridge released his famous twelve-second film *Buffalo Running* in 1883, the same year that William "Buffalo Bill" Cody debuted his novelty act Wild West show. By this time, the plains bison (buffalo) had been systematically slaughtered: from a pre-Columbian population of sixty million to a few thousand. The majestic animal, the livelihood of Indigenous peoples of the Great Plains—slicing from Canada's Alberta and Saskatchewan in the north, through ten US states, to the Mexican borderlands in the south—hovered on the verge of extinction.

Four years after the 1890 massacre of the Lakota (Sioux) at Wounded Knee, two of Edison's early, and now renowned, shorts depict five Lakota performers, all veterans of the Wild West show, performing the buffalo dance and the ghost dance. While Edison's cameras were rolling, the Indigenous population of the United States tracked that of their beloved bison, reaching its nadir at 237,000 from an estimated precontact population of 10 million. The origins of *The Vanishing American* began long before the 1925 silent film (based on the novel by the prolific Western author Zane Grey) premiered to critical acclaim—while perpetuating the misplaced American frontier spirit.

The stereotypical "noble savage" image of the Indian with feathered headdress and spotted horse was cemented into American folklore and frontier mythology by Hollywood and its enduring Western genre starring immortal silver screen cowboys such as Tom Mix, John Wayne, Clint Eastwood, and Yosemite Sam mercilessly brutalizing "uncivilized" Indians.* The epic television adventures of the Lone Ranger and his

---

* The award-winning, and highly recommended, documentary *Reel Injun: On the Trail of the Hollywood Indian* (2009), by Cree-Canadian filmmaker Neil Diamond, traces the history, portrayal, and stereotypes of Indigenous peoples in movies and on television.

Buffalo Bill's Wild West show: "A Congress of American Indians, representing various tribes, characters and peculiarities of the wily dusky warriors in scenes from actual life giving their weird war dances and picturesque style of horsemanship." Promotional poster, circa 1899. *(Library of Congress)*

pidgin-speaking Indian sidekick, Tonto, broadcast this caricature across 1950s baby boom suburbia.*

This misdirected interpretation diminishes the proactive strategies and collective agency of Indigenous peoples to promote and protect their own interests and agendas in the face of a cultural upheaval driven not only by the arrival of Europeans and disease but also by the entrance of the horse *and* their responses to it. Like the Chinese, numerous Indigenous nations coveted the horse as the great equalizer to defend their homelands from mounted invaders. Substitute the Xiongnu for American and Canadian settlers and escorting cavalry. In the process, the horse imposed its ethos on numerous Indigenous nations that enthusiastically adopted a nomadic, horse-based existence. Substitute Indo-Europeans, Scythians, and the Eurasian Steppe for Comanche, Lakota, and the Great Plains. Within our Centaurian Pact, change can be effected by *both* horses and humans.

---

* *Tonto* means "moron" or "fool" in Spanish, Italian, and Portuguese.

With headdress and horse: Oglala Lakota warrior Red Hawk, who fought alongside Crazy Horse at the Battle of the Little Bighorn, poses for famed photographer Edward S. Curtis in the Badlands of North Dakota, 1905. *(Library of Congress)*

Indigenous peoples were not sidelined spectators to the ongoing processes of colonization and the Columbian Exchange. They were active participants and ventured eagerly into the fur, bison, and horse trades. Both Europeans and horses "came not as an individual immigrant," says Crosby, "but as part of a grunting, lowing, neighing, crowing, chirping, snarling, buzzing, self-replicating and world-altering avalanche." It should come as no surprise that the lethal combination of guns, germs, steel, and horses also played out across the Great Plains.

On the vast prairies of North America, the closing years of the Columbian Exchange were caught on camera. The final frontier of the American West—the safety valve for outlaws and drifters and the playground for restless adventurers—captured the imagination of the urbane society. It was a strange and storied land inhabited by savage Indians, toiling settlers, stoic cavalrymen, exotic animals, and swaggering cowboys drinking, gambling, and brawling their way across dusty cattle drives, decadent boomtowns, and tumbleweed terrain.

Manifest Destiny: *American Progress* by John Gast, 1872. *(Library of Congress)*

The creative architects and enthralled audiences of these early films were trying to capture the august bison and wild Indian before they vanished forever from the American landscape. Theirs was a picturesque, but antiquated, worldview incompatible with American (and Canadian) progress and civilization—altruistic synonyms for imperialist capitalism and commercial expansion. "The fulfillment of our manifest destiny," wrote journalist John O'Sullivan in 1845, "is to overspread the continent allotted by Providence for the free development of our yearly multiplying millions."

The classic 1872 portrayal *American Progress* by painter John Gast contains all the prerequisite elements of Manifest Destiny and, by extension, this chapter. The backdrop is provided by the eastern modernity of New York, the Brooklyn Bridge, and chugging westbound trains giving way to a horse-drawn wagon train, snow-capped mountains, herds of feral horses, and a fading idyllic Indian tepee camp. A stampeding herd of bison being pursued by a mounted hunter exits the scene to the left, while wolves pick clean the bleached bones of their wasted kin. A despondent, half-naked Indigenous family, accompanied by dogs and horse-drawn travois, follow the bison into near extinction. Turning their feathered heads to look behind them—to *before*—these stragglers are being chased and replaced by a cattle-drawn Conestoga wagon, an Overland Mail carriage, the iron horse railway, a party of miners, fur traders, or fossil hunters, and a fenced-in pioneer family sod busting with a steel plow. Clutching a transcontinental telegraph line and a schoolbook, Columbia—the civilized, goddesslike female personification of the United States of America—gently guides the nation forward as she floats serenely toward the future.

Americans (and Canadians) fanned out across the western grasslands using the same domesticated horses, covered wagons, and languages that Indo-European migrants utilized six millennia earlier on the Eurasian Steppe. "Unless you live in the North American interior, it is easy to forget that the prairies of the Great Plains have the same land type and topography as many of the landscapes of Inner Asia," points out William Honeychurch. "People living in Kansas would probably feel right at home on the opposite side of the earth in the broad open spaces of Kazakhstan or eastern Mongolia, just as Mongolians might well see their northern steppes, forests, and mountains in the landscapes of Wyoming and North Dakota."

These trailblazing pioneers, however, faced an entirely different adversary and strategic situation than did Indo-Europeans, Columbus, Cortés, Pizarro, or even the early English and French settlers to the eastern shores of Canada and the United States. In an incredibly short time, Indigenous nations of the Great Plains acquired and mastered the horse. Like Indo-Europeans, they also frequently infused the horse into their personal names—none more hallowed than Crazy Horse. Mounted on their prized horses, they would fight, against overwhelming odds, to preserve their traditional lands and cultures from encroaching American progress.

Across European settler societies, an assortment of synchronous strategies was implemented to subjugate Indigenous populations: waging decisive military campaigns, destabilizing their political structures, undermining social conventions, eliminating identifiable cultural traits, creating economic dependency, drastically shifting demographics (which disease made certain), and expropriating Indigenous land. On the North American prairies, the horse played a shaping role in all elements of this paradigm and in the position of Indigenous peoples within it.

The introduction of horses (and guns) across the diverse Indigenous landscapes of the plains led to a complex and multidimensional ecological, economic, and sociocultural transformation—one that produced both positive and negative effects. Like previous equestrian cultures, including Indo-Europeans, Indigenous employment of horses can be broken down into five main categories: it provided a new food source (although not all nations ate horsemeat); bulk transport; long-distance transport; a fast and portable hunting podium; and a mobile, long-range fighting platform.

Relatively vacant and unexploited grasslands became commercially productive, valued for horse pasture and bison harvesting. These enterprises increased consumable territory across this environmental belt sixfold. The skyrocketing demand for furs and bison hides turned the plains into a battleground. Fierce competition erupted between increasingly combative and capitalist Indigenous nations.

Mirroring previous Eurasian models, the horse gave fully nomadic, mounted cultures, such as the Comanche and Lakota, a decisive military advantage over more sedentary and pastoral peoples. This enabled them to expand their range of influence, grazing and hunting territory, and their trading partners, which came to include Euro-American settlers, trappers, and fur traders who supplied firearms and ammunition.

Cyclical patterns of intensified economic exchange, horse raiding, bison reaping, and accrued wealth in horses upended long-standing Indigenous communal structures and social conventions. Traditional egalitarian customs and gender roles were undermined by the emergence of social classes and economic hierarchies. The complete cultural integration of the horse both enhanced and corrupted the lives and societies of Indigenous peoples.

It was originally hypothesized that escaped horses from the stocks of

THE EFFECTS OF THE INTRODUCTION OF THE HORSE TO NORTH AMERICA

UTILITY	PRIMARY IMPACTS	SECONDARY IMPACTS	FINAL RESULT
New Food Source	Prairie grasslands become viable and productive for subsistence and resources, including bison.	Wealth in horses upends economic, social, and cultural conventions, traditions, and mores.	Colonial capitalism undermines egalitarian principles, creating internal divisions, strife, and conflict.
Bulk Transport		The bison, horse, and fur trade (and theft) intensify in scope producing new political and community structures.	
Long-Distance Transport	Exploitable territory increases sixfold.		
Mobile Hunting Platform		Conflict over grazing and hunting lands (and resources) intensifies.	Colonial and national governments implement "divide and conquer" strategies to subjugate Indigenous Peoples.
Mobile War Platform	Nomadic horse-based Plains Nations possess military advantage over sedentary agricultural neighbors.	Military prowess and warrior societies command prestige and power.	

Consequences of the Columbian Exchange: Introduction of horses into North America. *(Jorde Matthews/Be Someone Design)*

Hernando de Soto and Francisco Vázquez de Coronado, who traversed the southern United States in the 1540s in search of Cíbola, were the first to be acquired and propagated by Indigenous peoples. Pedro de Castañeda de Nájera, the chronicler attached to Coronado's expedition, echoed Cortés in recognizing that "Horses are the most necessary things in the new country because they frighten the enemy most, and after God, to them belongs the victory." The Zuni of Arizona sent word to their allies that they had been attacked by "very fierce men who rode animals that ate people." The Acoma Pueblo rubbed Spanish horse sweat on their faces to absorb supernatural power.

Contrary to these early theories, the initial Spanish expeditions of Pánfilo de Narváez and Ponce de León to Florida, and those of Coronado and de Soto a few decades later, did not leave any permanent equine inheritance to Indigenous peoples. The lasting (re)introduction of horses north of the Rio Grande into the American Southwest and beyond was a combination of three factors: raiding, trading, and the Pueblo Revolt of 1680.* "It is one of those accidents of history," writes historian Bernard Mishkin in *Rank and Warfare Among the Plains Indians*, "that an instrument of Spanish expansion, the horse, was an important factor in barring further expansion to the same nation."

Daunted by the seemingly endless grasslands relatively devoid of resources, Spanish penetration past the main settlement at Santa Fe (established in 1607 in what is now New Mexico) into the Great American Desert was marginal at best. Where the Spanish saw a wasteland, the Apache, Comanche, Shoshone, Lakota—and eventually Americans—saw horse-powered opportunity. Punitive Spanish strikes and slave raids against the Pueblo, Apache, and Ute were dispatched from Santa Fe, which became the first radial focal point for the dissemination of horses.

Following the pattern established previously in South America and Mexico, trading and raiding were the two main arteries through which Indigenous peoples first obtained horses prior to the Pueblo Revolt. An

---

* During the period discussed, the area encompassing Arizona, New Mexico, and Texas is more accurately Northern New Spain (subsequently Mexico). The area was administered from Mexico City for a century longer than it has been from Washington, DC.

administrative dossier from New Mexico spanning the years 1606 through 1608, for example, mentions Apache and Navajo war parties harassing Spanish settlements and making off with horses. These marauders are described as "expert and able in the management of arms and horses," indicating that both skills had been acquired some time earlier.

While the Navajo embraced horses, they maintained a hybrid culture. A nomadic horse complex based on bison, like the Apache, Comanche, and Lakota, never materialized. The Navajo grew the three sisters of corn, beans, and squash (and peaches), and used their horses to tend to their sheep and trade their agricultural surplus and distinctive textiles and baskets. The Apache adopted a more advanced equestrian ethos than their Navajo cousins.*

Apache creation stories foretold that "It would come about that the broad earth would be covered with horses." The prophecy was correct, and the Apache were partly responsible for its realization by acting as a catalyst for distributing horses. Apache horse raids intensified between 1617 and 1622, furnishing all communities throughout New Mexico with an abundance of horses. Alonso de Benavides, a Spanish friar, reported in 1630: "Our herds have already propagated much there . . . herds of swine, mules and famous horses, particularly for military use." In 1637, for instance, a botched Spanish slave raid on the Apache and Ute resulted in the loss of numerous horses.

The demand for Indigenous slaves was fueled by the Spanish encomienda, or "feudal" system, in New Mexico. Local Indigenous peoples and rival prisoners served as "peasants" on massive cattle ranches. Each wealthy landowner was expected to pay taxes to the governor in Mexico City. While colonial law forbade Indigenous ranch hands from riding, they were immersed in the maintenance and breeding of horses.

This injunction was relaxed in 1621, allowing Indigenous cowboys to ride if they converted to Catholicism. It was a risky and potentially hazardous Faustian bargain. Across New Mexico, subjugated and mistreated Indigenous populations were trained in the art of horsemanship. Many

---

* Both the Apache and Navajo are Na-Dene Athabaskan speakers who migrated from northwestern Canada to the American Southwest between 1200 and 1400.

simply gathered a small herd and galloped home, where they acted as riding instructors for their people.

Numerous Pueblo joined Apache farming communities, bringing their horses with them. By the 1650s, the Pueblo were trading Spanish horses to the Apache in exchange for bison robes. Alonso de Posada, a missionary attached to the Pecos Pueblo, complained that the Apache were "running off day and night the horse herds of the Spaniards." In 1676, for example, the colonial administration urgently requested the delivery of a thousand horses to replace those stolen by the Apache the previous year. By this time, the Apache were openly trading captives for horses, and the Ute were using horses as pack and draft animals. The Pueblo Revolt, dubbed the "Great Horse Dispersal" by early scholars, merely supplemented established and emerging stocks.

After decades of brutal subjugation and prolonged drought, in 1680 under the direction of the enigmatic spiritual leader Po'pay, the Pueblo launched a coordinated overthrow of Spanish rule. Limping back to El Paso, the Spanish abandoned more than three thousand horses and the entire New Mexico territory before reoccupying the region twelve years later. "The opportunity to seize horses during this revolt was once thought to have been instrumental in their spread to other Native American groups in the Southwest and on the Plains," clarifies Peter Mitchell in *Horse Nations*. "No doubt some were acquired in this way. . . . However, it is now clear that the horse had already bolted free of the Spanish stable well before the Pueblo Revolt."

A lively inter-Indigenous horse trade spanned the borderlands of the Southern Plains. "The spread of the horse northward into the Southwest," notes anthropologist Jack Forbes, "was primarily an Indian phenomenon." The chronology, routes, and range of horse diffusion across the western United States and southwestern Canada have been painstakingly mapped through the detailed diaries and reports of countless fur traders, explorers, and colonial administrations, in addition to Indigenous traditions, archaeological evidence, and demographic trends. "The trade spread like twin vines up both sides of the Rockies. To the east, on the Great Plains, the Apache gave horses to the Pawnee, the Pawnee to the Arapaho, the Arapaho to the Kiowa, the Kiowa to the Mandan," maps out David Philipps. "Horses stolen from the Spanish on the western side of the mountains went first to the Ute, then to the Shoshone around 1700, then to the Nez Perce around 1730, then to the Crow and Blackfoot around 1740, and, finally, to the Cree around 1750. In 1790, trappers spotted horses with Spanish brands in Canada." The Apache, Comanche, and Shoshone were largely responsible for the dissemination of horses.

The Apache transition to a horse nation had commenced prior to the Pueblo Revolt, and they quickly became the dominant power on the Southern Plains. The Spanish referred to the Apache as *los farones* (the pharaohs), in homage to the swift Egyptian cavalry of the book of Exodus. "There is a nation which they call the Apacha which possesses and is owner of all the plains of Cibola," scorned de Posada. "The Indians of this nation are so arrogant, haughty and such boastful warriors that they are the common enemy of all nations who live below the northern region. They hold them as cowards. They have destroyed, ruined or driven most of them from their lands." Apache regional expansion and the corresponding exchange of captives for horses with Spanish mine and ranch owners introduced horses to neighboring Indigenous nations.

The vaunted French explorer and *voyageur* René-Robert Cavelier de La Salle reported in 1682 that the Apache were raiding and trading horses with the Pawnee, Wichita, Kiowa, and Caddo of Kansas, Oklahoma, and Texas. These nations then shuttled horses farther north to the Omaha, Arikara, Arapaho, Cheyenne, and Lakota. On the western side of the Rocky Mountains, the Apache traded horses with the Ute, who shuttled them north to their distant Shoshone kin in Wyoming/Idaho. Acting as the primary middleman, the Shoshone then dispersed horses across the northwest basin and plateau regions.

While the Apache adopted a significant equestrian element to their culture, they also maintained a strong agricultural component. Even with the infusion of horses, they were still tied to the land, and carefully limited their herds to preserve pasture and farming lands. They also possessed few guns, as the Spanish were maintaining a strict ban on their traffic to Indigenous peoples. These two attributes proved to be their undoing. The early Apache acquisition of the horse brought them dominance but also made the relatively sedentary nation the primary target of aggrieved rival neighbors and a rising nomadic power.

The Comanche, a splinter band of Shoshone, first acquired horses and a bison hunting template in the 1680s through trade and contact with the Ute to the south. "Nothing could have prepared the Comanche for what they would become when they first acquired horses in the late seventeenth century," writes Mitchell, "and yet within a hundred years they had carved out a dominant position in the politics and economics of the Southern Plains, treating Spanish governors as equals and later raiding to within 200 km [125 miles] of Mexico City. Likely the very first North American society to take fully to an equestrian life, they perfected the use of horses for hunting bison to the point where they invite serious consideration as pastoralists, rather than hunter-gatherers."

Apache slave raids sparked a series of tribal coalitions and spiraling wars. After a fleeting alliance with the southbound Comanche in the early 1700s, the Ute soon looked upon these distantly related but aggressive nomads as *kimantsi* ("enemy" or "foreigner"), a name adopted by the Spanish as Comanche. They are first mentioned around 1705, raiding Apache, Taos Pueblo, and Spanish settlements for horses. The Comanche quickly rose through the ranks, becoming the dominant mounted culture in the region. By the 1730s, they had driven the Apache to more arid lands on the fringes of the southwestern plains, where farming was more difficult, and their population declined. Their homeland continued to shrink, giving way to Comanche, Mexican, and, eventually, American territorial expansion.

By turning the endless sea of relatively uninhabited grasslands into a weapon, the Comanche and their horses became the undisputed lords of the plains. A French fur trader, Athanase de Mézières, observed in 1770: "They are so skilled in horsemanship that they have no equal, so daring that they never ask for or grant truces." Donning their distinctive bison-horned headdress, Comanche warriors were the Scythians or Xiongnu of the American Steppe, armed with lances, bow and arrow, leather shields, and layered rawhide armor (borrowed from the Apache) for both horse and rider.* "The Comanche were boasting in all seriousness that the horse was created by the Good Spirit for the particular benefit of the Comanches," scribbled a bison hunter, "and that the Comanches had introduced it to the whites." American cavalrymen acknowledged begrudgingly their Comanche enemy as "the finest light cavalry in the world."

With their horse-powered monopoly on the bison trade, between 1770 and 1840 Comanche hegemony dictated the larger ebb and flow of geopolitics across the plains. Centered in western Texas, they created a multifaceted, kinetic empire encompassing portions of New Mexico, Colorado, Kansas, Oklahoma, and northern Mexico. The Comancheria region housed roughly eight million bison, which became the heartbeat of subsistence, trade, and war in an era of increasing nomadism. The Comanche engaged in what historian at the University of Oxford and award-winning author of *The Comanche Empire* Pekka Hämäläinen refers to as "exchange-oriented equestrianism." Like the initial dispersion of horses from Santa Fe by the Pueblo, Ute, and Apache, Comancheria became the second radial focal point for horse diffusion across the West.

The Comanche secured and expanded their hunting grounds and commercial reach through war and by exporting horses, bison hides (and the meat of both), and captives. In return, they obtained firearms and ammunition through intermediary trade with more agrarian peoples on the eastern edge of the plains, including the Pawnee, Kiowa, Wichita, Caddo, and Quapaw. These nations had access to weaponry through French fur traders operating along the Mississippi River watershed and Acadian (Cajun) settlements in Louisiana. They also provided the Comanche with essential carbohydrates, including corn, to supplement a protein-heavy diet of bison and the occasional horse.

Plying their trade north to British and American trappers traversing the Missouri River valley, the Comanche brought an influx of horses to their Shoshone relatives. The Shoshone played a crucial role as intermediaries in the horse trade running from the Southwest to the Northern

---

* This remained the mounted warfare package on the plains until guns became more common (and more accurate) during the early- to mid-nineteenth century.

Plains and the plateau region of the Pacific Northwest. Around 1725, Saukamappee, a Blackfoot Piikani elder, told a French fur trader about a Shoshone raid on a neighboring village: "Our enemies the Snake Indians [Shoshone] and their allies had Misstutim (Big Dogs, that is Horses) on which they rode, swift as the Deer, on which they dashed at the Peeagans. . . . This news we did not well comprehend and it alarmed us, for we had no idea of Horses and could not make out what they were."

When Walla Walla and Yakama warriors from the Columbia River watershed of Washington and Oregon found their Shoshone enemy riding "some strange-looking animals [that] looked like elk or deer, only they had no horns," they chose not to attack; instead, they bartered all their possessions for the remarkable beasts. Strutting naked alongside a stallion and a mare, they proudly entered their villages to a silent reception from their stunned and speechless families.

In May 1804, at the behest of President Thomas Jefferson, Captain Meriwether Lewis and Second Lieutenant William Clark of the US Army led a small expedition northwest across the Great Plains and Rocky Mountains—all the way to the Pacific Ocean. Its mission, as outlined by Jefferson, was to find "the most direct and practicable water communication across this continent, for the purpose of commerce," while assessing the economic potential of the flora and fauna in the 828,000 square miles of new territory that the United States had recently acquired from France. The Louisiana Purchase, with a price tag of $15 million, doubled the size of the country overnight at less than three cents an acre.

During its two-and-a-half-year excursion, the Corps of Discovery Expedition encountered dozens of Indigenous nations. Lewis and Clark were notably impressed with the horse herds of the Shoshone, from whom they procured vital mounts in exchange for axes, knives, guns, ammunition, and clothing. They were even more amazed when the Shoshone told them of their Comanche cousins to the south, where, as Lewis recorded, "horses and mules are much more abundant than they are here." Both explorers reported that the horse was wholly integrated into the cultures and economics of the Pacific Northwest. Nez Perce warriors, for example, were described as having fifty to a hundred horses each, with communal herds reaching upwards of two thousand.

The Nez Perce even developed their own breed: the Appaloosa. According to Lewis, their horses were "pided [sic] with large spots of white irregularly scattered and intermixed with the black brown bay or some other dark color." He was alluding to their classic leopard-spotted coat markings. An Appaloosa fetched more than $600 in the affluent eastern horse markets at a time when an average horse sold for $15. With the surrender of Chief Joseph in 1877, the Nez Perce were forced from their ancestral lands to distant reservations. Their 1,100 remaining horses were killed or confiscated. The Appaloosa was forgotten until it was revived in the 1930s by a group of dedicated breeders and subsequently appointed the state horse of Idaho.

For some pastoral nations on the fringes of the plains, the decision to embrace the horse initiated a drastic cultural shift toward bison commercialism. Many of these nations also adopted plains-style tepees, travois, and horse-based medicine cults. "If horses expanded the horizons of those who acquired them, perhaps they did so most on the Columbia Plateau," notes Mitchell, "where many groups followed the Ute example in altering their annual subsistence cycle to cross the Continental Divide and spend the summer hunting bison on the plains. Flatheads, Cayuses, Kalispels, Nez Perces, and Northern and Eastern Shoshones all did this." This seasonal hunting practice was mirrored on the edges of the eastern plains by the Pawnee, Omaha, Osage, and Wichita nations, among others, who ventured into the grasslands for up to seven months to hunt bison while retaining their agrarian roots.

As a result, bison populations on both peripheries began to dwindle by the 1830s. Progressively deeper hunting incursions by these perimeter nations turned the heart of the prairies into a brutally contested war zone awash with the blood of bison, horses, and human beings.

This horse-bison complex also extended into Canada. Representing the Hudson's Bay Company on the northern reaches of the Great Plains in Alberta, Anthony Henday reported in 1757 that the Siksika, Kainai, and Piikani nations of the Blackfoot Confederacy were hunting bison on horses. The Blackfoot, Assiniboine, Plains Cree, and Plains Ojibwa

(Saulteaux) peoples, straddling the American-Canadian border, picked up their first horses through the Shoshone exchange as early as the 1720s. At the same time, their southern Lakota neighbors were initiating their transformation into the last wholly integrated Indigenous horse nation of the United States.

In the 1660s, the Lakota left their traditional homeland in central Minnesota, drifting west across the Missouri River to escape the onslaught of the Iroquois (Haudenosaunee) Confederacy. During the seventeenth century, the Iroquois exploded out of upstate New York in a sequence-repeating rampage across the Northeast and the Great Lakes region to secure beaver furs, which they bartered for British firearms in order to exact revenge on traditional enemies. These ruthless fur-trading and retaliatory conflicts, known collectively as the Beaver Wars, fractured long-standing Indigenous relations. Much like Xiongnu expansion, Iroquois aggression also initiated knock-on migrations by uprooting and dispersing numerous nations, including the Lakota.

Subsisting on corn, wild rice, and hunting, the Lakota obtained firearms from the French in exchange for beaver pelts during the 1680s, and their first horses around 1700 from the Arikara and Omaha. It was not until the late eighteenth century, however, that their conversion to a nomadic horse-based bison culture fully materialized.

Following the Comanche playbook, the Lakota, still relatively unscathed by disease, pounced on the weakness of sedentary agrarian nations that had been washed out by sickness and famine and were poor in horses and guns. "Hunting buffalo and fighting tribal enemies was an all-absorbing way of life around which the Lakota had created a beautifully intricate and self-contained culture," writes historian Nathaniel Philbrick in *The Last Stand: Custer, Sitting Bull, and the Battle of the Little Bighorn*.

In short order, the Lakota secured hegemony over the Northern Plains through horse raiding and bison trading. One by one, the Arapaho, Arikara, Cheyenne, Crow, Hidatsa, Omaha, Pawnee, and Ponca were defeated, fled, or became subordinate allies. During the 1830s, for instance, the Lakota seized the sacred Black Hills in South Dakota and assimilated the remaining Arapaho and northern Cheyenne peoples into their commercial-military complex. By the 1850s, the Lakota had become the northern equivalent of the Comanche. The strength of both imperial Indigenous powers rested on the profitable marriage between horses and the industrial harvesting of bison.

Prior to the horse, teams of Indigenous hunters would stampede bison over cliffs, as represented by the iconic Head-Smashed-In Buffalo Jump site in southern Alberta. Another, more dangerous, method—reminiscent of the great horse-kill site at Solutré in France—was to use chutes constructed of rock cairns or tree branches to funnel bison into ravines, box canyons, or premade log (or human) corrals called "buffalo pounds." All parts of the animal were used to make weapons, utensils, tools, clothing, rope, decorations, spiritual artifacts, musical instruments, pillows, blankets, tepee covers, manure fuel (*nik-nik*), candles, soap, cooking oil, and, of course, food.

On horseback, a hunter singled out an animal, rode alongside, and delivered a fatal shot with a personalized arrow before moving on to the next target. One mounted hunter could quickly bring down hundreds of bison. "Horses democratized access to this essential resource by reducing the necessity for communal drives," states Mitchell. "They also massively expanded the radius over which hunting could be undertaken by allowing larger quantities of dried meat and pemmican to be transported to campsites and reducing the need for the entire tribe to stay close to herds. Most hunting came to be done by riding in and shooting bison at close range, but bison were also killed using lances."*

The original herds were so thick that multiple arrows fired simultaneously from one draw could bring down several animals. The unofficial Pawnee record was three bison with one bowshot. To feed their bison-based economy and a population approaching 40,000, the Comanche harvested at least 240,000 to 320,000 animals every year, with annual ceiling estimates reaching 574,000. While overhunting, primarily to meet market demands, certainly played a role in taxing bison numbers, so too did spiraling ecological pressures triggered by the introduction of horses and cattle.

---

* Pemmican is a high-caloric and full-spectrum nutrition "jerky" of dried and pounded bison meat, fat, and mixed berries. It can be stored for months, providing crucial sustenance on the plains during the winter freeze or for long-distance travel during any season.

Buffalo Chase, Bulls Making Battle with Men and Horses: This image was produced by famed American adventurer and artist George Catlin on the Upper Missouri River in 1832. *(Smithsonian American Art Museum/Mrs. Joseph Harrison Jr.)*

Horses had been absent from North America long enough that from an environmental standpoint, they were as much an invasive species as cattle, smallpox, Europeans, and tumbleweed. The diet of horses, cattle, and bison overlapped by 80 percent. They also competed for critical, and relatively sparse, water resources. It was a simple equation: more horses and cattle on the plains meant less grass and water for bison. Remember that an average thousand-pound horse consumes fifteen to thirty pounds of grass per day in addition to six to ten gallons of water. A herd of five hundred will easily crop four acres of grass per day.

This also meant that communities had to be continuously on the move to secure grazing lands through seasonal migration or war. "Thus, like the colonizers themselves," contends historian and author Sandra Swart, "the equine 'invaders' were the instruments of extensive and long-term changes within both natural and social landscapes." As events unfolded, horses increasingly dictated policy on the plains. The battle for supremacy and survival was not just being played out among rival Indigenous peoples, settlers, and cavalrymen but also by cattle, horses, and bison. For Indigenous peoples, nature was pitting their two most important economic portfolios—horses and bison—against each other.

Increasing numbers of feral horse herds chased the plains from Mexico to Canada. In 1777 a Spanish friar, Juan Agustín Morfi, remarked in his memoirs that northern Mexico was so overrun with mustangs "that their trails make the country, utterly uninhabited by people, look as if it were the most populated in the world." A Texas Ranger patrolling the rolling country between the Rio Grande and the Nueces River in 1840 was stunned when he stumbled upon "a drove of mustangs so large that it took us fully an hour to pass it, although they were traveling at a rapid rate in a direction nearly opposite to the one we were going. As far as the eye could extend on a dead level prairie, nothing was visible except a dense mass of horses, and the trampling of their hooves sounded like the roar of the surf on a rocky coast."

Passing through this same stretch in 1846 during the Mexican-American War, Lieutenant Ulysses S. Grant, regarded as one of the most exceptional horsemen in US history, recorded, "As far as the eye could reach to our right, the herd extended. To the left, it extended equally. There was no estimating the number of animals in it; I have no idea that they could all have been corralled in the State of Rhode Island, or Delaware, at one time. If they had been, they would have been so thick that the pasturage would have given out the first day. People who saw the Southern herd of buffalo, fifteen or twenty years ago, can appreciate the size of the Texas band of wild horses in 1846."* Although hunters targeted these feral herds, the market for horsemeat and leather was immaterial when measured against that for the bison.†

One notable use for horsehide was the covering, or "skin," of baseballs. By the outbreak of the Civil War, the game was already being touted as America's "national pastime" or "national game."‡ Horsehide

---

* Grant's celebrity Civil War horse, Cincinnati, stood 17.2 hands.

† Herman Melville "toyed with the idea of narrating the hunt for a sacred white buffalo, or a sacred white stallion" before settling on a great white whale for his 1851 classic *Moby-Dick; or, the Whale*.

‡ The total equine population in the United States at the outbreak of the Civil War was about 7 million. Roughly 3 million served, suffering 1.5 million deaths.

was more supple and textured than cowhide, producing premier baseball wraps, and unhittable curveballs. On September 17, 1862, the bloodiest single day of combat in American military history, 15,500 spectators turned out to watch the Brooklyn Atlantics play Brooklyn Excelsior. At the same time, 250 miles away along Antietam Creek near Sharpsburg, Maryland, 22,717 Americans became casualties of the Civil War.

Feral horse estimates are admittedly problematic, but, by 1840, there were at least two million mustangs and a half million domesticated horses grazing the plains.* While Indigenous peoples supplemented their stocks from feral herds, the preference was to steal tamed (already broken and trained) horses from rival nations and settlers. Raiding also offered warriors the opportunity to gain status through bravery while complementing their personal wealth in horses—which, for the elite, climbed into the hundreds.

Domestic and feral cattle compounded the competition for essential and overstressed resources. When grass was limited, cows ate mesquite pods. The seeds, however, are indigestible, and cattle excrement transformed nutritious bison grasslands into mesquite-ridden scrub brush.† There were 2.4 million head of cattle ranging the Great Plains in 1880. Five years later, this number had more than tripled to 7.5 million, reaching 12.6 million by 1900. Cowboy cattle drives became a mainstay of the corporate West.

Between 1866 and 1886, for example, roughly 40,000 cowpunchers representing the ethnic mosaic of America drove more than 9 million head of cattle 600 to 800 miles north from Texas to Kansas railheads at Abilene, Dodge City, and Wichita, where they boarded trains for slaughterhouses in Chicago. The beef industry, and its secondary leather products, grew so large that during this period, 1.4 million square miles, or 44 percent of the United States, was devoted to some form of cattle production. The cowboy and his companion horse became a hallmark of American (and Canadian) society and culture.

---

* Estimates for the peak population of feral horses in the United States between 1850 and 1860 range from two million to five million.

† Sweet mesquite bean pods are like candy to horses. Overeating, however, can cause severe digestive tract problems and even death.

During this same twenty-year span, roughly a million feral horses were also rounded up from Texas alone and shipped east to Chicago for resale and distribution. "The cowboy rode to glory," muses historian Frank Dobie in *The Mustangs*, "but *horseboy* never became a name." These horses fetched an average of $55 each in 1880 and $72 a decade later. Chicago became the largest horse market in the world, with 180,000 animals trading hands every year. On a good day, auctioneers at the Union Stock Yards, the largest in the city with more than 4,000 stalls, sold one horse every minute. As settlers poured into the West, surging horse, cattle, and human populations—in combination with wanton commercial killing—heightened survival pressures on tumbling bison herds.

Although numbers in Texas and Oklahoma were slipping by the 1830s, the opening of several prominent overland trails precipitated their fall. By the 1860s, bison populations across the Great Plains were collapsing at an accelerated rate. When the first wagon train left Independence, Missouri, in 1836, a furrowed wheel track had been cleared to Fort Hall, Idaho, eventually reaching the fertile river valleys of Oregon. The ease of passage for Conestoga wagons or converted flatbed prairie schooners (billowing white canvas tops looked like ship sails) was improved continuously, making the journey progressively both faster and safer.

The Oregon Trail, extending from Independence to Oregon City, was traveled by more than four hundred thousand migrant settlers during its peak years, from 1846 to 1869. In a single week in 1863, more than eight thousand wagon trains lumbered and creaked through Omaha, Nebraska. Farther south, the Santa Fe, Old Spanish, and Gila Trails spanned the southwest corridors of settlement and trade. American progress was stabbing westward.

On May 10, 1869, when our old friend Leland Stanford, president of the Central Pacific Railroad Company of California, drove in the last ceremonial golden spike of the Transcontinental Railroad at Promontory Summit, Utah, Manifest Destiny was immediately modernized, and outmoded overland wagon trails were rapidly overtaken.* The opening

---

* The last spike of the cross-country Canadian Pacific Railway was driven in 1885.

of the Transcontinental Railroad in 1869, the same year that President Ulysses S. Grant heralded his idealistic "peace policy" for the turbulent frontier, punched western settlement into hyperdrive.

Between 1870 and 1915, American railroad track mileage increased from 53,000 to 410,000. Train companies also offered lavish, all-inclusive "iron horse" vacations to the West. The sporting package included the unlimited slaughter of bison with newfangled long-bore rifles, which could kill at a greater distance even for amateur assassins. The Sharps rifle quickly became the favorite of "sharpshooters" and buffalo hunters, who could now exterminate one hundred bison an hour. In the winter of 1872–73, more than four hundred thousand bison hides were shipped through Dodge City alone.

The American population of the West increased from 550,000 in 1860 to almost a million a decade later. Successive Homestead Acts—beginning with the most comprehensive in 1862—provided pioneers (including women) who had never borne arms against the US with an allotment of 160 acres of land after paying a modest filing fee and fulfilling five years of continuous occupation.*

By 1890, marking the official "closing of the frontier" by the US Census Bureau (meaning that westward migration was no longer recorded), more than 160 million acres (250,000 square miles) of Indigenous land (nearly 10 percent of the US), was divvied up between 1.6 million settlers. The result was an agricultural explosion.

As the West was integrated into the US economy, the amount of land being cultivated increased from 408 million acres on 2.7 million farms in 1870, to 956 million acres on 6.4 million farms by 1910. Over this same forty-year period, however, agricultural employment declined from 52 percent to 27 percent. The drastic reduction in human labor was the product of larger horse teams hauling technologically efficient plows, seed drills, rakes, threshers, and harvesters. Mechanical innovations

---

* Passed by Congress during the Civil War, in addition to promoting settlement and enterprise, it was also believed that the enticement of free land would attract abolitionists, "free-soilers," and former slaves, thereby curtailing the western expansion of slavery.

propelled the United States, and its growing population of stalwart horses, into an agricultural powerhouse.

The census of 1840, the first to register draft animals, counted 4.3 million horses and mules. By 1910, this total had risen sixfold to almost 28 million, roughly twice the rate of human population growth.* By 1900, there was one horse for every three people in America, compared with a one-to-five ratio only fifty years earlier. "Census returns show that horse populations had increased 12 percent between 1840 and 1850, and 51 percent between 1850 and 1860," reports Ann Norton Greene in *Horses at Work: Harnessing Power in Industrial America*. "In five western states— Ohio, Indiana, Illinois, Michigan, and Wisconsin—horse populations had grown 106 percent overall, while the growth in individual states ranged from 61 to 380 percent."

Behind a single horse, standard or cast-iron moldboard plows could till an acre a day. Both models were finicky, prone to breaking, and collected mucky soil on the blades, interrupting operating performance. Although the blacksmith John Deere invented his superior steel plow in 1837 based on a discarded saw blade, it did not gain widespread use until the 1850s, when advances in metallurgy slashed steel prices. Now commercially viable, Deere plows were being manufactured in the steelworks of Pittsburgh at a rate of thirty-four thousand annually. They were lighter and stronger than cast iron, polished to allow sticky soil to peel away, and held an edge to cut through the dense root systems of prairie grasses and thick sod.

Correspondingly, the horse-drawn corn planter that was developed and refined between 1853 and 1860 enabled farmers to plant twelve to twenty acres per day, or roughly twenty times more than by hoe. By this time, the self-rake reaper, pulled by a two-horse team, could harvest ten to fifteen acres of wheat per day. The reaper was so common by 1860, it harvested 70 percent of American wheat. On factory-modeled "bonanza farms," established in the mid-1870s, massive combines pulled by any-

---

* Horses employed in British agriculture saw a slight increase from 966,000 in 1870 to 1.14 million in 1910, highlighting the established farming patterns of the British Isles compared with the newly turned soils of the American West.

Industrial farming: An immense harvesting machine pulled by a thirty-two-horse team, Spokane, Washington, circa 1900. *(Library of Congress)*

where from ten to forty horses cut and threshed twenty-five to thirty-five acres a day.

In 1866, Americans farmed 15.4 million acres of wheat, producing an average of 9.9 bushels per acre. By 1900, these numbers had risen to 42.5 million acres, averaging 12.3 bushels per acre. Between 1850 and 1900, American wheat exports increased fortyfold, from 5 million bushels to over 200 million. With the influx of homesteaders and their horses, the Great Plains were plowed and kneaded into endless wheatfields and the breadbasket of the world.

With the dual purpose of exploiting untapped western lands and assimilating Indigenous peoples, the government sanctioned an outlandish program for domesticating large native animals. The *US Agricultural Report of 1851* contained a section on the potential of domesticating bison, musk oxen, mountain goats, moose, elk, caribou, and deer "with especial reference to the economic employment of several species, as beasts of burden or draught, as furnishing food of excellent quality, or as yielding valuable materials for the useful arts." If effective, the report concluded, "these Indians might become pastoral people. . . . Much of this continent, now desolate, and supporting a scanty and half-starved population, may become a populous region, filled with towns and villages and owing much of its prosperity to the employment of some of our own native animals in a state of domestication." While this preposterous scheme never materialized, the government did sponsor the importation of a temperamental creature to help secure the West.

With its formal surrender in 1848, Mexico ceded over five hundred thousand square miles (55 percent) of its territory to the United States. Secured through unprovoked aggression and thinly veiled imperialism, this massive land grab included California, Nevada, Utah, Arizona, New Mexico, most of Colorado, smaller portions of Wyoming, Kansas, Oklahoma, and, of course, Texas. Military bureaucrats, including Jefferson Davis, the secretary of war under President Franklin Pierce, proposed the idea of using camels for patrol and transport across this Great American Desert.

Davis, the future president of the Confederate States of America, envisioned camels as the "gunships of the desert," doubling as a "camel express" postal service. The *US Agricultural Report of 1853* fervently promoted the acquisition of camels as "a matter of much national importance." New York financiers established the American Camel Company to secure the lucrative, and presumably imminent, military and mail contracts. In 1855, at Davis's insistence, Congress authorized $30,000 for the establishment of the US Camel Corps.

The following year, the first trial run using Arabian and Bactrian camels was deployed in Texas. *Scientific American* magazine proclaimed that the expedition would deliver "a fair prospect of the Arabian camel becoming a regularly naturalized and valuable American citizen." This declaration was a chimera. The teamsters, accustomed to more docile and cooperative horses, could not transfer their skills to ornery and obstinate camels and grew increasingly frustrated—and violent—with their charges.

With the Civil War looming, the flagging experiment was abandoned in 1861. The camels ended up as orphaned wards of the Lone Star State. Most died of neglect, abuse, and disease, while a handful escaped into the desert. The last verified camel sighting was in 1875, although spurious claims lasted into the 1930s, as these fugitive creatures entered the folklore of the American West.

Effective communication across the United States, which fulfilled its Manifest Destiny at the expense of Mexico, was established with the legendary (although hardly unique or innovative) Pony Express and countless stagecoach lines. During the 1850s, for example, Texas counted at least thirty-one stagecoach companies. To feed the frenzy of the "forty-niner" gold rush, the California Stage Company employed 134 coaches and 1,000 horses across 28 daily lines covering 2,000 miles.

The Butterfield Overland Mail Company, financed by Wells Fargo, shuttled passengers and correspondence across a 2,795-mile U-shaped route (to skirt the mountains) from San Francisco in the West, across the Southern Plains, to its two eastern termini of Saint Louis and Memphis. "It is a glorious triumph for civilization and the Union," wrote President James Buchanan to John Butterfield following the first complete run in 1858. "Settlements will follow the course of the road, and the East and West will be bound together by a chain of living Americans that can never be broken."*

A reporter who survived the maiden voyage had a slightly different opinion: "Had I not just come over the route, I would be perfectly willing to go back, but I know what Hell is like. I've just had 24 days of it." The only saving grace was that at least it didn't take twenty-five days, the maximum deadline.

To fuse the fortunes of East and West, the federal government paid Butterfield $600,000 a year to deliver US mail across the entire itinerary. Passengers supplemented profit by paying $100 for a one-way trip. With the cramped coaches stopping only long enough to change out horses and drivers and take on supplies at designated depots, travelers were forced to remain on board for most of the trip. "Pulling up to a Butterfield stage station," remarks historian Gerald Ahnert, "was like making a NASCAR pit stop." Despite these measures, the twenty-five-day mail

---

* Butterfield, along with Henry Wells and William Fargo, founded the American Express Company in 1850 and Wells, Fargo & Company two years later.

delivery was considered slow even for the time. Utilizing mounted relay messengers like those of ancient empires was originally deemed too costly and vulnerable.

To improve distribution speeds and service the growing population of California, which was nearing four hundred thousand, William Russell and two associates set up the Pony Express in April 1860. Operated by the Central Overland California and Pikes Peak Express Company, it duplicated the same, now familiar, dispatch rider postal system unveiled by the Assyrians and Persians.

Leaving Sacramento, the 1,965-mile route bent slightly north (opposite the Butterfield), heading through Salt Lake City and Laramie, Wyoming, before reaching its eastern terminus at Saint Joseph, Missouri. A courier covered a leg of 75 to 100 miles, switching to a fresh horse every 10 to 15 miles at one of 190 stable stations, before transferring the mochila, a leather satchel containing twenty pounds of mail, to a new messenger.

A dashing fifteen-year-old named William Cody is purported to have made the longest nonstop ride. Joining the Pony Express on his way to the gold fields in California, he covered 322 miles on twenty-one horses in 21 hours and 40 minutes. Cody seemed to be everywhere during the glory days of the Wild West. He was a trapper and "fifty-niner" miner in Colorado, a teamster for the Union during the Civil War, and a Medal of Honor recipient. He served as a scout for the US Cavalry, Othniel Marsh, and Grand Duke Alexei Alexandrovich of Russia. He also slaughtered 4,282 bison during an eighteen-month period spanning 1867 and 1868 as part of a contest to supply meat to railway construction crews, earning him his immortal nickname Buffalo Bill.

At a time when a decent horse cost $40, those of the Pony Express ran upwards of $200 in order to ensure the fastest and most efficient mounts. Averaging ten days per trip, the Pony Express shaved fifteen days off the Butterfield delivery time. The three fastest runs (all seven days) brought word of Abraham Lincoln's presidential election victory in November 1860, his inaugural address in March 1861, and the opening clash of the Civil War at Fort Sumter the following month. The exorbitant cost to mail a half-ounce letter was $5 ($150 today), dropping to $1 by the end of its fleeting eighteen-month stint.

The first transcontinental telegraph system, completed by the Western Union Telegraph Company, sent its inaugural message on October 24, 1861, terminating the short gallop of both the Butterfield Company and the Pony Express. The bankrupt Overland defaulted to Wells Fargo, which consolidated numerous mail and stage lines under its brand. As the largest stagecoach operator during the 1860s and 1870s, Wells Fargo's iconic red-and-gold coaches delivered people, supplies, precious metals, and currency across the West.

After millennia shuttling messages across the greatest empires in history, dispatch riders and their hardy mounts gave way to progress, disappearing from global landscapes as obsolete relics of the past. Despite its extremely brief and bankrupt tenure, the Pony Express stands as an enduring symbol of the American frontier.

———

The reintroduction of the horse to the western United States was part of a larger convergence—or collision—of cultures. It should not come as a surprise how quickly the horse became the socioeconomic heartbeat of numerous Indigenous nations. We have seen similar equestrian models, from the Indo-Europeans, Assyrians, and Scythians to the Xiongnu, Huns, and Mongols. The benefits and transformational capabilities of the "Big Dog" for trade, travel, hunting, and war were too obvious to be overlooked.

It was a commercial and military machine that allowed for the exploitation of the vast untapped grasslands of the Great Plains. Think back to those remarkable and specialized anatomical features that the horse acquired during its evolutionary arms race with grass. "Horses, in a purely mechanical sense, served their masters as finely tuned organisms that converted inaccessible plant energy into tangible and instantly exploitable muscle power," explains Pekka Hämäläinen in *Lakota America: A New History of Indigenous Power.* "It was an extraordinary shortcut that redefined the limits of the possible."

Part of this wholesale equine transformation was the substitution of the heavily utilized travois dog with the horse. As the Piikani elder Saukamappee recounted, "[We went] to see a horse of which we had heard so much. . . . Numbers of us went to see him, and we all admired him. He put us in mind of a Stag that had lost his horns, and we did not know what name to give him. But as he was a slave to Man, like the dog, which carried our things, he [the horse] was named the Big Dog." To the Assiniboine, horses were Great Dog; to the Arapaho, Blackfoot, and Cheyenne, they were Sky Dog and Elk Dog; to the Lakota, they were Holy Dog and Medicine Dog; to the Comanche, Magic Dog; and Seven Dogs to the Tsuut'ina of Alberta.

The Siksika of Alberta/Saskatchewan retell the story of a warrior named Sits-in-the-Night, who led a raiding party south to steal Shoshone and Crow horses. When they arrived home with their fascinating new creatures, the inquisitive audience inched closer, only to run away every time the horses moved or snorted. Finally, one female elder, succumbing to her unbridled curiosity, bravely clutched the reins and declared, "Let's put a travois on one of these big dogs, just like we do on our small dogs."

Attached to a travois, a horse could pull four times more weight than a dog, and twice as far. Tepees constructed from forty buffalo skins replaced the standard six-skin lodge, facilitating larger political assemblies, combined bison hunts, and allied military camps. This transport capacity also led to a greater accumulation of material wealth and the damaging slide toward a tiered, mercantile socioeconomic system. "Most significant," says Hämäläinen, "horses opened a more direct way to tap energy. Dogs, Indians' only domesticated animals before horses, used the Plains' greatest energy source—the grasses—only indirectly, by consuming the meat of grass-eating animals their owners acquired for them, whereas horses drew their energy directly from the grasses, bringing their masters one step closer to the ultimate energy source." Horses embodied the harnessed energy of this solar power.

The reaction and perspective of the plains peoples to these strange new horses were shared by Aboriginal Australians and the Maori of New Zealand. When the earliest horses swam ashore to New Zealand from British boats, Maori onlookers, perhaps apocryphally, assumed they were *taniwha*: water monsters. Horses, however, were commonly referred to as *kuri* ("dog"), *kuri waha tangata* ("people-carrying dog"), or *hoiho*, a Maori rendering of the word *horse*.

Aboriginal Australians also reacted to horses as overgrown dingoes. "What sort of creature is this?" relates an Aboriginal oral tradition. "I wonder where this animal has come from, it's so big." Others believed, like the Purépecha of Mexico, that horses could talk. Some thought the horse and rider were one centaur-like creature. In the arid desert of the interior—which makes up the bulk of Australia, as it is the driest continent on the planet—Aboriginal peoples were bewildered by how much water these "giant kangaroos" consumed. Aboriginal rock art depicts horses with a distinctly marsupial silhouette.

While possessing a unique assortment of animals, including marsupials, Australia was devoid of any purely earthbound placental mammals save rodents, which first washed up roughly six million years ago. (Platypuses and echidnas lay eggs.) Humans arrived around sixty thousand years ago, with dogs coming ashore by 2000 BCE.* On the eve of permanent British settlement in 1788, the total Aboriginal population stood between 550,000 and 750,000, divided into 554 distinct nations of semi-nomadic hunter-gatherers.

Australia followed the predictable progression of colonization, albeit in more rapid succession than other settler states. While the end results were the same, the British invasion of Indigenous Australia was more analogous to the rapid conquests of Cortés and Pizarro over the Aztec and Inca than it was to the more prolonged Spanish or American overthrow of the Mapuche, Apache, Comanche, and Lakota.

Europeans, disease, and horses swiftly occupied Australia. For Aboriginal peoples, there was no lag time to recover between incessant epidemics and the influx of Europeans set against the backdrop of the Frontier Wars: festering, low-intensity skirmishes and massacres spanning the century after contact. For the First Peoples of Australia, it was an immediate deluge of displacement, disease, and death. With the Federation of Australia in 1901, the Aboriginal population reached its nadir at roughly 94,000 survivors, scratching out a bleak existence on isolated reserves or in reprehensible state-sponsored "Stolen Generation" settlements and boarding schools.

---

* Over 80 percent of Australia's mammals, reptiles, frogs, and flowering plants are distinct to Australia, along with nearly half its birds and almost all its freshwater fish.

When the original seven horses struggled ashore with the First Fleet of 1,400 convict-settlers at Botany Bay (just south of modern Sydney) in 1788, Australia kicked off its own Columbian Exchange with the core-imports-turned-commercial-exports of cattle, sheep (wool), and wheat.* Shipments of horses—usually picked up from the Cape of Good Hope in South Africa, where both trekking Boer settlers and horses were first landed by the Dutch East India Company (VOC) in 1652—arrived with each new deposit of free or felon immigrants.†

Feral horses, known as brumbies, spread across the island faster than Europeans themselves. These invasive animals were quickly regarded as pests. They munched and trampled farm fields and consumed valuable water and grass vital for the economic enterprises of cattle and wool sheep. Brumbies, complained an early settler, made a habit of blazing past and spurring tame horses to run off with them, "leaving their owners to chew the cud of mortification."‡ Referred to as "a very weed among animals," by 1830, feral herds spanned the entire continent.

Within eighty years of the first landing, beachhead farming and ranching settlements ringed the entire island, and its north-south axis had been established. With their knowledge of local environments, many Aboriginal peoples certainly made use of the horse and engaged in equine activities. Skilled Aboriginal horsemen found employment on sheep and cattle stations and as mounted scouts in police forces, both of which also doubled as the spearheads of Euro-Australian penetration into the interior.

There was not, however, a comprehensive horse-culture conversion, even regionally, like those in the Americas. "The rapidity of pastoral expansion," stresses acclaimed historian Henry Reynolds in *The Other Side of the Frontier: Aboriginal Resistance to the European Invasion of Australia*, "precluded the possibility of a gradual acquisition of the techniques of horsemanship." In places like the desert interior, where British incursion

---

* Donkeys were introduced to Australia in 1793 from India.

† The administrator of the VOC, Jan van Riebeeck, proclaimed a year later, in 1653, that for his colonial outpost to survive and succeed, horses were "as necessary as bread in our mouths."

‡ While the origin of the term *brumby* is debated, it likely stems from the Aboriginal Bidjara word *baroomby*, meaning "wild," similar to the word *mustang* from the Spanish *mesteño* (also "wild").

was somewhat delayed, the harsh landscape prevented Aboriginal peoples from adopting horses. "Aboriginal Australians, in sum," adds Peter Mitchell, "had little time and few, if any, places in which equestrian societies could develop *and* be maintained."

Similarly, New Zealand was devoid of all mammals (save bats) when its first peoples, the Maori, arrived around 1300 from the Cook Islands of Polynesia. The Maori quickly adopted a mixed subsistence of farming, fishing, and hunting-gathering. By 1870, a century after British captain James Cook's troubled visit, New Zealand was home to 80,000 horses, 400,000 cattle, 9 million sheep, and 300,000 Pākehā (non-Maori) peoples. The Maori population numbered around 47,000, down from an estimated 110,000 prior to contact. By this time, disease and intense conflict, including the Musket Wars (1814–33), which pitted Maori against Maori, and the British-versus-Maori New Zealand Wars (1845–72), had run their usual course of colonization.

The first horses to New Zealand landed at the Bay of Islands in 1814. Maori *rangatira* (chieftains) immediately obtained horses as gifts and bribes, and they quickly spread through Maori communities across the North Island. Initially a status symbol, as horses became more common, they were used for standard traction occupations. Like Indigenous cultures in Australia, sub-Saharan Africa, Southeast Asia, Indonesia, and the tropics of the Americas, the Maori did not morph into an equine culture. This conversion was simply not compatible with the landscapes and climate of New Zealand or their cultural roots as Pacific Islanders.

The naïve Eurocentric (or perhaps equine-centric) notion of the horse as the sole beneficial and uplifting element of the Columbian Exchange for Indigenous peoples—fitting apologetically into narratives otherwise filled with disease, desolation, and death—is not accurate. According to Pekka Hämäläinen:

> Virtually all modern histories portray the rise of the Plains Indian horse culture as a straightforward success story. According to this view, horses spread northward from the

Spanish Southwest, repeatedly creating a frontier of fresh possibility, opening for each tribe in its path a new era of unforeseen wealth, power, and security. . . . That success story has a bleak undercurrent that went largely unnoticed until recently, when ground-breaking studies shed light on the harmful effects of horses on Plains Indians socioeconomic systems and the environment. Horses did bring new possibilities, prosperity, and power to Plains Indians, but they also brought destabilization, dispossession, and destruction.

The horse was a double-edged sword, and as the Xiongnu found out, ultimately bit the hand that feeds. The Cheyenne creator deity Maheo cautioned about this exact duality when contemplating the adoption of a horse culture: "If you have horses, everything will be changed for you forever. You will have to move around a lot to find pasture for your horses. You will have fights with other tribes, who will want your pasture land or the places you hunt. Think before you decide."

This prudent counsel was echoed by a Kainai war chief with the memorable name Buffalo Bull's Back Fat, who warned his people in the early nineteenth century: "Don't put all your wealth in horses. If all your horses are taken from you one night by the enemy, they won't come back to you. You will be destitute. So be prepared. Build up supplies of fine, clean clothing, good weapons, sacred bundles, and other valuable goods. Then, if some enemy takes all your horses, you can use your other possessions to obtain the horses you need." A wise, still-pertinent lesson to always diversify your portfolio.

The adoption of the horse had severe and enduring repercussions, contributing directly to the ruin of plains nations. The introduction of invasive horses upended long-standing and intricately evolved grassland ecologies, contributing directly to the collapse of bison populations. It also undermined traditional military, economic, egalitarian, gender, and social conventions. These were replaced with bison-based export capitalism; divisive social, gender, and economic hierarchies; and constant intertribal raiding and warfare. Paternalistic American and Canadian departments of Indian Affairs pounced on this discord as potent instruments of divide-and-conquer subjugation.

The southern horse nations, for example, created the conditions for their own eventual downfall. On the eve of the US invasion of Mexico in April 1846, the borderlands of Texas and northern Mexico had been largely vacated. Unremitting Comanche, Kiowa, and Apache raids had driven settlers from these regions. "Scarcely has a hacienda or rancho on the frontier been unvisited, and everywhere the people have been killed or captured," recorded an eyewitness to the devastation. "The roads are impassable, all traffic is stopped, the ranchos barricaded, and the inhabitants afraid to venture out of their doors." As Hämäläinen points out, "The destruction proved critical during the Mexican-American War; through their power politics, Comanches and Kiowa had inadvertently paved the way for the United States' takeover of the Southwest."

These militaristic equestrian nations now faced the twin threats of dwindling bison herds and the new combative power of the United States, which was hell-bent on westward commercial expansion. "Indeed," states Mitchell, "the destabilization that Comanches—and also Apaches—wrought there after Mexico's independence arguably paved the way for the American conquest of Arizona and New Mexico in 1846 by weakening the Mexican state's infrastructure across its northern provinces and undermining its citizens' allegiance. . . . American takeover of the Southwest and of Texas in the mid-1840s replaced two weak states (Mexico and Texas) with a much stronger and more determined one, equipped, as it turned out, with an aggressive policy of expansionary settlement for both ranchers and farmers." Eventually, every nation, Indigenous or otherwise, buckled under the rolling weight of American economic imperialism.

Indigenous peoples eagerly entered this capitalist economy by actively supplying bison hides and other furs in exchange for firearms and other commodities. "The fur trade exposed them to market fluctuations and alcohol, and threatened to reduce them to debt and dependency," explains Hämäläinen. "The real threat of market penetration was at once subtler and more profound: the fur trade helped change the relatively egalitarian tribes into highly stratified rank societies." Horses circulated as the common currency within Indigenous economic, political, and social contracts.

Marriage practices serve as a poignant example of horse wealth dislodging traditional mores by establishing imposed social hierarchies. Men who acquired more horses could harvest more bison to trade for more goods to obtain more horses. Some individual men owned upwards of three hundred or four hundred horses, creating a concentration of capital. These men could afford higher bridewealth—a dowry presented to a woman's family to transfer membership of a woman and her children from her family or clan to that of the husband.

Fathers were more than happy to offload their young daughters, as relentless raiding and protracted Horse Wars incessantly sheared male populations. It is estimated that females comprised 65 percent to 75 percent of numerous plains nations, leading to large, polygynous households for horse-wealthy men. Red Cloud, the famed Lakota leader and friend of Othniel Marsh, for example, paid twelve high-end horses for his first wife, Pretty Owl. By the time he was twenty-four, he had acquired enough horses to afford six wives.

Breaking with traditional matriarchal structures and defined gender norms, extra wives were essentially slaves or servants who maintained laborious horse herds and processed bison hides and meat. Unlike primary wives (usually the first three), these women had no social or political standing. They were dressed in tattered clothes to mark their inferior station, had no possessions, and were frequently controlled by violence. Since many girls married immediately upon the onset of menstruation, they had a far greater risk of dying in childbirth. "Exploited, controlled, and hoarded by male elites," remarks Hämäläinen, "the extra wives were considered less companions than instruments of production." These women enriched horse herds and bison exports through their labor and the fruits of their loins, which included the collection of bridewealth from female offspring. "Some men," recounted Comanche elder Post Oak Jim, "loved their horses more than they loved their wives."

These connections were not lost on contemporary fur traders. "It is a fine sight to see one of those big men among the Blackfeet," wrote

Charles Larpenteur in 1860, "who has two or three lodges, five or six wives, twenty or thirty children, and fifty to a hundred horses." He goes on to mention that a Kainai chief, Seen From Afar, possessed "10 wives and 100 horses." James Schultz, a fur trader who lived among the Piikani of Montana from 1879 to 1882, wrote, "Horses were the tribal wealth, and one who owned a large herd of them held a position only to be compared to that of our multimillionaires. There were individuals who owned from one hundred to three and four hundred." Chief Many Horses, for example, owned about five hundred horses.

Males with few or no horses had difficulty securing wives (and bison and horses). These deprived men were also perceived as cowardly and craven, since horses were procured through raiding.* They were marginalized from the hierarchies of tribal politics and power, signaling a clear shift away from customary democratic and collective principles. "Those with few horses found themselves materially and socially impoverished in societies that were becoming less egalitarian," states Mitchell, "while in many cases horses also became essential for bridewealth payments when men sought wives.... Horses favoured the development of social inequalities."

The Kiowa, for example, had an explicitly tiered power structure based on the number of horses a man possessed: elite (dozens or hundreds of horses), middle class (ten to twenty animals), and poor (no horses and had to work, hunt, and raid for others). For the Crow nation, the recalculated social status of a wife was determined by how many horses were remunerated to her father. Fines and punishments were also exacted through horses. These cultural deviations created institutional castes of wealth and prestige, leading to internal dissent and rivalry. American and Canadian authorities then effectively utilized this dissension as a tool of divide and conquer, resembling the Chinese strategy that splintered and subjugated the Xiongnu.

This cultural upheaval was concurrent with devastating disease. "Played out in a highly labile system of shifting alliances, horses' contribution to the eventual restriction of all Plains peoples to reservations should not

---

* The battle scars of horses (always painted red) and the "pat handprint" on the right hip were tributes bestowed upon mounts who brought their masters home unscathed. These equine battle honors were also visual reminders of the bravery of their riders.

be underestimated," writes Mitchell. "Nor should that of epidemic disease." As with the Black Death, the horse accelerated the delivery of pathogens across the plains. Disease found a thriving market for transmission among the horse, bison, and fur trade.

A smallpox outbreak between 1779 and 1782, for example, claimed half the Comanche population—perhaps as many as fifteen thousand to twenty thousand people. The same virus dispatched another four thousand in 1816. This was just the opening deluge on a generational merry-go-round of disease and death.

The Great Plains smallpox epidemic of 1836 to 1840 rippled through the Missouri River delta before seeping south, killing upwards of 20,000 (with some estimates reaching 35,000) Indigenous people. "I Keep no a/c of the dead," wrote Francis Chardon of the American Fur Company, "as they die so fast it is impossible." The Mandan were virtually annihilated, with a mere 125 scarred survivors remaining from a population of 3,600. Other nations, including the Hidatsa, Crow, Arikara, Blackfoot, Pawnee, and Assiniboine, suffered death rates between 35 percent and 65 percent of their total populations. "The destroying angel has visited the unfortunate sons of the wilderness with terrors never before known," wrote an English fur trader, "and has converted the extensive hunting grounds, as well as the peaceful settlements of these tribes, into desolate and boundless cemeteries."

Eight years later, a second round of smallpox complemented by cholera once again halved the Comanche and Kiowa populations. Miners flocking to the Colorado gold rush set off a series of overlapping epidemics of smallpox, measles, scarlet fever, whooping cough (pertussis), and cholera between 1861 and 1867 that sliced through the Ute, Comanche, Kiowa, Cheyenne, Arapaho, and Lakota.

In summary, between 1733 and 1870, an epidemic of smallpox (often escorted by another pathogen) ripped across the plains every ten to twenty years. Disease was a leading factor in the demise of waning Indigenous populations and their ability to resist escalating US military campaigns and the surge of land-hungry settlers spurred by the Homestead Acts. The rapid and intense American occupation of the West also overwhelmed exhausted bison populations.

Strategic state-sponsored slaughter by assassins such as Buffalo Bill to starve Indigenous peoples, eroding and shrinking habitat, competition for resources with horses and cattle, sport shooting, and overhunting all contributed to the rapid extermination of the bison. During the punitive Indian Wars in the wake of the Civil War, American forces also systematically impounded or butchered Indigenous horse herds. This severely undermined their ability to hunt, trade, and fight. By the official closing of the frontier in 1890, coinciding with the massacre of Lakota ghost dancers at Wounded Knee, bison estimates in the United States ranged from only three hundred to five hundred, with another fifteen hundred to two thousand sheltering in Canada.*

The wanton eradication of the bison undercut the nomadic subsistence and socioeconomic base of the plains nations. "That made the Euro-American military takeover virtually effortless," asserts Pekka Hämäläinen. "Exhausted by starvation, disease, and decades of fighting, the northern tribes could rally only weak resistance against the encroaching Americans and Canadians. By 1877, after only a few fights with the U.S. Army, all northern tribes were confined to reservations on both sides of the forty-ninth parallel."

While the resistance of nomadic horse nations such as the Apache, Comanche, and Lakota lasted longer than that of their sedentary farming counterparts, eventually they all shared the same fate. To save their starving and destitute people, leaders had little choice but to sign treaties, surrender their lands, and accept the reservation. Indigenous peoples entered the destructive era of assimilation, poverty, disease, prejudice, and cultural decay.

During the intermittent Apache Wars between 1849 and 1886, famed leaders Mangas Coloradas, his son-in-law Cochise, Victorio, and Geronimo confronted settler incursion and both Mexican and American military forces. With the surrender of Geronimo and his handful of

---

* The Canadian North-West Mounted Police, now recognized as the Royal Canadian Mounted Police (RCMP), was established in 1873 as a paramilitary police force to patrol the western frontier and US border. Its primary objectives were to maintain peaceful relations between Indigenous peoples and settlers, and to apprehend unlawful and aggressive American bison hunters and whiskey traders.

followers in 1886, however, the remaining Apache were consigned to reservations scattered across New Mexico, Arizona, and Oklahoma.

Ravaged by epidemics, the Comanche population followed the trend of their precious bison, plunging from forty thousand during the 1770s, to twenty thousand in the 1820s, to fewer than five thousand on the eve of the Civil War. When the Antelope band under Quanah Parker finally capitulated at Fort Sill, Oklahoma, in 1875, only fifteen hundred horses were left to surrender from a herd that had numbered over fifteen thousand a decade earlier. Following the Comanche Wars (1836–77) with the US Army, the total number of Comanche and Kiowa registered on reservations in Indian Territory barely reached three thousand. Other plains nations, such as the Nez Perce, Ute, Cheyenne, Arapaho, and Lakota, fought to preserve their fading nomadic lifestyles against settlers and soldiers on the final frontier that had been pushed to its limits.

After a series of battles in a long-running conflict with the US Army, remembered for names like Red Cloud, Black Kettle, Crazy Horse, Sitting Bull, and Custer, the Lakota and their Cheyenne and Arapaho allies finally lost the Sioux Wars (1854–1890). "A cold wind blew across the prairie when the last buffalo fell," lamented Sitting Bull, "a death wind for my people." The last of the horse nations, which, as a collective, lasted no more than two hundred years, joined the Apache, Comanche, Shoshone, and Crow on reservations under the paternalistic watch of the US Bureau of Indian Affairs.

To facilitate agricultural production on so-called idle Indian lands, the Dawes Severalty Act of 1887 allotted select parcels of reservation land to Indigenous men. The remainder was auctioned off to settlers, creating a checkerboard pattern of tribal and non-tribal lands. When two million acres of Indigenous land were made available in Oklahoma on April 22, 1889, more than fifty thousand settlers raced to claim plots— *on this single day*—epitomizing the "land rushes" of the West. As a result of the Dawes Act and subsequent land grabs, by 1900, Indigenous peoples had lost 66 percent of territory that had been in their possession only fifteen years earlier.

To solve the "Indian Problem" and promote assimilation, Indigenous children were removed forcibly from their remaining lands and shipped to callous Indian boarding schools that cruelly intended to "kill the Indian and save the man." In Canada, cadaverous and grisly Indian Residential Schools carried out this same inhumane mandate. Thousands of stolen children and traumatized generations of permanently scarred survivors, many coping with substance abuse problems, are the painful legacies of these "civilizing" and "enlightened" policies.*

In 1928 Lewis Meriam of the US Department of the Interior submitted an exhaustive and meticulously researched nine-hundred-page assessment on the condition of Indigenous peoples in the United States. The venerated mounts of the once-mighty plains nations received token mention under a paragraph entitled "Worthless Horses." As you might expect, the *Meriam Report*, as it came to be known, also presented a stark picture of deprivation and despair in the wake of the Dawes Act. In 1934 the commissioner of the Bureau of Indian Affairs, John Collier, described Indigenous society as "physically, religiously, socially, and aesthetically shattered, dismembered, directionless."

The horse was partly responsible for this heartrending portrayal and tragic situation. As the Columbian Exchange rolled to a close across the reconfigured and bison-less prairies, Indigenous peoples, besieged by poverty and prejudice, attempted to preserve what was left of their beloved lands and fading cultures. America was now *their* New World.

As an advocate for both Indigenous peoples and women, Buffalo Bill Cody handsomely employed Geronimo, Chief Joseph, Red Cloud, Sitting Bull, Calamity Jane, and Annie Oakley in his Wild West show. "What we want to do is give women even more liberty than they have," he argued. "Let them do any kind of work they see fit, and if they do it as well as men, give them the same pay." He described Indigenous peoples as "the former foe, present friend, the American" and reasoned that

---

* In both the US and Canada (as well as Australia), Indigenous children suffered unconscionable emotional, physical, and sexual abuse at the hands of the teachers, pastors, nuns, and priests supervising these institutions. In 1919, for example, 389 Indian Residential and Industrial schools were operating in Canada, primarily in the West, accommodating 12,413 children. Between 1894 and 1908, mortality rates in Western Canadian schools ranged from 35 percent to 60 percent over a five-year period (meaning that five years after entry, 35 percent to 60 percent of children had died of disease, neglect, starvation, abuse, or suicide). The conditions in the twenty-five federally funded American schools housing more than 6,000 kids in 1902 mirrored those in Canada.

"The former foe, present friend, the American": Sitting Bull and William "Buffalo Bill" Cody, 1885. *(Library of Congress)*

"every Indian outbreak [war] that I have ever known has resulted from broken promises and broken treaties by the government." Cody's life and legacy transcend centuries. He has also been the subject of over twenty-five movies, beginning with the 1926 silent film *With Buffalo Bill on the U.P. Trail.*

When Othniel Marsh, who had once hired a young William Cody as a scout during the Bone Wars, died in 1899 at his home near Yale University, the world was in flux. The preeminent animal that Marsh and Thomas Henry Huxley had painstakingly traced back to the fifty-seven-million-year-old *Hyracotherium* had been reintroduced to its original home on native land.

The modern horse gave birth to Hollywood, thanks to a wager by Leland Stanford and his galloping filly Sallie Gardner. For better or for worse, it also changed the course of history across its evolutionary cradle of the Great Plains and scripted the fates of the subjects of the first movies: the American bison, Indigenous peoples, and pioneering settlers. By the closing of the frontier at the turn of the twentieth century, the future of American progress was electric with opportunity, industry, and invention.

Horses powered farming across the planet, feeding a surging global population. They carried these same jingoistic masses across the blood-soaked battlefields and unsurpassed carnage of two world wars. In bustling metropolises such as London, Paris, and New York, horses were the lifeblood of commerce, communication, construction, and public transport, while providing the backbone for postal delivery, police forces, firefighters, ambulances, and other services.

In New York, horses shuttled spectators on carriages and omnibuses past designer boutiques, extravagant restaurants, swanky athletic clubs, the Polo Grounds baseball stadium—and massive piles of festering manure—to silent movies flashing the last glimpses of the vanishing frontier. The screens twitched with scenes from once upon a time in the West with a cast of characters from a bygone, antiquated era that would soon include the horse. Its final curtain call, between 1870 and 1920, however, was also *the* Age of the Horse.

CHAPTER 15

Spiritual Machines
The Supremacy of the Horse

New York City was sinking in shit. The metropolis and its three and a half million residents were being swallowed by the surging contamination and accompanying communal health problems produced by their indispensable personal, public, and commercial transportation. Instead of greenhouse gases and climate change, in 1894 the mounting pollution crisis took the shape of unimaginable piles of rancid manure produced by more than two hundred thousand urban horses and mules, prompting *The New York Times* to brand it "Stable City."

While the "Great Manure Crisis of 1894" was gaining steam, the Chicago World's Fair—convened in 1893 to honor the four-hundredth anniversary of the arrival of Columbus—was celebrating American progress and civilization. The sanitary splendor of the modern urban era was symbolized by its aesthetic White City centerpiece. The architects of the fair were so concerned about maintaining their facades (literally made from white "stucco" staff) of "America the Beautiful" that they declined to host Buffalo Bill's Wild West show. Spurned but not defeated, he set up camp immediately outside the official grounds, raking in massive amounts of revenue without giving a cut to the fair. While the World's Columbian Exposition, as it was formally called, promoted the impression that "thine alabaster cities gleam," the putrid sights and sordid smells of reality in Chicago, Paris, London, Toronto, and New York told a very different story.

One commentator complained that the squalid streets of Gotham

The White City: Two hardy Michigan horses hauling a mountain of logs bound for the construction of the Chicago World's Fair, February 1893. *(Library of Congress)*

were "literally carpeted with a warm, brown matting of comminuted horse dropping, smelling to heaven and destined in no inconsiderable part to be scattered in a fine dust in all directions, laden with countless millions of disease-breeding germs."* A typical urban traction horse produced roughly 30 pounds of manure and 1.5 gallons (12.5 pounds) of urine per day. By the time of the Manure Crisis, New York City received upwards of 3,000 tons of manure a day!† "With the exception of a few thoroughfares," complained a newspaper editorial, "all the streets are one mass of reeking, disgusting filth, which in some places is piled to such a

---

* Gotham, a nickname for New York City, was first used in 1807 by author Washington Irving, known for the short story "The Legend of Sleepy Hollow," with its enduring character the Headless Horseman.

† This is roughly the *combined* weight of your favorite NFL football team, the Statue of Liberty, one Blue Whale, the total amount of gold mined in South Africa in a year, and all the garbage produced in Baltimore in a day.

height as to render them almost impassable." Urban centers around the world were facing the same shitty situation.

The forty-five largest US cities had an average density of 426 horses per square mile. Milwaukee, home to 350,000 people, topped the list at 709. Its 12,500 horses produced 185 tons of feces a day. In 1894 the 8,500 equine Torontonians tendered 128 tons of excrement per day, while the 6 million residents of London were bombarded by a daily output of 4,500 tons from 300,000 horses. Urban English working horses were producing more than 10 million tons of manure every year.

Street sweeping during the intensity of working-day traffic was next to impossible. Nightly sanitary squads generally shoveled equine waste off to the side, where it stacked up and lined the streets and buildings like snowbanks. "Many scraps of waste land in the poorer quarters of towns were turned into vast dung heaps," explains acclaimed English social historian Francis Thompson, "considerably aggravating the squalor, stench, and unhealthiness of such parts of the urban environment." Vacant lots, public parks, and alleyways accumulated heaping piles of manure reaching sixty feet.

Futile fleets of wagons were used to cart out the waste. Of course, these were pulled by horses—who produced the very ordure they were picking up! The Boston City Scavenger Department, for example, employed over 500 horses pulling 120 wagons, 14 cesspool cleaners, 9 street sweepers, 6 watering carts, and 175 carcass and manure sleds. In London 1,500 horses dubbed the Heavy Brigade were tasked with the nightly duty of hauling off dung as they donated their own. Small amounts were shipped and sold to local farmers, but mass transport outside the immediate rural ring was neither feasible nor cost-effective. Moreover, farms produced their own livestock manure, and urban supply far outpaced rural demand.

According to Thompson, the scenes of mid-nineteenth-century London portrayed by novelist Charles Dickens in *Oliver Twist* and other tales represented an "'excremental vision' of society as a gigantic choked up sewer, of a preoccupation with ordure, filth, and bowel movements." Battling its own Manure Crisis in 1894, *The Times* of London predicted that "In 50 years, every street in London will be buried under nine feet of

The Great Manure Crisis: Sidewalks separate pedestrians and brownstone row houses from towering piles of rancid manure lining Varick Street, New York City, 1895. *(Museum of the City of New York)*

manure." The cities of the world were being submerged under mountains of manure, and their residents were at the mercy of escorting disease.

During spring thaws and rainy seasons, streets turned into thick sludges of mud, manure, urine, and other refuse that seeped into basements. New Yorkers grumbled that in a few more years, the accumulated manure would be so high that the fourth story of buildings would become the first floor. "Streets in the lower part of the city," observed *Scientific American*, "are completely blocked three or four days of the week." The trendy New York brownstone row houses with their elegant staircase-stoops climbing to second-story parlors were purposefully built to rise above the streaming seas of horse manure and urban muck.

Passing vehicles, one Londoner pointed out, "would fling sheets of such soup—where not intercepted by trousers or skirts—completely across the pavement." Ankle boots were worn by all classes as shields from this slurry of excrement, while the affluent wore long, ankle-length "splashguard" outer garments. These styles were not dictated solely by a priggish, puritanical society to stifle sexual arousal at the sight of an uncovered ankle, as has often been assumed. The reasons for these modifications were far more pragmatic and based on good ol'-fashioned horse sense.

During dry spells, reported one New Yorker, "any little puff of wind filled the air with powdered horse manure that settled on the passerby and had to be wiped from the eyes and lips," creating the perfect storm for transmitting disease. A health official in Rochester, New York, for example, calculated in 1900 that the "15,000 horses in that city produced enough manure in a year to make a pile covering an acre of ground 175 ft [feet] high and breeding sixteen billion flies." The US Department of Agriculture's Bureau of Entomology estimated in 1895 that horse manure provided the breeding grounds for 95 percent of the urban fly population.

Other disease-vectoring rodents and insects were attracted to the dung heaps, igniting deadly seasonal epidemics of malaria, yellow fever, and typhus. These clouds of insects were snacked on by swelling flocks of urban birds—especially proliferating swarms of English sparrows, who also splashed their poop across urban spaces and faces. This simmering zoological stew of fecal matter also fed rampant outbreaks of lethal water-borne afflictions, including cholera, typhoid, dysentery, giardia, *E. coli*, and assorted parasitic worms.

In 1894, as manure piled up across fourteen inner-city dump sites, New York City health inspectors reported higher rates of infectious disease "in dwellings and schools within fifty feet of stables than in remoter locations." The Broadway and Seventh Avenue Street Railway stable, the largest in the world, housed 2,500 horses across four floors. A fire insurance map of Philadelphia marks 175 individual stables within only 130 city blocks.

Adding to this atmosphere of urban disease and decay were thousands of dead horses. The average working life of a city horse was a fleeting five to seven years. Across urban landscapes, injured, lame, and incapacitated horses were shot and left to rot in the streets. "Dead horses were extremely unwieldly," remarks transportation historian Eric Morris.

"As a result, street cleaners often waited for the corpses to putrefy so they could be more easily sawed into pieces and carted off." Between 1880 and 1910, an annual average of 26,000 dead horses were removed from the streets of London, 15,000 in New York, and 12,000 in Chicago, which maintained its position as the premier horse market for feral and farmed horses shipped in from the West.

Between 1887 and 1897, the New York branch of the American Society for the Prevention of Cruelty to Animals (ASPCA), founded in 1866 primarily to care for unwanted horses, put down between 1,700 and 8,000 horses a year. The ASPCA and the French Society for the Protection of Animals, created twenty years earlier, were both inspired by their pioneering British counterpart, the Royal Society for the Prevention of Cruelty to Animals, established in 1824 (the first of its kind in any country) by Member of Parliament Richard "Humanity Dick" Martin.

The Irish politician had previously introduced the first British animal anti-cruelty bill. It was ratified in 1822 based on English philosopher and social reformer Jeremy Bentham's assertion in *An Introduction to the Principles of Morals and Legislation* (1789) that "The question is not, Can they *reason?* nor, Can they *talk?* but, Can they *suffer?* Why should the law refuse its protection to any sensitive being? The time will come when humanity will extend its mantle over everything which breathes." Similar laws were passed in New York in 1828 and Massachusetts in 1835, and, by 1907, every state had anti-cruelty statutes on the books. Comparable laws were invested in the Canadian Criminal Code in 1892.

Other national philanthropic organizations sprang up, including Our Dumb Friends League (later renamed Blue Cross Fund), raised in 1897 to care for working horses in London, and the American Humane Society, with its mission to petition for the welfare of urban horses—and children. To announce its founding in 1877, free copies of the freshly released book *Black Beauty: His Grooms and Companions—The Autobiography of a Horse* were handed out to children to promote inner-city literacy as well as English author Anna Sewell's goal "to induce kindness, sympathy, and an understanding treatment of horses." With its effective use of anthropomorphism and "first-person" narration by the eponymous

Black Beauty, the novel mobilized the rhetoric of slavery in the vein of Harriet Beecher Stowe's *Uncle Tom's Cabin* (1852) to appeal to the masses in its plea for the compassionate treatment of horses.

The instant international bestseller recounts the trials and tribulations of Black Beauty from his carefree days as a foal with his mother on a serene English estate through his hardships and misery as a cab horse on the tough streets of London. Black Beauty eventually finds peace and tranquility in retirement on country pastures reminiscent of the rehabilitated cavalry horses from the Battle of Waterloo charging on the meadows of Sir Astley Cooper's manor. Most urban horses had no such happy endings.

Dead and bloated horses left rotting in the streets and those that perished in distress from traffic accidents, overwork, exhaustion, or disease were discarded at industrial rendering plants (and canneries) that historian at Northeastern University Clay McShane labels "the organic equivalent of a scrap iron processor." In 1895 one horse carcass produced a profit of $24, equivalent to roughly $1,000 today. Appearing in the 1850s, these large-scale recycling centers turned horse carrion into a surprisingly thick catalog of secondary products, similar to those developed from the bison by Indigenous peoples.

Shaved hair was used for blankets and as stuffing for cushions. Hides were turned into highly valued cordovan leather. Hooves were boiled to extract oil used for glues and gelatin. Leg bones were fashioned into knife handles and combs. Ribs and skulls were treated to remove oils and then burned for bootblack polish and filters in sugar refineries. The by-products from this process were the chief sources of ammonium carbonate, insecticides, and phosphorus for match tips. Horsemeat was sold for human consumption at "pork shops" and, more frequently, ground into pet food. Fats were skimmed for candles and soaps. The remaining vat sludge, or "soup," was used as fertilizer or pounded with potash to produce substances used in dyes and poisons.

These processing facilities, usually erected in the heart of urban centers, emitted an unimaginable, reeking stench from the butchery, boiling, burning, and bilge waste. Most of the rendering vats were outdoors for fear of explosions and the prohibitive cost of indoor closed boilers,

Destined for the rendering plant: Kids playing in the filthy street beside a dead workhorse, with milk wagons in the background, New York City, circa 1900. *(Library of Congress)*

or autoclaves. Of course, in death, many horses were also dumped unceremoniously into waterways or eaten by their owners—or pets.

It is hard to believe that just over a century ago horses were literally everywhere and used for everything. They were the preeminent beast of burden. "Between 1870 and 1900, the number of horses in American cities grew fourfold, while the human population merely doubled," reports Tom Standage in *A Brief History of Motion: From the Wheel, to the Car, to What Comes Next.* "By the turn of the century, there was one horse for every ten people in Britain, and one for every four in the United States."

I want you to stop reading and pause for a moment after you mentally substitute horses for every aspect of our current motorized and mechanical farming, manufacturing, transport, distribution, services, war, and trade.

It is a mind-numbing exercise. "Horses were ubiquitous, working in cities, towns, and factories, on farms and frontiers, on streets and roads, alongside canals, around forts, ports, and railroad depots," stresses Ann Norton Greene in *Horses at Work*. "Horses were particularly dense around cities. Census returns would confirm the visible evidence that the horse population had grown substantially, and that Americans consumed more power from horses than from any other source."

The total US horse population peaked in 1915 at twenty-five million, with another five and a half million mules. Sustaining this massive national herd consumed roughly 30 percent of the country's total crop area. It is fair to say that between 1880 and 1920, horses ruled the world, and humans depended on them more than at any other time in our history.

Ratio of Horses to Humans in US Cities, 1900*

City	Population	Ratio Horse to Human
Kansas City	214,000	1 to 7.4
Minneapolis-Saint Paul	366,000	1 to 9.3
Los Angeles	102,000	1 to 12.7
Denver	134,000	1 to 14.7
Memphis	102,000	1 to 17.0
Saint Louis	575,000	1 to 17.5
Buffalo	352,000	1 to 18.5
San Francisco	343,000	1 to 20.1
Chicago	1.69 million	1 to 22.9
Pittsburgh	322,000	1 to 23.0
Cincinnati	326,000	1 to 23.3
Toronto (Canada)	210,000	1 to 25.0
Philadelphia	1.29 million	1 to 25.3
New York	3.44 million	1 to 26.4

At the turn of the century, American cities had upwards of five million horses performing every function and job imaginable. They were the

---

* McShane and Tarr, *The Horse in the City: Living Machines in the Nineteenth Century*, 16; US Census Bureau, 1900.

heartbeat of commerce and transportation. A bird's-eye view of the planet would show the urban centers of humanity crawling with horses—and their fly-ridden manure.

Horses pulling omnibuses, railcars, wagons, and carts hurried, hurtled, and crashed across rush-hour traffic on busy commercial streets darkened by the first skyscrapers and clattered down the quiet picket-fence-lined cobblestones of suburbia. They shuttled goods and passengers to and from railway stations and ports. Horse teams hauled building materials to, and removed debris from, construction sites, as cities expanded outward *and* upward during the unprecedented growth of the Gilded—or Gelded—Age.

Soon after their first appearance in Chicago in 1885, skyscrapers dominated the skylines of major American centers.* The iconic triangular Flatiron building anchoring the south side of Madison Square in Manhattan, for example, opened its doors in 1902, followed seven years later by the neighboring Metropolitan Life Insurance Company tower. "Humans could not have built nor lived in the giant, wealth-generating metropolises," notes McShane, "without horses."

Cities reverberated with the deafening sounds of horse-related occupations and infrastructure, including blacksmiths, farriers, wheelwrights, tanners, drivers, carters, breeders, breakers, knackers, teamsters, hostlers, veterinarians, groomers, saddlers, stables, markets, canneries, rendering plants, and carriage, coach, and cab makers. Horses literally drove people and profits.

By the mid-seventeenth century, scheduled stagecoach services carrying up to eight passengers appeared across Europe. An English travel almanac from 1784 cautioned that "the number of carriages is so great in Paris that one can scarcely withdraw from the burden they cause," while warning also of the danger they posed to pedestrians on the "narrow, crowded, dirty streets." Elaborately painted and decorated twelve- to eighteen-passenger omnibuses drawn by teams of two to six horses were shuttling people across Paris, London, New York, Philadelphia, and

---

* The ten-story, 138-foot-tall Home Insurance Building opened in Chicago in 1885.

Preparing the foundations of the future: Horse teams hauling materials to and from a skyscraper construction site at the intersection of Front and Yonge Streets, Toronto, 1903. *(City of Toronto Archives)*

Boston by the 1830s. Swanky omnibuses appeared in both Toronto and Montreal in 1849. By the 1850s, most urban centers across the Americas and Europe had adopted public omnibus transportation.*

In 1850, for instance, the seven hundred omnibuses operating in New York—dubbed the City of Omnibuses—averaged 120,000 passengers per day, while those in Paris handled 95,000. Reminiscent of contemporary subway travel, *The New York Herald* quipped, "Modern martyrdom may be succinctly defined as riding in a New York omnibus." In London, the omnibuses were so packed that people took to riding on the roof, giving rise to the renowned red double-decker bus.

An advertisement for the debut of the Baltimore Omnibus Company in 1844 provides a concise overview of this new public transportation: "They are quite handsome affairs, well fitted up, richly decorated, drawn by good horses and we believe driven by careful drivers. . . . In other cities, in addition to the general convenience, they have tended greatly

---

* *Omnibus* means "for all" in Latin, as in *justitia omnibus* ("justice for all").

to enhance the value of property in the outskirts of the City, enabling persons to reside at a distance from their places of work." Omnibuses created the first suburban rings around modern cities. They were not, however, the only form of horse-driven public transportation on the manure-plastered roads of the creeping urban sprawl.

Horse-drawn railcars, or trams, in downtown corridors across North America and Europe gained traction and popularity by the 1860s. With less friction and more concentrated power, railcars accommodating 20 people moved at eight miles per hour, roughly twice as fast as omnibuses. New York, of course, started the trend in the United States in 1832, with Toronto and Montreal following suit in Canada in 1861. By 1890, there were 5,800 miles of tramline (supported by 250,000 horses and 37,500 employees) in US cities alone, servicing 400 million passengers a year. Ridership in New York, for example, increased from 23 million in 1857 to 161 million by 1880.

In the twenty-eight American cities with populations over one hundred thousand in 1890, residents rode public transportation an average of 172 times per year. The typical New Yorker clocked in a whopping 297 annual trips. Across the country, horses were pulling 27 percent more passengers than they did a decade earlier, as urban populations soared and routes extended farther into suburbia.

Joining the fleets of mass public transportation were smaller public cabs, private carriages and carts, commercial wagons, and postal, police, and fire teams.* Horses, and all manner of vehicles, clogged the gridlocked streets. Adding to the chaotic traffic congestion was the bicycle, or "dandy horse." The introduction of the "safety bicycle," with its chain-drive gear transmission and smaller, pneumatic tires, in the late 1880s kicked off a decade-long bicycle craze. It did nothing, however, to curb horse traffic. Or manure.

All this transportation, accommodating an increasing number of people on the move, meant big business for vehicle makers. The eighty manufacturers in Cincinnati, for instance, cranked out 130,000 horse-drawn carriages in 1890, at a time when there were more than 13,800

---

* *Cab* is short for the French *cabriolet*, a light, two-wheeled carriage drawn by a single horse.

Spiritual machines: A vast array of horse-drawn vehicles carrying wool, coal, ice, ale, fish, furniture, people, and other products cross Pyrmont Bridge in Sydney, Australia, 1894. On the far right, a young man in a white shirt is sweeping the gutter. As late as the 1930s, the city paid boys, nicknamed "sparrow starvers," to collect manure to sell as fertilizer. (Sydney Museum of Applied Arts and Sciences Powerhouse Collection)

American companies producing some form of horse-drawn transport. The Studebaker Company of South Bend, Indiana, built 75,000 that year, and the Kentucky Wagon Manufacturing Company in Louisville, 90,000. Abbot, Downing & Co. of Concord, New Hampshire, was so successful in constructing durable stagecoaches that their product cornered the market and was forever known as the Concord coach.

Companies adorned their wagons with trademark logos, and their horses became part of their brand marketing. What would Budweiser beer be without its recognizable Clydesdales, or Coors without its Belgians? The H. J. Heinz Company was known for its white wagons and contrasting black Percheron horses. Dairy companies preferred milky white horses to match their products and promote the allure of purity. Horses were adorned with clanging bells and flashy outfits to draw attention. The Levi Strauss & Co. two-horse patch, debuting in 1886, depicts two horses attempting to pull apart a pair of jeans, demonstrating its strength and durability. Without knowing, Strauss invoked the historical connection between horses and the creation of pants three thousand years earlier by Indo-European riders of the Tarim Basin.

The fabrication of specialized vehicles demanding different horsepower strengths was comparable to today. Large-haul express companies preferred heavy coach horses averaging 1,450 pounds, while smaller delivery and buggy services used standard horses between 1,000 and 1,200 pounds. Heavy draft horses were considered too slow, lumbering, and clumsy for the quick pace and mind-rattling chaos of city driving.

The first firehorses, for example, were hastily drafted into service in 1832 by the Fire Department of the City of New York (FDNY) during a cholera epidemic as emergency substitutes for dead, sick, and dying fire crews. The horses proved so invaluable in saving time, property, and lives that they were promoted to full-time employees, pulling custom fire trucks. American firehorses were larger than their transportation counterparts, standing sixteen to seventeen hands and weighing 1,500 to 1,600 pounds on average.

In 1893, the year prior to the Great Manure Crisis, W. J. Gordon published his statistical findings in *The Horse-World of London*, providing an exhaustive representation of urban horses across the global order. According to the prolific author, the city had 2,210 omnibuses operated by 11,000 employees and 22,000 horses. Another 10,000 horses pulled 1,000 railcars across 135 miles of tramline. Over the course of the year, these trollies carried 1.9 million passengers more than 21 million miles.

London also registered 11,297 cabs pulled by 22,000 horses. Gordon also discovered that of the 15,000 licensed cabmen, more than 2,000 were convicted of various driving infractions, including cruelty to animals, drunkenness, loitering, and the catchall of "willful misbehaviour." Many were repeat offenders.

The city also stabled over six thousand commercial horses, six hundred Royal Mail horses, three thousand brewery and distillery horses, and three thousand more supplying London with eight million tons of coal annually. Another twenty-six thousand found themselves on dinner plates during 1893. Thousands of additional horses serviced the massive fleet of private coaches, carts, and carriages.

In total, roughly three hundred thousand horses drove the personal and economic demands of a bustling city of six million people. Over the course of the year, these steadfast horses also bequeathed 1.6 million tons of manure and 685,000 tons of urine to the streets of London.

Predictably, transportation was at a standstill, prompting the Royal Commission on London Traffic to investigate solutions to the gridlock. From 1870 to 1900, the average speed of travel in London dipped by 25 percent. (The average speed of cars in central London today is equivalent to that of a horse-drawn carriage in the 1890s: eight miles per hour.) The overflowing traffic generated head-hammering noise pollution from the clamoring din of thousands of clip-clopping shod horses and clattering iron-rimmed or hobnailed wheels. This provided the incessant background rhythm for a competing chorus of whinnying, snorting, squealing, and the cruel cracking of whips and bellowing commands of chauffeurs

Gridlock: Rush-hour traffic on the streets of Philadelphia, 1897. *(US National Archives)*

and horse drivers. And just like today, there were also hazardous slowpokes, reckless speeders, infuriating tailgaters, and volatile road-ragers.

Horse traffic, however, was far more chaotic and dangerous because of the dual unpredictability of humans *and* horses. The complete horse-drawn vehicle package was also stretched in disjointed sections of the horse team, traces, and vehicle, presenting a longer, rambling target for accidents and entanglements. Suddenly halting a team of horses pulling an enormous amount of weight was quite another issue altogether. Accidents were frequent and often lethal.

Across the United States between 1890 and 1900, for example, there were 750,000 annual horse-drawn traffic accidents causing injury or death. Serious traffic injuries were ten times greater than modern automobile levels, with death rates almost double. In 1900, for instance, horse accidents killed one out of every seventeen thousand New Yorkers. A century later, car accidents killed one out of every thirty thousand. Cities attempted to control horse traffic and its associated noise with bylaws.

Boston forbade horse-drawn traffic from passing courthouses so that judges and juries could hear the proceedings and testimony. Doctors across the world claimed that the incessant and unrelenting racket was the cause of a growing number of "nervous disorders." Like modern noise-reducing barriers erected along interstates and highways, in high-volume neighborhoods, people wrapped their houses in straw bales to muffle the commotion. Some cities limited teams to four horses.

Posted speed limits were common: San Francisco, 10 mph; Chicago and Detroit, 6 mph; and New York, 5 mph for commercial vehicles and 8 mph for light passenger vehicles. Speeds, however, could not be measured, nor could limits be enforced. The volume and corresponding danger generally controlled the pace of traffic. It was the unofficial duty of all drivers to even out and smooth pack the roads by keeping out of the ruts, which damaged wheels and were potentially dangerous to horses' fragile legs. Signs with slogans such as "Keep out of the ruts and save the road" were posted along well-traveled routes.

As horse-drawn transportation surged, streets were widened where possible, and layers of packed-gravel macadam roads became standard by the 1850s. With the advent of buoyed American petroleum production in the 1890s, these gave way to smoother asphalt, making the fast-approaching shift to automobiles relatively painless. Sidewalks to separate the flow of horse traffic and manure from doorways, stoops, and pedestrians also became commonplace by the 1850s.

In the immediate aftermath of the Confederate surrender on April 9, 1865, ending the Civil War, General Grant was asked about a possible political future. He responded shyly that his only political ambition was to return home to Galena, Illinois, and petition his local council for sidewalks. Three years later, he was president of the United States.

As a result of these escalating transportation concerns and the nagging Manure Crisis, New York hosted the first international urban planning conference in 1898. The ten-day agenda revolved around three inseparable and inescapable issues: horses, traffic, and manure. "Horses," writes Tom Standage, "had become both indispensable and unsustainable." After only three days, the meeting was adjourned. None of the delegates or scientists could see any workable solutions to the insurmountable problems and human dependence on horses. Hopefully our efforts to tackle greenhouse gas emissions will last longer than a weekend.

Human reliance on horses, and the need for an alternative form of power, were never more evident than during the Great Epizootic of 1872. "Imagine an equestrian health disaster that crippled all of America, halted the government in Washington DC, stopped the ships in New York, burned Boston to the ground, and forced the cavalry to fight the Apaches on foot," evokes writer and avid rider CuChullaine O'Reilly. "The outbreak is known as the most destructive recorded episode of equine influenza in history." In October 1872 a highly transmissible equine strain of influenza A—spread mainly through coughing and close contact, causing severe respiratory infection—literally went viral.

With stunning speed, it swept down from Toronto, Canada, coursing through the international arteries of trade and travel across the United States and Mexico, to Central America and the Caribbean. Within three days of the first symptoms in Toronto, all streetcar horses making up the bulk of the city's 2,500 horses were infected. Roughly 65 percent of the horses in New York City and 95 percent in Rochester fell ill.

On November 9 the Great Boston Fire raged for sixteen hours, as firefighters and civilian volunteers struggled to pull vehicles, inching forward at a crawling pace all for the want of a few skilled horses. More than half of the city's seventy-five firehorses were dead or unfit for service, coughing uncontrollably in their stables as the blaze spread through the downtown business district. The fire claimed thirty lives, including twelve firefighters, and consumed sixty-five acres and 776 buildings, causing the equivalent of $1.6 billion in damages.

Farther west in Arizona, the US Cavalry was handicapped with sick and dying horses as it tried to corral Cochise and his Apache warriors. "There was still another source of discomfort which should not be overlooked," recorded Captain John Gregory Bourke. "At that time the peculiar disease known as the epizoötic made its appearance in the United States, and reached Arizona, crippling the resources of the Department in horses and mules; we had to abandon our animals, and take our rations and blanket upon our backs, and do the best we could." Humanity had taken horses for granted, and, without them, the pandemic paralyzed most aspects of modern society.*

Transportation, trade, and commerce ground to a halt across North American cities as the virus ripped through dense horse populations, infecting upwards of 80 percent. The supply chain backed up and collapsed in the face of mass horse shortages. Produce and meat piled up and rotted in the static ports dotting the Great Lakes, Erie Canal, and Atlantic Ocean. Crates and shipping containers stuffed with merchandise sat motionless in eerily quiet train stations. Farmers ceased slaughter, and factories shuttered.

There was talk of economic recession, which came the following year with the Panic of 1873. Banks failed, and the value of silver plunged. The bulging wealth of the Second Industrial Revolution and post–Civil War speculative bubble burst into a global twenty-three-year Long

---

* Highly effective vaccines are now available to combat most strains of equine influenza.

Depression, which helped foment the economic triggers of the First World War.*

As dazed and confused populations were confronted with the harsh realities and immediate repercussions of the equine epizootic, in October 1872 *The Nation* magazine reminded its readers of "The Position of the Horse in Modern Society":

> Our talk has been for so many years of the railroad and steamboat and telegraph, as the great "agents of progress," that we have come almost totally to overlook the fact that our dependence on the horse has grown almost *pari passu* with our dependence on steam. We have opened up great lines of steam communication all over the country, but they have to be fed with goods and passengers by horses. We have covered the ocean with great steamers, but they can neither load nor discharge their cargoes without horses. We have collected at the mouths of our great rivers and at the intersections of our railroads vast bodies of people, covering miles on miles of area with their dwellings and factories, but have left them wholly dependent for their intramural travel and for their regular supplies of food and clothing on horses. More than this, we have within the last few years made horse labor an almost essential condition of the protection of our great cities from fire. . . . We have come to think of him as a machine. . . . [T]hey are the wheels of our great social

---

* During this severe economic depression from 1873 to 1896, the British lost ground to Germany (which amassed the most dynamic economic and industrial growth seen in the past two hundred years) in most commercial portfolios. Between 1870 and 1912 Britain's total share of global trade was slashed from 32 percent to 14 percent, iron output fell from 50 percent of the world's share to 12 percent, while copper decreased from 32 percent to 13 percent. Even coal production, the lifeblood of Pax Britannica and the propulsion behind the assembly of her mighty empire, was threatened. By 1910, German coal output equaled that of the United Kingdom. German steel production increased tenfold between 1880 and 1900. At the outbreak of war in 1914, Germany was smelting twice the amount of pig iron as British foundries. Domestic agriculture accounted for 95 percent of German consumption, whereas Britain imported 58 percent of its daily per capita caloric intake.

machine, the stoppage of which means widespread injury to all classes and conditions of persons, injury to commerce, to agriculture, to trade, to social life.

The author warns candidly that "[W]e are now for the first time forcibly reminded that a plague might break out among horses, as plagues have broken out among men, which would sweep them away by the hundred or thousand every day, and which would momentarily baffle science. . . . [T]he sudden loss of horse labor would totally disorganize our industry and our commerce, and would plunge social life into disorder." Sound familiar?

During its relatively short six-month outbreak, the mortality rate from equine influenza hovered somewhere between 3 percent and 5 percent, with urban levels reaching 10 percent. Mules, donkeys, and captive zebras appear to have been hit especially hard. "Its lasting effect, however, was that in the various fields where horses served," notes historian Jonathan Levin in *Where Have All the Horses Gone?: How Advancing Technology Swept American Horses from the Road, the Farm, the Range and the Battlefield,* "it hastened the search for an alternative to the horse, whose traditional dependability was put in question."

Just as the 1973–74 oil crisis spawned domestic production (and Japanese imports) of smaller, more fuel-efficient compact cars, a century earlier the Great Epizootic of 1872 helped stimulate the search for cleaner, less infectious, and more reliable forms of mechanical transportation to replace the faithful, but mortal, horse.

In 1260, as the Mongols were penetrating the Levant, the eccentric English philosopher and friar Roger Bacon—who, as mentioned, was the first European to record the formula for gunpowder—eerily predicted a future dominated by mechanization: "Machines may be made by which the largest ships, with only one man steering them, will be moved faster than if they were filled with rowers; wagons may be built which will move with incredible speed and without the aid of beasts; flying machines can be constructed . . . machines will make it possible to go to the bottom of seas and rivers."

His contemporaries, who already regarded him as a heretical wizard, must have thought Bacon had gone completely mad. Six centuries later, his seemingly insane ideas were slowly brought to life in what computer scientist and futurist Ray Kurzweil labeled "the age of spiritual machines."

The 5,500-year-long supremacy of the human-horse dyad, however, did not come to an abrupt or unforeseen end. The Horse Age was not overthrown suddenly by a mechanical coup or *Matrix*-like insurrection. This plodding "dehorsification," as Russian writer Isaac Babel called it, and overlapping handoff to an automated era was gradual. It took almost a century and two calamitous world wars to complete the transfer of power.

Horses reached their pinnacle in global output between 1900 and 1915, just as the age of mechanization lurched forward. They unwittingly encouraged their own demise, which for the often overworked and abused horses was a blessing in disguise. Horses helped build the railways, subways, electric trams, war machines, and automobiles of the modern world. They were essential in hauling materials and debris to and from construction sites. Horses provided transport and fuel to the factories that manufactured the engines that forced them into a long-overdue and well-deserved retirement.

The slow death of the living machine breathed life into those twisted from metal and forged in the fires of the steel mills of Andrew Carnegie and the petroleum rigs of John D. Rockefeller. By 1890, both US industrialists had consolidated their respective monopolies on the two most important global commodities of the Second Industrial Revolution, which lasted from 1870 to 1914.

During this period, coinciding with the last land grabs of colonization, or "Imperial Scramble," international trade increased fourfold, and the world's steel output rose from five hundred thousand tons to sixty million tons, half of it fabricated in the United States. Steel, petroleum, and rubber, not manure, lined the unsoiled road to modernization and its immaculate ideals of the White City. The machine, not the horse, would carry the torch of progress into a new future for humankind.

American Modernization During the Gilded Age, 1870 to 1920*

	1870	1900	1920
Farms (millions)	2.7	5.7	6.4
Farmed Land (million acres)	408	841	956
Wheat (million bushels)	254	599	843
Workforce Agricultural (millions)	53	38	27
Workforce Industrial (millions)	28	32	44
Steel Production (million metric tons)	0.8	11.2	46
Oil Production (barrels/million)	5.1	63.5	441.7
Railway Track (miles)	53,000	259,000	407,000
Gross National Product ($ billion)	7.4	18.7	91.5
Life Expectancy at Birth (years)	42	47	54

The steam engine patented by Scottish inventor James Watt in 1769 had numerous precursors dating back almost a century to the genius Dutch polymath Christiaan Huygens, who also made groundbreaking contributions to optics, mechanics, astronomy, and motion a decade before Isaac Newton. To compare the output, or rate of work, performed by steam engines and draft horses, Watt devised the measurement of horsepower. Although there are now different standards and definitions of horsepower, it is basically a unit measurement of power or output from an engine or motor.†

Building on this tradition, in 1804 British mechanical engineer and inventor Richard Trevithick deployed the first steam locomotive. In 1820, more than fifty years before Mark Twain narrated the shenanigans

---

* US Bureau of Labor Statistics; US Census Bureau; US Department of Agriculture; US Energy Information Administration; Eric Foner, Kathleen DuVal, and Lisa McGirr, *Give Me Liberty!: An American History*, 7th ed., vol. 2 (New York: W.W. Norton & Co., Seagull, 2022), 608.

† Today there are two common definitions of horsepower: mechanical (or imperial), which is 745.7 watts, and metric, which is 735.5 watts. One human can exert 1.2 horsepower; a single horse, 15 horsepower.

of Tom Sawyer and Huckleberry Finn, seventy-five steamboats navigated the Mississippi River. When Twain published *The Adventures of Huckleberry Finn* (often called the Great American Novel) in 1885 more than six hundred were paddling the interior rivers of the United States, transforming the tenets of American progress.

Down on the farm, however, massive, plodding, expensive, and relatively inefficient steam tractors could not outwork the horse and never really caught on. By 1910, only seventy thousand sat, predominantly idle, in American barns. Over time, the internal combustion engine, not steam propulsion, unseated the horse from its commanding reign.

In 1876 German engineer Nicolaus Otto built the first petroleum-powered, compressed-charge internal combustion engine. Despite standing more than seven feet tall, it generated only limited power. Over the next two decades, numerous contributors made successive improvements in reducing size and weight while increasing horsepower. Gottlieb Daimler, for instance, installed a small engine on a bicycle in 1885, creating the Reitwagen (riding car), or first motorcycle. A year later Carl Benz patented his Motorwagen (a derivative of Greek and Latin meaning "movable self"), which is considered the birth of the modern automobile. In the English world, however, "horseless carriage" dictated the lexicon during these early years.*

Health professionals and scientists hailed the automobile as a sanitary, safe, and efficient form of transportation. After all, "Cleanliness," as the English novelist Aldous Huxley stressed in *Brave New World*, "is next to fordliness." In Huxley's dystopian World State (plied with the Indo-European-inspired mind-control drug soma), Henry Ford is a messianic-like father figure. The calendar notates years in AF (Anno Ford), with the history-altering 1908 marking year zero. I am currently writing in the year 115 AF.

Although he grew up on a farm in Michigan, young Henry never really took a shine to horses ever since a spooked colt dragged him around with

---

* The word *car* evolved from the Proto-Indo-European *$k\text{ṛsós}$* (wagon) and *$k^w\text{ék}^w\text{los}$* (wheel or circle).

his foot caught in the stirrup. "I never had any particular love for the farm," he recalled. "It was the mother on the farm I loved." In 1875, when he was twelve, Ford watched an enormous steam-engine tractor lumber across a field, remarking later that "it was the first vehicle other than horse drawn that I had ever seen." He quickly determined, however, that "steam was not suitable for light vehicles, [as] the boiler was dangerous." After working as an engineer for Thomas Edison for nine years, in 1899 Ford set out on his own to revolutionize the auto industry.

In year zero—1908—he did just that. The cheap, mass-produced Model T horseless carriage changed the course of history. The "Tin Lizzie" was marketed as the Everyman car, with slogans such as "Even You Can Afford a Ford." Henry flipped the switch on the assembly line, and we all know what happened next.

By 1920, the Ford Motor Company was cranking out 2 million cars per year (up from 12,000 in 1910), and the Model T accounted for 57 percent of global market share. Conversely, only 90 manufacturers in America produced horse-drawn vehicles, down from 13,800 just thirty years earlier. While the internal combustion engine fitted to cars, tractors, and other wheeled vehicles was the main catalyst behind the leisurely departure of the horse, other technologies, including electric railcars and subways, also contributed to its urban demise.

The Metropolitan Railway in London opened the very first subterranean line in 1863, followed by the first true "deep-level tube," or subway, in 1890. Budapest and Glasgow opened their first operational lines in 1896, and Boston unveiled the first American underground the following year. Built with the muscle of 1,500 horses, the New York City subway system opened its doors in 1904, accommodating 127,381 passengers on its first day. By 1919, Paris, Berlin, Athens, Philadelphia, Hamburg, Buenos Aires, and Madrid had joined the growing club of cities with subterranean travel. Modernization was tightening its perimeter on the last roundup of natural horsepower.

In 1890 the American Street Railway Association conducted a series of comparative tests between horse-drawn and electrified trams. The data revealed that horses cost 3.72 cents per car mile compared with 2.37

The slow death of the living machine: A modern tramline juxtaposes a team of horses dragging a hearse through the thick sludge of mud and manure on St. Clair Avenue, Toronto, 1908. *(City of Toronto Archives)*

cents per electric tram mile. "The report undoubtedly sped the disappearance of the horse," states McShane, "since the presidents of virtually every large street railway were at the meeting." Between 1888 and 1902, 97 percent of American horse-drawn tramlines and streetcars had been electrified, replacing over a hundred thousand traction horses. The rails of Toronto and Montreal were converted by 1894, and those of the United Kingdom went electric between 1901 and 1903. At the outbreak of war in 1914, London boasted the largest electric streetcar system in the world, and its conductors-turned-soldiers were traded for "conductorettes" doing their bit for the war effort.

While the conversion to electric trams lowered the cost per mile by 36 percent, automobile ownership produced even greater savings. In 1915 it was 38 percent cheaper to own and operate a car for five years (the average working life of an urban horse) than a horse and passenger wagon. Registered motor vehicles in France increased from 146,000 in 1915, to 677,000 in 1925, to 2.3 million by 1935, while those in Germany rose from 93,000, to 350,000, to 2 million over the same ten-year increments.

The Urbanization of Motorized America, 1880 to 1950*

Year	Urban Population as Percent	Cities with 100,000+	Vehicle Registrations
1880	28.2	20	
1890	35.1	28	
1900	39.6	38	8,000
1910	45.6	50	468,500
1920	51.2	68	9,239,161
1930	56.1	93	26,749,853
1940	56.5	92	32,453,233
1950	64.0	100+	49,161,691

The convenience and associated industry of automobiles sped up the progression of suburbanization, encroaching on the habitat of the metropolitan workhorse. "Moreover, like many inhabitants of America's rapidly growing urban centers," writes Richard Bak in *Henry and Edsel: The Creation of the Ford Empire*, "Henry was personally affronted by the ubiquitous horse manure and urine." The automobile magnate did his part to clean up the streets while saving his Ford Motor Company, headquartered in Detroit, some overhead expenses.

Initial Ford factories in both the United States and England used manure as a supplemental fuel source, burning through two thousand pounds a week. These early "eco-friendly" practices lasted until 1939, when more efficient energy sources dropped in price. Horse droppings were used to manufacture the very machines that would depose the steadfast animal. Or framed another way: one contamination was used to produce another contaminator. Transportation has always had emission problems.

The motorized solutions that saved us from horse pollution in the twentieth century are now imperiling us with their own pollution in the twenty-first century. "In our attitudes about transportation and mobility, in our relationships to automobiles, and in our language of movement

---

* US Census Bureau; US Department of Transportation; Federal Highway Administration (FHWA) Research Library

and power, horses are still with us," observes Ann Norton Greene. "We must understand their role as the important, prime movers in the nineteenth century as we face the energy challenges of the twenty-first."

Like cyclical fashion trends, hairstyles, and pants, what is old is now new. Toyota has been developing technology to "recycle," or convert, the methane in manure into hydrogen to power fuel cells for electric cars. The goal is to be able to run a car for a year from the annual excrement of one animal. Numerous companies and pilot projects are also studying the potential of using this same manure synthesis as a large-scale biofuel alternative for industry.

However, in a 2002 scientific report published by the European Molecular Biology Organization, the author concedes that "the massive costs involved in establishing the infrastructure required to produce, store and distribute hydrogen means that its utilisation is still years away, despite the fact that car manufacturers in both Europe and the USA are already at an advanced state in the development of hydrogen-powered vehicles." More than twenty years later, this statement holds true.

The original shift from horse to car also completely overhauled our urban and rural infrastructure. Between 1910 and 1930, the number of workhorses powering urban America shrank by more than 90 percent to a mere three hundred thousand. "The horse," Ford eulogized unsentimentally in 1929, "is gone." As horses went, so too did their vast economic and visible presence.

The last horse-pulled engine of the FDNY, for example, made its final run in 1922 in front of thousands of spectators gathered to catch one last nostalgic glimpse of the horse era. That same year, fifty thousand Michiganders lined the streets of Motor City to bid farewell to its last horse-drawn fire brigade as it dashed into the history books down the car-parked streets of the automotive capital of the world. Horse-related stables, transit stops, water troughs, professions, and stacks of manure were traded for petrol pumps, repair shops, parking lots, bus stops, and other infrastructure and occupations supporting the surging number of automobiles and specialized vehicles on European and North American roads.

Changing of the guard: A horse-drawn firetruck (with water tower) races past a parked automobile, Washington, DC, 1913. *(Library of Congress)*

The US Federal Aid Road Act of 1916, or Good Roads Act, was the first legislation of its kind, providing $75 million for countrywide construction as well as subsidies for state and municipal projects. Further legislation to create a national transportation grid, including the Interstate Highway System, accelerated between 1921 and 1956 to meet the public and commercial demands of the motorized—and increasingly suburban—age. While the shift to cars was expedited by the fact that they used the same roads as horse-pulled vehicles, it is easy to forget that they shared these crowded streets for decades.

Growing up in Canada during the *Leave It to Beaver* baby boom generation of the 1950s, my parents remember the milkman with his horse and wagon disrupting their high-stakes road hockey games. For door-to-door deliveries, horses were still preferred. They memorized their routes by landmarks and knew instinctively when to halt and proceed without guidance. While the horse advanced the wagon unattended, the milkman walked back and forth from house to house, swapping full bottles for empties without having to get in and out of a car to continuously move it. "I recall that when I was a boy playing on the street in

New York, our game had to be interrupted every little while to let a horse and wagon go by—milk wagon, iceman, vegetable man, or old clothes man," echoes Jonathan Levin. "We took it for granted and resumed our game, usually punch ball or stick ball, after its passing." By the time "Jerry Mathers as the Beaver" aired its series finale in 1963, the milkman was just another memory of bygone North Americana.

Like its urban brethren, the farm horse also began its slow trot to redundancy as tractors and other agrarian machines sank their teeth into mass-produced crops. Families, neighbors, and entire communities gathered in great excitement to welcome tractors and cars to rural America. "The arrival of the Model T in hinterland cities like Omaha and Denver," notes Rice University historian Douglas Brinkley in *Wheels for the World: Henry Ford, His Company, and a Century of Progress, 1903–2003*, "was an event as eagerly anticipated as a Billy Sunday evangelical revival or Buffalo Bill's Wild West Show." So too was the arrival of the Fordson tractor in 1917.

Having cornered the automobile market, Henry Ford sought to apply the same manufacturing and marketing principles to tractors. "I suspected that much might be done in a better way," he said, evoking his childhood on the farm. "I have followed many a weary mile behind a plow, and I know the drudgery of it." The first gas-powered tractor was the Hart-Parr, introduced in 1903. Its immense size and cost, crawling speed, and mechanical issues made it impractical for most farmers, and it sold poorly in its first year—a total of fifteen. One of these forlorn fourteen-thousand-pound behemoths is on display at the Smithsonian Institution's National Museum of American History in Washington, DC. It is immediately obvious why this machine was a commercial failure.

When the Ford Fordson, a lightweight, relatively small, and inexpensive gas tractor, was unveiled in 1917, there were fewer than eighty thousand tractors on US farms, many of them idle or inoperable steam engines. Although mechanically compromised, by 1921, three hundred thousand Fordson tractors were owned by American farmers. It quickly forced other manufacturers to drop out of the competition. Two years later, Ford controlled 76 percent of American tractor sales.

In 1924 the International Harvester Company one-upped Ford by releasing the Farmall, a sturdier "all-purpose" (hence the name) tractor that could plow, furrow, *and* cultivate. Set on a raised chassis, it could drive across most fields without damaging the crops. When the Fordson was scrapped from US production in 1928, the Farmall comprised 60 percent of the tractor market.

Continuous improvements to more economical and technical farming machines allied to innovations in hybridization, fertilizers, and pesticides instigated the era of industrial farming known as the Green (or Third) Agricultural Revolution. Like its Neolithic and medieval predecessors, this explosive agrarian transformation also led to unbridled—and unprecedented, by comparison—population growth. The number of people on the planet jumped from 1.9 billion in 1920 to 7.8 billion in 2020. An International Harvester advertisement extolled the virtues of this new age, declaring that "Every Farm Is a Factory."

Horses, Mules, Tractors, Trucks, and Cars on American Farms, 1870 to 1950 (millions)*

Year	Horses	Mules	Equine Total	Tractors	Trucks	Cars
1870	7.6	1.2	8.8			
1880	10.9	1.9	12.8			
1890	15.7	2.3	18.0			
1900	17.9	3.1	21.0			
1910	20.0	4.2	24.2			
1915	21.5	5.1	26.6			
1920	20.1	5.7	25.8	0.25	0.14	2.1
1930	13.7	5.4	19.1	0.92	0.91	4.1
1940	10.4	4.0	14.4	1.6	1.0	4.1
1950	5.4	2.2	7.6	3.6	2.2	4.2

Globally, between 1910 and 1970, working farm horses declined by more than 90 percent, while human labor fell by 70 percent. Between 1920 and 1945, in the aggregate, each tractor displaced four working

---

* US Department of Agriculture, Census 1954. The number of donkeys employed on farms peaked in 1910 at a mere 106,000. This table does not include specialized motorized farming equipment such as combines, bailers, brooders, grinders, pickers, and milking machines.

horses. "Horses, you see, belong to the vanished agricultural past," wrote the futuristic novelist George Orwell in 1937, "and all sentiment for the past carries with it a vague smell of heresy."

In the United States, the tractor put twenty million farm horses out of work but freed up eighty-eight million acres of land to yield human consumables rather than horse feed. This equates to roughly 25 percent of the nation's entire crop area. Across the United Kingdom and its commonwealth of empire, including Canada and Australia, some fifteen million acres shifted away from horse fodder.

As a result, the average consumer now had access to a far greater selection of cheaper foods than ever before. The advent of mechanized farming within an accessible motorized society produced changes to everyday diets. In 1947, for example, more than seven million transport trucks hauled foodstuffs across an expanding grid of US roadways. The consumption of American table staples—potatoes, turnips, onions, carrots, corn, and cabbage—which were grown locally and easily stored in cold cellars, dropped by 25 percent between 1920 and 1945. The shelves of neighborhood supermarkets began to resemble those of my local Safeway.

Tractors by Continent, 1920 to 1990 (thousands)*

Region	1920	1930	1939	1950	1961	1970	1980	1990
N. and C. America	294	1,030	1,576	4,220	5,326	6,038	5,606	5,841
S. America			17	70	297	465	880	1,186
Europe		130	270	990	3,698	6,077	8,454	10,356
Soviet Union		78	440	430	1,212	1,978	2,646	2,609
Asia				35	200	783	3,475	5,599
Africa			6	95	235	334	439	532
Oceania				142	351	428	427	403
WORLD TOTAL				5,552	11,318	16,102	21,932	26,526

With its hegemony over steel and petroleum production, the relatively rapid American ascent to mechanized industrial agriculture was

---

* From Giovanni Federico, *Feeding the World*, 48.

somewhat of an anomaly. "Agricultural motomechanization began to develop in the interwar period," state Marcel Mazoyer and Laurence Roudart in *A History of World Agriculture: From the Neolithic Age to the Current Crisis*. "But it is necessary to emphasize that, in 1945, animal traction was still overwhelmingly prominent in most industrialized countries, and motomechanization was deployed in all these countries only after World War Two."

While international aggression stimulated mechanization on and off the battlefield, for its deadly encore, the horse endured two final whirlwinds of war. The volatile German kaiser from 1888 to 1918, Wilhelm II, who was prone to mercurial outbursts and crippling depression, maintained that mechanization was "a temporary phenomenon," insisting that the future still belonged to the horse. Only after their immeasurable, involuntary sacrifice during the World Wars were horses consigned globally to what former US secretary of state Condoleezza Rice has called "the roadkill of history."

The twentieth century marked both the rise and fall of the Horse Age. As Ulrich Raulff explains: "The last century of the era of the horse witnessed not only the exodus of the horse from human history, but also its historical climax: never before had humanity been as heavily dependent on horses as when Benz and Daimler's first internal combustion engines began rattling away. . . . [T]his changeover was a considerably drawn-out process: the two world wars prompted a mercilessly heavy reliance on horses and it was only after the middle of the century that traction power was cheap enough to lead to a dramatic decline in horse numbers." After five millennia of unbroken immortal combat, the last military hoorah of the horse was, unfortunately, also the most destructive.

CHAPTER 16

The Final Draft
War, Mechanization, and Medicine

The eloquent diary entry of Second Lieutenant Arnold Gyde serves as a timeless example for millions of other universal soldiers who echoed his heartbreak and sorrow regarding their own generation of warhorses. Gyde, who trudged to the western front with the British Expeditionary Force (BEF) in 1914, presents a vivid picture of the pity of war across all conflicts through his own carnage-stained memories of the Great War for Civilization:

> Horses have been slaughtered by the score. They looked like toy horses, nursery things of wood. Their faces were so unreal, their expressions so glassy. They lay in such odd postures, with their hoofs sticking so stiffly in the air. It seemed as if they were toys, and were lying just as children had upset them. Even their dimensions seemed absurd. Their bodies had swollen to tremendous sizes, destroying the symmetry of life, confirming the illusion of unreality.
>
> The sight of these carcasses burning in the sun, with buzzing myriads of flies scintillating duskily over their unshod hides, excited a pity that was almost as deep as pity for slain human beings. After all, men came to the war with few illusions and a very complete knowledge of the price to be paid. They knew why they were there, what they were doing, and what they might expect. They could be buoyed up by victory, downcast by defeat. Above all, they had a Cause, something

to fight for, and if Fate should so decree, something to die for. But these horses were different: they could neither know nor understand these things. Poor, dumb animals, a few weeks ago they had been drawing their carts, eating their oats, and grazing contently in their fields. And then suddenly they were seized by masters they did not know, raced away to places foreign to them, made to draw loads too great for them, tended irregularly, or not at all, and when their strength failed, and they could no longer do their work, a bullet through the brain ended their misery. Their lot was almost worse than the soldiers'!

It seemed an added indictment of war that these wretched animals should be flung into that vortex of slaughter.

The equine comprehension of war is innocently articulated by Black Beauty: "'Do you know what they fought about?' I asked. 'No,' he said, 'that is more than a horse can understand, but the enemy must have been awfully wicked people, if it was right to go all that way over the sea on purpose to kill them.'"

Between 1914 and 1918, an unsurpassed international herd of more than sixteen million horses, donkeys, and mules were conscripted to fight in the First World War, suffering eight million killed. It remains the bloodiest conflict for horses in the history of warfare. The human butcher's bill was equally horrific, bested only by its encore enflamed just two decades later. Of the mind-boggling sixty-eight million men mobilized, roughly ten million died and another twenty-eight million were wounded. Approximately six million civilians suffered war-related deaths. All of this might be more than anyone, horse or human, can understand.

During the industrial slaughter of the World Wars, however, our undying martial union and shared mortal fate became more entwined than ever. Both horses and humans were little more than cannon fodder; statistics on a ledger of death.

The First World War was the natural battlefield extension of the transition from horse to machine that was already unfolding across cities and farms. It was a conflict that transpired in the midst of a global identity crisis. The Great War was waged during a historical epoch that straddled the modern motorized era and a waning horse-powered existence.

While aging generals clung to the pageantry and tradition of cavalry, outside of the Levant it was somewhat of an anachronism in a relatively stagnant conflict dominated by trench warfare, barbed wire, poison gas, grenades, machine guns, and artillery. The gallant mounted charge was an archaic relic from a lingering age of feudalism, aristocratic chivalry, and an increasingly spurned social hierarchy.

Only 15 percent to 20 percent of Great War horses were earmarked for "riding purposes." These mounts were retained for dispatch delivery and rear-echelon duties, as cavalrymen increasingly fought dismounted as conventional trench-variety infantry. There were, however, isolated and modest cavalry actions highlighted by the limited achievements of Russian horsemen in the early years of the war and by the Canadian Cavalry Brigade in 1918.

General Edmund Allenby's brilliant maneuvers at Megiddo in September 1918, and his subsequent capture of Amman, Damascus, and Aleppo, were really the last successful large-scale cavalry operation. These triumphs received acclaim in part because they evinced an elegant and chivalrous, if not exaggerated, contrast to the bloodbath unfolding in the futile trenches of the western front. "His success gave a polish to the old shield of cavalry, letting it simmer in the desert sun," notes Ulrich Raulff. "Most onlookers were intoxicated by the image of the old aristocratic cavalrymen gracing the same ground in the theatre of war that had once been ridden by the splendid cavalries of Alexander the Great and Napoleon."

From its ancient Assyrian origins to its sighing exhale under Allenby, after more than 2,600 years of faithful service, the cavalry horse was finally put to pasture. The Horse Age was coming to an explosive close.

While cavalry was generally retired from the battlefield, horses and mules maintained their preeminent position as the principal means of transport for all belligerents. "The horse as a traction engine," states Raulff, "experienced a sinister boom." At its peak strength of 388,000 personnel in July 1918, for example, the ratio of animal to mechanical transport in the Canadian Expeditionary Force was twenty-two to one.

Although the First World War witnessed the introduction of trains, planes, trucks, tanks, motorcycles, and armored cars, it was still a war of horses. "Machines were everywhere on the battlefield," emphasizes military historian Andrew Iarocci, "but none could match the speed and agility of horses. At least as much as previous conflicts, the First World War was driven by horsepower." They supplied the invaluable, yet unsung, grunt work, towing artillery, massive shells, and fuel for these newfangled machines. They lugged wagons loaded with timber, armaments, ammunition, rations, water, and every other conceivable article of war. Horses hauled ambulances full of wounded soldiers to field hospitals, passing their equine comrades headed in the opposite direction shuttling reinforcements to the front.

The nature of combat and the sheer unprecedented scale of the war necessitated the mass mobilization and immense contribution of transport horses. A division during the Franco-Prussian War (1870–71), for example, required roughly 50 tons of supplies per day. By 1916, on the western front, this figure had tripled to 150 tons (and to 650 tons by 1944).

Put another way, at Waterloo in 1815, Napoléon's 246 artillery pieces fired a total of 24,500 rounds. Over three days at Gettysburg in 1863, the Union launched 32,814 rounds from 372 artillery guns. The Prussians fired 33,200 rounds at the French throughout the Battle of Sedan (northeastern France) in 1870. During the two-week artillery barrage preceding the 1917 Battle of Messines, on the western front, almost 3,000 Allied guns fired *3.75 million* shells, prompting Major General Charles Harington to remark on the eve of battle: "Gentlemen, we may not make history tomorrow, but we shall certainly change the geography." As industrial war advanced, horses still supplied the weaponry to accelerate the slaughter.

But horses had their limitations. "That the old modes of transport were inadequate to handle the demands of modern war," remarks military theorist Martin van Creveld, "is demonstrated by the permanently fixed lines of trenches that were the hallmark of World War I." The trenches themselves were a repercussion of a protracted stalemate along the front.

As a war of movement and the fruitless outflanking campaigns of the summer and fall of 1914 gave way to winter, the soldiers on the western front dug in facing one another across no-man's-land held only by the rotting dead. The 435-mile-long trench line, snaking like an angry scar from the Swiss Alps to the Belgian coast on the North Sea, remained relatively static until the spring of 1918. Stalemate and attrition remained the order of the day.

The carnage of the Battle of the Somme, from July to November 1916, reached unprecedented proportions, epitomizing the futile struggle unfolding on the cratered, decomposing battlefields of Europe. In total, in less than five months, one and a half million men were killed, wounded, or missing in a space covering no more than twelve miles of trench line and an Allied axis-of-advance reaching a paltry six miles.

Paul Bäumer, the German protagonist narrator of Erich Maria Remarque's enduring literary masterpiece *All Quiet on the Western Front* (1928), provides a human face for these statistics:

> We see men living with their skulls blown open; we see soldiers run with their two feet cut off, they stagger on their splintered stumps into the next shell-hole . . . another goes to the dressing station and over his clasped hands bulge his intestines; we see men without mouths, without jaws, without faces; we find one man who has held the artery of his arm in his teeth for two hours in order not to bleed to death. The sun goes down, night comes, the shells whine, life is at an end.
>
> Still, the little piece of convulsed earth in which we lie is held. We have yielded no more than a few hundred yards of it as a prize to the enemy. But on every yard there lies a dead man.

On these disemboweled, meat-grinding battlefields bathed in the stench of poison gas, death, and decay, horses faced all the same dangers, ghoulish horrors, and unimaginable suffering. Private Christopher Massie of the Royal Army Medical Corps summed up the respect most troops held for their horses. "He is a soldier. . . . He is a mate of ours—one of us. A Tommy." Soldiers of all stripes wrote empathetically about horses who became central figures in their war experience.

It is beyond any words of mine to attempt to convey with any emotional integrity just how much they cared for their equine comrades. As famed British soldier-poet Siegfried Sassoon stated bluntly, "I dislike the idea of a lot of good horses being killed and wounded." The forthright compassion and profound sorrow soldiers felt for these innocent animals permeate their letters, diaries, and memoirs. In many instances, the insidious sights, smells, and sounds of dead and wounded horses haunted them a great deal more than those of humans.

While it is too lengthy to quote in its entirety, one of the most poignant narratives in *All Quiet on the Western Front* is the vivid, ethereal description of the tortured shrieking of mutilated horses. "It is not men," laments Bäumer, "they could not cry so terribly . . . It's unendurable. It is the moaning of the world, it is the martyred creation, wild with anguish, filled with terror, and groaning. . . . We sit down and hold our ears. But this appalling noise, these groans and screams penetrate, they penetrate everywhere." His comrade and friend, Detering, a tenderhearted, homesick German farmer, mournfully agrees: "I tell you it is the vilest baseness to use horses in the war." This was a common refrain. "To me," wrote Private David Polley, "one of the beastliest things of the whole war was the way animals had to suffer. . . . [M]any a gallant horse or mule who had his entrails torn out by a lump of shell was finer in every way than some of the human creatures he was serving."

It is estimated that only 20 percent to 25 percent of the 8 million equine fatalities were caused by shell and shot. For example, of the 256,000 horses and mules lost by the British Expeditionary Force on the western front, only 58,000 resulted from enemy fire. In total, the British employed 1.2 million horses across all theaters of war, suffering 484,000 dead—one horse for every two men.* The horse and mule count of the American Expeditionary Force (AEF) during its brief foray into the war, beginning in 1917, was 68,000 killed out of 243,000. The French and Germans each lost roughly a million horses from respective totals of 1.5 million. Russian horse estimates are crude but hover around 2.5 million,

---

* By comparison, 326,000 of the 494,000 British horses deployed to South Africa during the Boer War (1899–1902) died.

with death rates comparable to those of the French and Germans, at 65 percent to 70 percent. Most horses died of disease and starvation, or were euthanized for shell shock (PTSD), burns, lameness (primarily from stepping on nails), trench foot, and blindness, blisters, and respiratory distress caused by poison gas.

One of the most communal memories was of mired horses sinking slowly into the endless depths of thick, swallowing mud. "Tales got around of men slipping from the duckboard paths in darkness and sinking inch by inch, in some cases to drown in the liquid sea of mud around them," remembered Lieutenant Reginald Hancock, a veterinary officer with the Royal Field Artillery. "Some of my battery horses suffered thus and either drowned slowly by inches or had to be shot before they did so. . . . I was summoned to half a dozen horses just visible above the ground. There was nothing to do but shoot them."

A British private named Sydney Smith shared a similarly nightmarish scene: "I had the terrible experience to witness three horses and six men disappear completely under the mud. It was a sight that will live forever in my memory, the cries of the trapped soldiers were indescribable, as they struggled to free themselves. The last horse went to a muddy grave, keeping his nostrils above the slush until the last second. A spurt of mud told me it was all over."

Some lucky horses cheated death, like that of Driver Percival Glock, who spent twelve exhausting hours extricating his equine friend from the suffocating grip of a sludge-filled shell hole. Others received proper medical attention. Over the course of the war, 2.56 million horses and mules passed through veterinary hospitals, with nearly 2 million returned to duty.

Of the total 725,216 horses treated by the British Army Veterinary Corps, for example, 529,064 (73 percent) were saved. Major General Sir John Moore, the director of veterinary services, praised the horse as "a weapon in the hands of the Allies that went a long way towards the fall of the enemy." By the summer of 1918, Moore had 133 equine-specific hospitals and aid stations run by 17,200 personnel under his command in France and Belgium.

A young officer in charge of fifty horses complained to Major

"He is a soldier. He is a mate of ours—one of us": British soldiers try to rescue a horse (and water cart) from a muddy grave during the Battle of Passchendaele, Belgium, August 1917. *(National Army Museum/Alamy Stock Photo)*

General Moore about the poor quality of oats. "Well, sir," he explained, lobbying for his voiceless equine comrades, "they're so small they get stuck in the horse's teeth," to which his unamused superior barked, "That's tough. You'd better indent for a supply of toothpicks."

The 5.95 million tons of British horse fodder sent to the western front outweighed all shipments of ammunition (5.27 million tons). Horse fodder represented the largest cargo unloaded at French ports and was, in fact, the single most demanded logistical item of the entire war. At the outbreak of hostilities in August 1914, no one could have envisioned or predicted the enormous equine armies—and accompanying medical and provisioning infrastructure—that would eventually dominate the war effort.

When Britain and her empire mobilized for war, the military was custodian to a mere 19,000 horses and 6,000 mules. Within two weeks, the BEF had acquired 165,000 horses. Like Joey, the equine hero of the First World War children's novel *War Horse* (1982), later adapted for the big screen by Steven Spielberg, most were appropriated from civilians through the legal auspices of the Impressment of Horses and Horse-Drawn Vehicles in Time of National Emergency Act.*

Although the United States officially remained neutral until April 1917, it served as the primary Allied stockyard, shipping 1.35 million horses and 270,000 mules to Europe throughout the war. Canada supplied another 57,000 horses and 18,000 mules, in addition to its own military herd of over 100,000. An international collection of horses and mules, not machines, were the primary engines of war.

For instance, when the British Expeditionary Force first engaged the Germans on August 23, 1914, near Mons, Belgium, it had just 827 motor cars—of which 747 were donated by civilian benefactors—and only 15 motorcycles. On the final day of combat, Armistice Day, November 11, 1918, the operational motorized British transport fleet (excluding tanks and armored cars), totaled only 122,000, while France steered another 70,000 vehicles. During its six months of major combat operations beginning in June 1918, the American Expeditionary Force utilized 105,000 trucks and 4,000 planes. More importantly, the United States supplied a whopping 80 percent of all Allied petroleum demands. The Germans, by contrast, produced only 20 of their bulky A7V tanks, and had only 23,000 operational transport vehicles and 2,270 aircraft in the field at the Armistice. For all nations, however, these mechanized elements were subordinate to the enlistment of organic horsepower.

At the outbreak of war, Germany impressed civilian horses and quickly raised 615,000. Requisition from the home front continued while the Imperial army also seized horses from occupied territory, including 375,000 from France. Germany also looked outside of Europe to supplement its herds.

Robert Skinner, the US consul general in Berlin, was quoted in *The New York Times Magazine* in May 1914 expressing concern that covert German agents were advertising "in certain American newspapers for 500 American thoroughbreds, 1,000 more or less pedigreed horses, and 1,000 draught horses for artillery use." Most likely, this was typical pro-

---

* The British purchased ten thousand Fordson tractors to help offset the agricultural handicap of requisitioned horses.

paganda attempting to shift paranoid American public opinion toward Britain and France as war loomed on the horizon. There was, however, genuine German espionage on US soil during the war to subvert the vital shipments of horses bound for Allied stables and active fronts.

Anton Dilger, born in Virginia to German immigrants, earned his doctor of medicine from the University of Heidelberg in 1912. With a specialty in animal cellular biology and in vitro tissue cultures, in October 1915 he returned to the United States, concealing four glass vials containing lethal contents.

Teaming up with his brother Carl, he rented an unassuming cottage-turned-laboratory six miles from the White House, where he proceeded to culture the causative bacterial agents of anthrax and glanders.* Dr. Dilger, whose father had won the Medal of Honor for his Civil War heroics as a Union officer at the Battle of Chancellorsville in 1863, spearheaded a German equine biological warfare program in the United States. With orders from the German general staff in Berlin, Dilger was hell-bent on crippling the supply chain of Allied horses.

Baltimore stevedores, recruited and handsomely paid by German spies, were given ampoules with instructions to methodically "paint" the liquid bacterium in the nostrils of horses awaiting shipment in the dockyards. These saboteurs reported later that they delivered the pathogen through injection, edible sugar cubes, and water troughs. After abandoning the American program in August 1916, Dilger continued his covert operations under various aliases in Mexico and then Spain, where he died at age thirty-four of influenza in October 1918.

The roving 1918–19 flu epidemic, abetted by cramped and squalid trench conditions and repatriation centers harboring soldiers awaiting return to all four corners of the globe, infected more than five hundred million people and killed seventy-five million to one hundred million worldwide—five times more than the war that helped it go viral.

This altogether strange and surreptitious episode of equine biological warfare was not uncovered until 1930, when the dockworkers, who went unpunished, emerged from secrecy. They produced detailed testimony of

---

* The First World War marked the first use of anthrax as a biological weapon.

Dilger, his macabre machinations, and their collusion in his nefarious plot to assassinate Allied horse herds. Given the amateur delivery methods of his unviable cultures, the intrigues of Dilger and his accomplices failed to produce the desired epizootic. The Germans would have to win the war of horses another way. As the conflict progressed, however, this was an increasingly gloomy prospect.

In an autobiography written after the war, the German quartermaster in chief General Erich Ludendorff reflected, "Our horses were getting worse and worse, and remounts came forward slowly. We had to make lorries to replace horse transport, although here, too, we were met with difficulties in the supply of materials. . . . The casualties in horses were extraordinarily high, and the import from neutral countries was hardly worth reckoning. The homeland and the occupied districts could not cover the losses."

He also understood imperatively that "We could not carry on the war in the west without the horses from the Ukraine." In 1918 the Germans commenced rapid invasions into the Baltic states, southern Russia, and Ukraine. Although they secured 140,000 horses from the birthplace of domestication on the Pontic-Caspian Steppe, these were not enough to turn the tide.

An Allied naval blockade was finally beginning to strangle German imports, and the German war machine was quickly and literally grinding to a halt. Domestic grain yields for 1917 decreased by 48 percent from 1914, while the potato harvest fell by 15 percent. A dire lack of steel, petroleum, fodder, and other crucial war resources severely hampered German industrial and military capabilities. Harsh rationing of all commodities on the home front, including horses, created desperate conditions and undermined the morale of a steadily starving population. Lastly, dwindling German manpower could not endure the arrival of legitimate American combat power in the spring of 1918 and the vast potential of Yankee motor pools and waves of freshly drafted doughboys. In November the guns fell quiet on the western front—for this war anyway.

In *War Horse*, the iconic Joey and his adolescent owner, Albert, return to England. "Both of us were received like conquering heroes," says Albert, "but we both knew that the real heroes had not come home, that they were lying out in France." In reality, only 62,000 equine veterans like Joey ever made it back home.

While a handful were reunited with their former owners, most were auctioned off. That they fetched an average of £37 each—equivalent to two years' base pay of a private in the BEF—reflects the critical shortage of horses on the home front. Most British horses, however, remained in Europe in some form: 56,000 were put down, 61,000 "unserviceables" were sold for meat, while another 320,000 were auctioned to local farmers. Of the 136,000 Australian horses, only *one* made it back down under.*

Horses were not the only animals that had a front row seat to the slaughter of the Great War. At its operational peak in August 1917, for example, the total British animal army consisted of 591,000 horses, 213,000 mules, 47,000 camels, 11,000 oxen, and thousands of messenger dogs and pigeons. At this time, there were more than 3 million horses on the western front alone.

In addition to these invaluable service animals, military units and individual troops kept a zoological "Tommy's Ark" of pets and mascots, including tigers, lions, bears, birds, cats, foxes, monkeys, pigs, goats, and every other creature imaginable. The most famous of these mascots was none other than the naïve, slow-witted, but optimistic and steadfast Winnie-the-Pooh.

The character is based on a Canadian black bear cub smuggled across the Atlantic by Lieutenant Harry Colebourn of the Royal Canadian Army Veterinary Corps. Winnie, named after his hometown of Winnipeg, loved to chase horses and became a cherished member of the Fort Garry Horse—one of the few regiments to remain mounted cavalry throughout the war. When Colebourn deployed to the front in 1915, he donated his girl Winnie to the London Zoo, where she became a favorite attraction of author Alan Alexander "A. A." Milne and his son, Christopher Robin. Winnie and Colebourn, who survived the war, are

---

* As a symbolic gesture, Sandy, belonging to Major General Sir William Bridges, who was killed at Gallipoli in 1915, was the sole horse to return to Australia. To limit transportation and quarantine costs, of the 13,000 surviving Aussie service horses, 2,000 were put down and 11,000 disembarked in India.

Winnie-the-Pooh: Lieutenant Harry Colebourn of the Royal Canadian Army Veterinary Corps with Winnie and a Fort Garry horse on Salisbury Plain, England, 1914. *(Lindsay Mattick/Colebourn Family)*

commemorated by numerous plaques and statues, including those at the London and Winnipeg Zoos.

Tributes honoring warhorses from all conflicts dot the planet, including: the humbly inspiring Boer War Horse Memorial in Port Elizabeth, South Africa; three versions of the American Civil War Unbridled Veterans; those honoring the highly decorated Korean War horse Sergeant Reckless of the US Marine Corps; and horse-specific memorials in London, Australia's Canberra and Melbourne, and Ottawa, Canada.

Unveiled at Hyde Park, London, in 2004, the beautifully somber and stirring Animals in War Memorial states simply: "They had no choice. . . . [T]hey all played a vital role in every region of the world in the cause of human freedom. Their contribution must never be forgotten." Horses made other monumental lifesaving donations to the war effort—and humanity—that largely went unnoticed. Millions of people, including many of us today, owe our lives directly to the horse.

Amid the sickening bloodshed and twisted wreckage of the First World War, medical teams enlisted heroic horses to produce miracle vaccines for the severe bacterial infections tetanus and diphtheria, which had fatality rates between 10 percent and 20 percent.* Prior to mass im-

---

* Diphtheria is caused by a bacterium that produces a toxin that forms a tough, thick membrane that sticks to the nose, throat, and mouth. Eventually, the narrowing of the airways causes lethal respiratory distress by "asphyxiating the victim," earning it the ghastly

They had no choice: Animals in War Memorial, Hyde Park, London. *(Jeffrey Blackler/Alamy Stock Photo)*

munization programs beginning in the United States and Canada in the mid-1920s, diphtheria was the number one killer of children under fourteen years of age. Roughly 20 percent of those under five (and over forty) who contracted the disease died.

German Nobel Prize winner Emil von Behring, known as "the savior of children," isolated the antitoxin, or antibody, serum for both diseases from horses during the 1890s. As a result of his groundbreaking experiments, serums from specifically farmed horses were collected and used to successfully inoculate humans against both tetanus and diphtheria. By 1900, the mortality rates in New York City had been cut in half. The First World War intensified ongoing research and marked a watershed in military medicine.

In preceding conflicts, roughly 65 percent of all fatalities were caused by disease. During the Great War, this was reduced to 20–25 percent,

---

nickname "the strangling angel of children." Tetanus, known as "lockjaw," is caused by bacterial spores that produce a toxin that causes muscle spasms and paralysis, or "locking." These convulsions can be so severe that they break bones and restrict breathing, leading to death.

including the 1918 influenza epidemic. Given that the bacterial spores that cause tetanus are commonly found in soil and manure and enter the body through cuts or punctures from contaminated objects, bullets, bayonets, and shrapnel were potent delivery systems. Noble steeds harboring lifesaving serum came to the rescue of frontline soldiers.

The ungainly, tousled little horse Brick Top (specimen T#17), at Connaught Laboratories in Toronto, for example, was hailed in 1918 as a real warhorse for having supplied enough antibody serum to treat fifteen thousand soldiers during his four years of service. Thanks to this unlikely hero, and his comrades, the rate of tetanus infection among wounded Canadian soldiers was reduced to a trifling 0.1 percent. Brick Top's American counterpart, "Old Dan the Retired Fire Horse," was given credit for saving one hundred thousand of Uncle Sam's finest, while Jim, a former milk wagon horse, produced eight gallons of diphtheria vaccine.

The most renowned and courageous event associated with diphtheria was not of horses but of dogs during a lethal outbreak in Nome, Alaska, in January 1925. Battling whiteout blizzard conditions and temperatures hitting -62°F (-52°C), a relay of 20 mushers and 150 sled dogs raced across 674 miles of tundra in five and a half days to successfully deliver lifesaving horse-produced vaccines to dying children. Balto, the final lead dog, became the enduring cultural emblem of the Great Race of Mercy.

Humanity owes the horse an immense debt of gratitude for its unheralded contribution to these miraculous lifesaving vaccines. Infants now receive the all-too-easy single-shot combination DTaP, or Tdap, (diphtheria, tetanus, and pertussis), sparing them from unimaginable suffering and death. As a result of mass vaccination, these diseases, among others, including polio, measles, mumps, and rubella, have largely been eradicated across the world. Since 1980, only twenty thousand cases of diphtheria and fifty thousand cases of tetanus are reported annually, predominantly in underdeveloped, unvaccinated pockets of our planet. Horses continue to offer additional medical assistance.

Specific antivenoms for deadly snake and spider bites are also produced by horses. The estrogen-rich urine from pregnant mares is used to

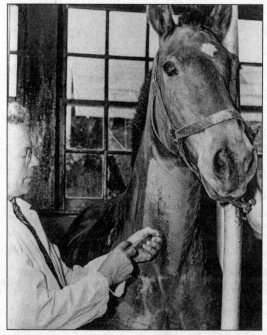

Lifesaving donations: A noble steed at a diphtheria serum and vaccine farm, Otisville, New York, 1944.
(Library of Congress)

manufacture hormone drugs to ease the passage of menopause and for birth control, although this practice has come under scrutiny in recent years. Currently only Premarin (*pre*gnant *mar*es ur*in*e) and Prempro estrogen replacement therapy medications remain on the market and are used annually by over fifteen million women worldwide, including nine million Americans.

From medicine and industry to weapons and carnage, the First World War was a turning point in human history. The conflict produced both the climax of military horse transport and the denouement of cavalry. Technology, from wristwatches and telephones to tanks and chemical weapons, forever changed the nature of warfare. For industrializing nations, the interwar years witnessed an uptick in urban and rural mechanization. The horse was progressively losing its pinnacle position as the driver of human society both on and off the battlefield. This technological evolution, however, was an unequal one.

The earlier and committed American adoption of a motorized society transferred to a fully mechanized military—although not without contention. As Brigadier General George Van Horn Moseley, commander of the US First Cavalry Division, understated in a memo to his superiors in 1929: "When the cowboy down here is herding cattle in a Ford, we must realize that the world has undergone a change." The importance of this timely jump-start, including the correlated infrastructure of steel, petroleum, factories, engineers, and mechanics, cannot be overstated when international events dictated the mobilization of a sophisticated military-industrial complex.

When Japan awoke the American sleeping giant with its preemptive strike on Pearl Harbor on December 7, 1941, the war, which had been raging in Europe for more than two years, entered a new industrial phase of motorization and mass production. Horses could not compete with the unprecedented intensity and output of America's total war economy.

The rest of the world, including Germany, lagged far behind the United States in all aspects of mechanization. For all the historical bluster about Adolf Hitler's Volkswagen "People's Car," Germany was predominantly a horse-drawn nation. It lacked the industrial foundation, proficiencies, and familiarity conducive to producing *and* maintaining a comprehensive mechanized military.

In 1937, as the world inched closer to war, the United States produced 4.8 million vehicles, while the combined output of the eventual Axis Powers of Germany, Italy, and Japan was a mere 428,000. The number of cars per 1,000 people is a revealing measure of this marked disparity: United States, 200; Britain, 48; Germany, 16; Japan, 1. German strategists and propagandists were acutely aware of their motorized shortcomings prior to the war.

On the eve of the invasion of Poland in September 1939, the army transport chief, Colonel Rudolf Gercke, notified his peers on the German general staff that "as regards transport, Germany is at the moment not ready for war." That was a furtive, obsequious way of informing the führer that only 14 of the 102 German divisions were mechanized. "Such an assessment could only reinforce the assumption that literal horsepower was going to be crucial for the army's war effort," writes military historian at Western Carolina University David Dorondo in *Riders of the Apocalypse: German Cavalry and Modern Warfare, 1870–1945*. "If and when a war between Germany and her neighbors became a war of attrition, as indeed it did as of 1942, the matter would only become more acute."

Gercke was not alone in his grim assessment of German capabilities. General Georg Thomas, director of the Defence Economy and Armament Office, circulated a detailed memorandum in late August 1939 concluding that Germany "could not last through a war on the grounds of its war economy." His evaluation proved correct.

Albert Speer, the minister of armaments and war production, could seemingly never find a way to sidestep Nazi red tape, meddlesome incompetency, and sycophantic political bickering in order to establish an effective war-based economy. He warned Hitler bluntly that his Reich would ultimately lose the struggle because of industrial mismanagement and an "arthritic organizational system" contrasted by the "organizationally simple methods" utilized by the Allies to mass-produce machines.

At one point, for example, the Germans were developing 425 different types of aircraft. While Nazi scientists built innovative jets and rockets by the end of the war, they utterly failed to deliver desperately needed frontline weapons in bulk to match the assembly-line production of Russian T-34 tanks and American Sherman tanks, B-24 bombers, and General Motors two-and-a-half-ton (the original "deuce-and-a-half") supply trucks.

To compensate for the deficit of motorized transport, the Germans tried their hand at genetically engineering the perfect military horse by combining the brute strength of heavy draft breeds with the spirited temperament and speed of Thoroughbreds. Under the leadership of hippologist Gustav Rau, experimentation to produce this equine master race was headquartered at the Army Remount Depot at Grabau-Schonboken, supplied by smaller "studding stations." Indicative of the confusion and competition that plagued the German war economy, the Nazi Schutzstaffel (SS) also established a branch devoted to equine eugenics.

Both programs met with dismal failure, in part because the select stallions became so overtaxed and drained sexually, they could no longer sire offspring. It seems the Nazi production of horses was marred by the same industrial inefficiency and bureaucratic bungling as that of trucks, tanks, planes, and other critical weapons of war.

Hitler drastically overestimated his own strength, or was perhaps masking his own insecurities, by mocking American economic power. "What is America," he asked his cabinet rhetorically, "but millionaires, beauty queens, stupid records, and Hollywood?" Of course, he reserved a special loathing for the ruthless, despotic premier of the Soviet Union, Joseph Stalin, and his Communist regime. In June 1941, on the eve of his Operation Barbarossa invasion of the Soviet Union, Hitler boasted confidently that "Bolshevism will collapse like a pack of cards." On both fronts, American industry and Soviet resolve exposed Hitler's malignant narcissism. In the final count, German horses, American machines, and Russian manpower won the war.

For the Allied nations—primarily the United States—who were ahead of the motor curve, the unrivaled presence of horses on the German front lines facilitated the overthrow of the Third Reich, which feigned mechanized superiority at the outbreak of war. Nazi blitzkrieg was just another bluff based on Hitler's Big Lie. It was carefully crafted propaganda to hide the true vulnerabilities of an anemic and backward German military equipped for the *last* war while attempting to fight the next. The Nazis had far more ponies than panzers. Blitzkrieg, as it turns out, was fueled not by oil but by oats.

The German army that goose-stepped and galloped through Poland, Denmark, Norway, Belgium, the Netherlands, and France in 1939 and 1940 was only 15 percent mechanized at best. As motorized casualties mounted and German industry faltered, the horse was increasingly conscripted into service. By 1944, more than 90 percent of the German military relied on hooves for transport. "Too often, the average person thinks of the German army as a mechanized juggernaut, having been totally deceived by marvellous German war footage showing German armoured columns rampaging across the Russian Steppes, or Erwin Rommel dashing off into the desert at the head of his panzers,"

"The German Mazeppa. 'Away! Away! My breath was gone, I saw not where he hurried on. . . .'": This cartoon is based on the 1819 poem *Mazeppa* by Lord Byron, loosely recounting the popular legend of Ivan Mazepa (1639–1709), a Ukrainian military leader. A youthful Mazepa has a love affair with a married countess while serving as a page at the court of the Polish king John II Casimir Vasa. On discovering the affair, the incensed count punishes Mazepa by tying him naked to a feral horse. The bulk of the poem describes the traumatic journey of the hero strapped to the runaway horse, heading into the great unknown. The mustached horse is Hitler (who had become chancellor of Germany in January 1933), leading a strapped and helpless Germany on a traumatic journey into the great unknown future—which came to pass as the most devastating war in human history, complete with genocidal holocaust and nuclear weapons. (Punch *magazine, July 1933*)

stresses military historian R. L. DiNardo, author of the thoroughly researched *Mechanized Juggernaut or Military Anachronism?: Horses and the German Army of WWII*. "The harsh reality, however, is that the basic means of transport in the German army was well known to Alexander, Hannibal, Caesar, Gustavus, Marlborough, Frederick, and Napoleon: namely, the horse."

The fundamental impression of the German army as some unbeatable, mechanized monster moving with lightning speed and imposing strength persists across general society and academia alike. Horse-based reality, however, reveals an entirely conflicting truth. The fact that this mechanized blitzkrieg myth persists more than seventy-five years later is a credit to the Nazi minister of propaganda, Joseph Goebbels, and his cunningly crafted and mind-spinning Big Lie disinformation campaign. His calculated manipulation of the ignorant masses (and of our notions of blitzkrieg) was nothing short of genius and bestowed Hitler, an Austrian corporal and political neophyte, with cult-of-personality authoritarian power.

Prior to and during the war, Goebbels released newsreel films to domestic and international audiences showcasing the sophisticated, mechanized components of the German military. This was done clearly to impress the German people and intimidate the awestruck opposition. Later clips were infused with wartime footage showing the beast of blitzkrieg in all its terrifying splendor. The screens flashed with Stuka dive bombers shrieking their mounted ram-air sirens, thundering tanks crashing through enemy fortifications, and motorized quick-strike infantry hurling stick grenades, hammering rounds from machine guns, and hurdling trenches. Horses, however, are conspicuously absent.

This footage was in turn used as propaganda by the Allies, including legendary American filmmaker Frank Capra, commissioned by the War Department to actually *promote* the myth of a German "mechanized juggernaut" to scare people into wartime action, stoke the war effort on the home front, and encourage enlistment.

These enduring propaganda clips spin-doctored by Goebbels are still spliced into the modern documentaries we watch today. There is perhaps a touch of irony to the fact that Goebbels used the medium of film, forged by the horse, to hide their overwhelming presence in the German armed forces, or Wehrmacht.

In truth, Germany, which was home to 3.8 million horses at the outbreak of the Second World War, employed over a million more of these animals than in the previous war. Roughly 2.7 million horses, including 1.21 million confiscated from occupied territories, were thrust into military service, suffering 1.7 million dead. "At the beginning of the Second World War," writes German historian Heinz Meyer, "an infantry division possessed more than twice as many horses as an equivalent division in the First World War. The greater number of heavy weapons and the more widespread use of such equipment necessitated this increase in the supply of horses. In the nonmotorized troops of the First World War, there was one horse for every seven men; in the Second World War, it was more like one horse for every four soldiers." The supply tonnage for an average combat division increased 77 percent from that of its Great War counterpart. Horses were the main mode of transport for the German army. The Wehrmacht began the war with 590,000 horses, many arrogated during Hitler's prewar Lebensraum land grabs.

Although the invasion of Poland was a lopsided campaign, during the four weeks of fighting Germany lost 50 percent of its transport vehicles and three hundred tanks. The famous, and favorite, history class knight's tale of brave Polish cavalry armed with lances defending their homeland against evil Nazi panzers on the opening day of the war was, again, more the product of Reich propaganda than reality. In this version, the Poles were ready to defy the modern military age and overwhelming German firepower from horseback.

What occurred on September 1, 1939, was a chance encounter between a disoriented and ensnared cavalry regiment and an encircling German mechanized force. Horse and rider were cut to pieces in minutes as they attempted to escape by charging through gaps in the constricting German line. Goebbels seized on the tragic episode to portray the horse-drawn Nazi army as some mechanized Goliath. Widely publicized in strictly controlled media, the event was concocted to convince the world—notably Britain and France—of the invincibility of the führer's war machine and that resistance was indeed futile.

As the fighting intensified, the critical demand for vehicles always outran the Reich's production capacity. "No matter how skillful the leadership of the German Army was in operational art and on the battlefield, combat competence could not make up for the fact that the great bulk of the army marched to war with its equipment drawn by horses," emphasizes renowned military historian and prolific author Williamson Murray. "The disparity between the armored, motorized elite and the regular, plodding infantry became more glaring as the war continued. No level of military competence could bridge that gap." The smokescreen shrouding the charade of blitzkrieg was the overwhelming German concentration of their meager mechanized forces.

By necessity, limited air, armor, and motorized elements were assembled in a few elite divisions at the tip of the offensive wedge to maximize striking power, rather than being wastefully dispersed piecemeal across horse-drawn formations. The bulk of the lumbering follow-on forces relied on the Great War transport of horses, railways, and boots. "These units would prove decisive in the early campaigns of the war yet also created two very different armies, 'one fast and mobile and the other slow and plodding,'" writes DiNardo. "While this dual nature did not become a major factor in Poland and France . . . it did play a big part in the ultimate failure of Operation Barbarossa. . . . The defects in Germany's army, economy, and approach to war finally became evident."

Over short distances, as with the opening campaigns in Poland, western Europe, and the Balkans, this dichotomy did not present itself as a glaring problem or military handicap. The battle spaces in these areas of operation were not prohibitive, allowing the trailing infantry and supply columns to stay reasonably tethered to the forward mechanized units. These quick victories were fought over relatively narrow fronts where German invasion forces could maximize their advantages against unprepared and comparatively ill-equipped enemies. "The army that was to shake the West was a scavenger," notes Van Creveld. "Its hopes for victory rested in part on the capture and utilization of French, Dutch, and Belgian vehicles." Dependence on horses, or unreliable foreign transport, generally did not hinder the execution of these early operations.

It should be noted, however, that as the invasion of France came to a close in June 1940—after just six weeks—the armored spearhead had outpaced the strained horse-drawn infantry and logistical units, which struggled to deliver vital petrol and ammunition. During the advance on Dunkirk, France, in late May, horses were succumbing to exhaustion even with four thousand equine reinforcements arriving weekly from Poland.

According to Jilly Cooper in *Animals in War*, "Many of the horses pulling ammunition and galloping about at Dunkirk were ironically seen to have British First World War brandings." These were no doubt those (now old) horses auctioned off to local farmers in 1918 and impounded by the Germans in 1940. The German offensive ground to a halt (the reasons are still debated), allowing four hundred thousand British and French troops stranded on the beach at Dunkirk to miraculously slip away across the Channel back to England aboard a motley flotilla of military and civilian watercraft.

The German occupation of the Low Countries and France—notably the famed breeding grounds of William the Conqueror in Normandy—bolstered depleted transport stocks. Hitler's horses, however, never got the chance to embark on a copycat invasion of England. The führer had ordered an overambitious amphibious assault, dubbed Operation Sea Lion, to begin in late August, with an initial wave of 170,000 men and 57,500 horses.

The German Luftwaffe (air force) would first relentlessly pummel British cities in what became known as the Battle of Britain, or the Blitz. London alone withstood fifty-seven consecutive nights of devastating bombing raids. The Royal Air Force (RAF), despite being outnumbered, heroically repelled enemy planes, depleting Reich Marshal Hermann Göring's Luftwaffe to the point where, by the end of October, with winter approaching, Hitler had no choice but to postpone Operation Sea Lion indefinitely.* It would never be launched.

With the British Isles isolated and relatively impotent, in the spring of 1941, Hitler turned his overconfident, haughty gaze eastward. Gambling away his good fortune in a game of Russian roulette, the inherent

---

* It should be noted that more than 20 percent of RAF pilots during the Battle of Britain came from allied countries: Canada, Poland, New Zealand, Australia, Czechoslovakia, Belgium, South Africa, and even nine airmen from the still neutral United States.

limitations of the führer's horse-drawn military became starkly exposed. The myth of blitzkrieg, and successive German army groups, evaporated at the gates of Leningrad, Moscow, the Caucasus, and Stalingrad.

The massive invasion force that stood primed on the Soviet frontier consisted of 3,000,000 men, 500,000 wheeled vehicles of all types, 3,350 tanks, 2,000 aircraft, and upwards of 750,000 horses. The seemingly respectable number of vehicles, however, betrays their tactical value. The Germans used more than two thousand different types of vehicles, many of them antiquated French, Belgian, and Czech designs with nonexistent spare parts, creating a bewildering nightmare for supply and maintenance crews. Broken-down vehicles were immediately cannibalized for parts to keep others alive for the time being.

Extreme distances, barren steppes, harsh climate, negligible infrastructure, and a determined enemy cordially welcomed the Germans to the Soviet Union. "It was generally recognized," explains Van Creveld, "that the Russians would have to be defeated within the first 500 km (300 miles) if they were to be defeated at all." Protracted warfare over boundless spaces favored the defending Soviet forces. German transport horses proved to be a liability in what Soviet strategist and general Mikhail "the Red Napoléon" Tukhachevsky coined the "deep battle." They restricted striking range by acting as the brakes on the lead mechanized elements, directly impacting the operational conduct of the German offensive.

Instead of effecting sweeping pincer movements across a wide front, the Germans were relegated to conducting small encirclements to keep the horse-trotting supply "tail" attached to the leading mechanized "teeth." They could not cast a large enough net to catch the Russians, who, like the ancient Scythians, methodically withdrew, exacting a scorched-earth policy, slowly sucking their pursuing enemy farther into the endless steppes. For the Germans, this was an immediate strategic miscalculation in a campaign that was designed to encircle and destroy the bulk of Soviet forces before the debilitating winter freeze. Like Napoléon and his ill-fated 1812 overture, Hitler also failed to fully appreciate the vast expanses, severe weather, and iron-forged will of the Russian people.

The ineffectiveness of German horse transport was compounded by the complete absence of modern infrastructure. "The Russian campaign, with its structural conditions, belongs back in the horse age," conceded German historian and veteran of the eastern front Reinhart Koselleck. "It could not be won with horses, and certainly not without them." These structural conditions included only fifty-one thousand miles of incompatible railway across the entire Soviet Union.

Behind the mechanized vanguard, inadequate numbers of *Eisenbahntruppen* (railway troops) construction crews feverishly converted the Russian rails to the German gauge. It would, however, never be enough to make up for the lack of basic motor transport pushing forward the needs of the front line. Of the 850,000 miles of "road," only 18 percent met the standard European definition. Most were little more than muddy wagon tracks with two wheel ruts melting endlessly into the steppe.

Conscripted Russian *panje* horses heightened the logistical impasse. Although hardy and resilient, at thirteen to fourteen hands, they were too small to pull the standard German 105-millimeter howitzers, heavy steel transport wagons, and winter sleds. Customarily hauled as a single unit with gun, two-wheeled limber, and ammunition, the four-ton 105-millimeter howitzer required a traction team of six heavy draft horses. The more cumbersome 150-millimeter howitzer, which was transported in pieces, required eighteen horses. Only when the Germans produced a lighter, but lower volume, wagon toward the end of 1942 or utilized humble peasant carts did these stout horses become serviceable. The lack of suitable roads, viscous mud, and weather conditions made any form of transportation, including horses, a sluggish and perilous undertaking.

By the onset of winter in 1941, the undefeated German war machine was floundering in the mire. Instead of transferring supplies, horses were now dredging and freeing trucks from engulfing sludge. "Even the smallest vehicles," recalled a German soldier, "were often only loosened from the spot by a team of four [horses]." In November, for example, the Germans lost 5,996 supply trucks, while Reich factories churned out only half that number, at 2,752. The capacity of the substitute horse-pulled sleds was far below combat requirements. Of the 500,000 vehicles that left the line of departure in the spring, only 75,000, a mere 15 percent, were still operational.

Amid the Soviet counteroffensives pushing west from Moscow in the bitterly cold winter of 1941–42, more than 180,000 German horses—20 percent—died of combat, starvation, disease, and hypothermia. "Of course, difficulties were the greatest during the winter of 1941–42, the first winter of the campaign against Russia," wrote a cadre of senior German officers in their postwar report *Horses in the Russian Campaign*, commissioned by the US Army Historical Division. "The supply systems had yet to be organized; the wide-gauge Russian railroads had to be converted to the narrower German gauge. At the beginning of the winter, no fodder inventories of any kind were on hand. . . . The high mortality rate of horses could not be checked. One thousand horses perished of exhaustion every day. One could say that they had literally starved to death. Under these conditions, large numbers of horses could no longer be kept at the front. . . . Thousands of horses contracted heart and lung diseases; they either wasted away or died suddenly. The loyal horse toiled in the harness until it staggered or fell." By March 1942, the Germans had lost 265,000 horses.

Even the 200,000 horse reinforcements brought in from Germany and occupied Europe for the rejuvenated southern campaign in the

The Big Lie of blitzkrieg: A team of German transport horses flounder in a quagmire wasteland on the Russian front near Kursk, March 1942. *(Das Bundesarchiv)*

spring of 1942 to seize the critical oil of the Caucasus did little to change the strategic situation. "The capture of the Caucasus would kill two birds with one stone," writes military historian Richard Overy in his analytical masterpiece *Why the Allies Won*: "The Soviet armies would be deprived of the oil needed to fight, and Germany would capture the oil she required to combat Britain and the United States."

But when the Germans reached the northern Caucasus oil fields of Maykop, which produced two million tons annually, they found the wells, refineries, and stockpiles torched by the retreating Russians. German forces were halted only eighty miles west of Grozny—a city with an annual petroleum production that exceeded all German supplies. Farther east, the primary target of Baku, and its twenty million annual tons of oil (three times Germany's yearly consumption), also remained out of reach. Indeed, Hitler was right when he had prophesied to his general staff in August 1942, "Unless we get the Baku oil, the war is lost."

Still clinging to delusions of invincibility, by this time, Hitler was out of touch with the situation on the ground. He demanded of his generals: "It must not happen that, by advancing too quickly and too far, armored and motorized formations lose connection with the infantry following them; or that they lose the opportunity of supporting the hard-pressed, forward-fighting infantry by direct attacks on the rear of the encircled Russians." To accomplish this, another 400,000 horses and a paltry 59,000 trucks reinforced a German army of more than 3 million men. Throughout 1943, an additional 380,000 horses were funneled into the maelstrom.

However, no amount of conscripted Nazi horses could outpace the arsenal of mass-produced trucks, jeeps, tanks, and planes pouring off assembly lines at Ford, General Motors, Chrysler, Boeing, Lockheed, and Martin within the unparalleled military-industrial complex of America and its allies.

The Red Army, which mobilized 3.5 million horses during the Second World War—roughly a million more than in the prior war—was becoming increasingly mechanized, thanks to a retooling of domestic production in the spring of 1942 and the influx of American imports. The dynamic economic, manufacturing, and military strengths that the United States brought to bear within the Grand Alliance (to use a

Churchillian phrase) of the UK, USSR, and USA was above and beyond what German horses—or anything else on the planet, for that matter—could handle. "There is only one thing worse than fighting with allies," British prime minister Winston Churchill remarked, "and that is fighting without them!"

The Tehran Conference in late 1943 brought the big three Allied leaders—Churchill, Joseph Stalin, and US president Franklin D. Roosevelt—to Iran to coordinate military strategy. During a dinner for Churchill's sixty-ninth birthday, Stalin rose to propose one of his numerous toasts: "This is a war of engines and octanes. I drink to the American auto industry and the American oil industry." Stalin had reason to rejoice.

United States factories churned out more than 2.3 million transport vehicles of all classes during the war, 25 percent of which eventually carried the insignia of his Red Army. Meanwhile, US oil companies, buoyed by the great boom in the West beginning in the mid-1920s, pumped out six billion of the seven billion barrels of petroleum used by the Allies. During the war, roughly 90 percent of global output and 96 percent of refining capabilities were under Allied control.

American industry outfitted thirty-five million Soviet troops with the necessary tools to repulse the Nazi onslaught. "It was only the American and British [and Canadian] armed forces that could tactically resupply their troops in the line through motor transport alone, and then thanks to the unique productive capacity of the American oil industry and automobile plants," declares John Keegan in *A History of Warfare*. "So ample, indeed, were American resources that they sufficed not only to supply the US army and navy with all the trucks and fuel they required but to equip the Red Army also with 395,883 trucks and 2,700,000 tons of gasoline, thus providing the means, as the Soviets themselves freely admitted later, by which it advanced from Stalingrad to Berlin."

Uncle Sam also supplied Stalin's armies with another 151,000 light transport vehicles, 77,900 jeeps, 956,000 miles of telephone cable, 380,000 field phones, and 35,000 complete radio stations. "The American automotive industry not only made the horse redundant for all practical purposes in the U.S. Army," stresses David Dorondo, "it also helped complete the motorization and mechanization of the British ground forces, and it did a great deal to put Stalin's legions on wheels and tracks."

Canada, too, mass-produced modern armaments, including 1.1 million soldiers, 815,729 transport vehicles (more than double the output of Germany), 50,000 tanks, 43,552 artillery guns, 16,431 aircraft, 1,140 capital ships, and 1.7 million small arms. During the final years of the war, Canadian factories were cranking out 525,000 artillery shells and 25 million rifle cartridges *per week*. Canadian mines provided the Allies with a bounty of aluminum (40 percent), nickel (75 percent), asbestos (75 percent), zinc (20 percent), lead (15 percent), and copper (12 percent).

Possessing a cornucopia of natural resources, Canada also supplied massive amounts of iron ore, uranium, timber, wheat, corn, beef, and pork. By 1943, Canadian wheat yields even outpaced wartime demand. Farmers used the surplus to fatten hog numbers, so that by the following year, 7.4 million had been butchered. Other Commonwealth or allied nations, including India, Australia, New Zealand, and South Africa, also contributed significant agricultural and industrial turnouts in addition to military manpower. Hitler was not as fortunate in his choice of friends.

Total Wartime Armaments Production of the Major Powers, 1939 to 1945*

	Aircraft	Tanks	Military Trucks	Artillery	Major Vessels
USA	324,840	88,479	2,382,311	224,874	8,812
USSR	158,218	105,232	197,100	485,648	161
UK	131,549	30,396	480,943	39,800	1,156
Canada	16,431	50,000	815,729	43,552	1,140
Germany	117,881	61,700	345,914	159,147	954
Japan	79,123	4,771	165,945	13,350	589
Italy	9,890	2,908	83,000	7,200	43

Germany, like its allies Italy and Japan, was handcuffed by a lack of oil, steel, and manufacturing capacity to fuel its military machine. Of

---

* Richard Overy, *Why the Allies Won*, 331–32; Government of Canada, Veterans Affairs; Canadian War Museum; Library and Archives Canada.

the twenty essential resources required to wage mechanized war, including steel (iron), petroleum, rubber, coal, lead, copper, and aluminum, the Reich was almost entirely deficient. The imposed German dependence on horses, however, was due to a dire shortage of the twin towers of industrialization: steel and petroleum.

Axis nations controlled a mere 3 percent of oil output and 4 percent of refining. "The Allies had long regarded oil as the German Achilles' heel," stresses Overy. "It would be wrong to argue that oil determined the outcome of the war on its own, though there could scarcely have been a resource more vital to waging modern combat." Imports of oil from the Soviet Union and iron ore from Sweden dried up at the outbreak of war. Scrounging from lesser allies, domestic production, and synthetic ersatz alternatives lagged well behind what was needed to support the demands of protracted war.

Field Marshal Erwin Rommel, the legendary "Desert Fox," was so hindered by inadequate fuel supplies during his North African campaign that he lamented to his wife, "Shortage of petrol. It's enough to make one weep." His Afrika Korps reached El Alamein in Egypt in July 1942 with no fuel, a mere nineteen serviceable tanks, and an ammunition depot 1,100 miles to the rear in Tripoli, having driven back and forth across the vast—but undiscovered—oil fields of Libya.

The seventy-five Italian divisions, like their German counterparts, were also primarily horse drawn and were well below their paper numbers in both men and materials. The handful of so-called elite mechanized divisions were poorly equipped, lacked modern heavy tanks, and possessed only 350 second-rate transport vehicles. Italian industry was anemic, producing only 83,000 transport vehicles, 2,908 tanks, 7,200 artillery guns, 9,890 aircraft, and 43 capital ships over the course of the war.

In the Pacific theater, the Japanese had one transport vehicle for every 49 soldiers, whereas the American ratio was one to thirteen. "For every American soldier in the Pacific, there were 4 tons of supplies," notes Overy; "for every Japanese, a mere 2 pounds." The paucity of mechanization hampered the Axis Powers in all theaters of war, from Europe and North Africa to the Caucasus and Okinawa.

The German reliance on horses had numerous secondary repercussions that destabilized the war effort. Strategic planners were acutely aware that during the First World War, the appropriation of horses drastically undercut domestic agriculture, and they were determined not to repeat this famine-inducing misstep. Bread is as important to victory as bullets.

War exigencies, however, eventually overrode prewar sensibilities. Almost 40 percent of Wehrmacht horses were drafted from German farms, and agrarian output stagnated as a result. Between 1939 and 1944, total German yields of wheat decreased by 23 percent, barley by 39 percent, rye by 11 percent, and oats, vital for horses, by 28 percent. The Reich sought to counter domestic shortages by mercilessly seizing horses and harvests from subjugated peoples.

Soviet horse populations were devastated by both German and Russian roundups. The total crashed from 21 million in 1939 to 7.8 million in 1945, a mind-boggling war levy of 63 percent. According to the late Alexander Nove, an acclaimed Russian-born historian, of the 11.6 million horses in occupied Soviet territory, roughly 7 million were "killed or taken away."* As civilian populations succumbed to starvation, many horses were eaten. Between the loss of plow horses, which suppressed the ability to farm, and the German confiscation of fractional crops, famine gripped hardest in the Netherlands, Poland, and the Soviet Union.

No arrogation or occupation policies, however, could alleviate the gathering disparity between Allied and Axis agricultural and armament production. In 1943 alone, the Allies produced 151,000 aircraft to an Axis count of 43,000. By the following year, Allied air superiority and intensified bombing raids led to a 31 percent reduction in German aircraft assembly and a 35 percent decrease in vehicle and tank manufacturing.

---

* During the interwar years, Stalin's forced agricultural collectivization and the ensuing terror-famine of 1930 to 1933, killing upwards of 9 million people, also devastated the Soviet horse population by 47 percent, falling from 32 million in 1928 to 17 million in 1933. There is an argument that the famine in Ukraine (Holodomor), resulting in 3.5 million to 4 million deaths, represents genocide to erase Ukrainian national identity. With the Russian invasion of 2022, this historic event has become a hot topic.

Throughout the war, German division strengths were methodically reduced to counter the grave lack of motorized transport, among other dwindling munitions, progressively dulling sharp-end combat power.

Average Forward German Division Strengths, 1939 to 1945

Year	Personnel (All Ranks)	150mm Artillery	105mm Artillery	Trucks	Horse-Drawn Vehicles	Horses
1939–1940	17,734	12	36	942	1,133	5,375
1944–1945	9,069	9	30	370	1,375	3,177
Difference %	-48.9	-25.0	-16.7	-60.7	+17.6	-40.9

German forces embattled in the Soviet Union were also increasingly tied to umbilical cords of supply, which horses—and Hermann Göring's ham-fisted, sycophantic airlift promises to regain lost standing with Hitler—simply could not sustain. "The German invasion of the Soviet Union," concludes DiNardo, "failed in part because of the German Army's heavy reliance on the horse for transport." Field Marshal Friedrich Paulus and his ill-equipped, starving, and freezing Sixth Army, and its horses, found this out the hard way at Stalingrad.

During the winter campaign of 1942–43, the Sixth Army trapped at Stalingrad required a daily resupply of six hundred tons. The most it received, and only on three occasions, was three hundred tons. The situation became more desperate and untenable by the day.

In late November, when the Soviets surrounded the city, the Germans had only 25,000 horses left. Roughly 179,000 of their counterparts had been killed in combat, succumbed to starvation or disease, or were devoured by ravenous German soldiers picking clean their bleached bones with frostbitten fingers. "Here one realised particularly clearly what the last days of Stalingrad had been to so many of the Germans," wrote eyewitness Alexander Werth, a British war correspondent. "In the porch lay the skeleton of a horse, with only a few scraps of meat still clinging to its ribs. Then we came into the yard. Here lay more horses' skeletons, and to the right, there was an enormous, horrible cesspool—fortunately, frozen solid. . . . No one doubted that this was *the* turning point in World War II."

The turning point: A Junkers Ju 52 transport plane drops supplies to an awaiting horse-drawn wagon outside Stalingrad, January 1943. *(Das Bundesarchiv)*

On January 19, 1943, Paulus informed Hitler that "the last horses have been eaten up." Ignoring his führer's order to fight to the death, the newly promoted field marshal formally surrendered on February 2. "A curious odor will stick to this campaign," wrote a German prisoner of war, "this mixture of fire, sweat, and horse corpses."

The bloodiest clash in history claimed a staggering two million casualties. "The Germans lost some twenty divisions, thirteen of which were infantry. Vast equipment losses were incurred, and massive losses in motor transport could not be replaced by German industry," write R. L. DiNardo and Austin Bay in their article "Horse-Drawn Transport in the German Army" for the *Journal of Contemporary History*. "The end result was that in 1943 the German army was even more dependent upon horses." By January 1944, German motor transport was down 50 percent from its peak only a year earlier.

Stalingrad was, unequivocally, *the* decisive battle of the war. Hitler's chief of staff, Field Marshal Wilhelm Keitel, clearly understood that at Stalingrad, the führer had "played [his] last trump, and lost." Soviet propaganda branded the event as its own version of Hannibal's brilliant double-flanking, or pincer, cavalry encirclement and annihilation of the Roman legions at Cannae in 216 BCE during the Second Punic War. Following the valiant Soviet stand at Stalingrad, losing became a habit for Hitler's retreating horse-powered armies. Victory was now out of reach.

After its mauling on the eastern front by the Soviet steamroller at Stalingrad, Kursk, Belgorod-Kharkov, and Kyiv, Germany now also faced an unprecedented Allied invasion force assembling across the Channel in England. "By 1944, American and British [and Canadian] forces were fully motorized," explains Overy. "German ground forces relied on horse transport. Neither the quality nor quantity of German technical resources was sufficient to prevent the dissipation of the myth of German invincibility."

By June 6, 1944, when 160,000 Allied troops landed at Normandy, France, only 10 percent of the entire German military was mechanized. A large proportion of the 1.2 million horses fielded by the Germans in 1944 were the impounded Russian *panje* breed. The stark reality was not lost on the soldiers, who began cynically referring to their panzer divisions as "*panje* divisions." Propaganda could no longer mask the truth.

The lack of German mechanization also lent itself perfectly to Allied air superiority during the Normandy campaign. The Germans were forced to move only short distances at night. In retreat, horse-drawn artillery had to be removed from the line early to avoid capture, creating a huge disparity in firepower. As the Allies struck out from the beaches, fleeing German horse-transport columns were caught in the open and mercilessly strafed and bombed.

By the end of August, the American, British, and Canadian breakout from Normandy cost the Germans roughly 290,000 troops, 1,900 tanks, 20,000 vehicles, and 30,000 to 40,000 horses. The horrible scenes of equine devastation were extreme enough to even evoke public pity from an old warhorse who had competed in equestrian events at the 1912 Olympics: US Lieutenant General George Patton.

Although fully mechanized, the western Allies did use select, and largely locally procured, horse and mule trains in North Africa; Sicily, Italy; the Philippines; and Burma. Of the 26,403 horses acquired domestically by the US military, only 49 ever left the country—a far cry from the 1.35 million shipped overseas during the previous war. Likewise, of the 34,000 homegrown mules, only 7,800 saw combat duty.

In 1944–45 across the entire Pacific theater, for example, the Allies employed just 6,758 horses and 23,595 mules, primarily along the mountainous Burma-China-India corridor. US forces also hired 4,000 hardy mules to deliver ammunition in the rugged terrain of Italy. "They served without a word of complaint or lack of courage," remembered Sergeant Edward Rock Jr., who served in Burma. "Mules fell in battle, mortally wounded, and we shed tears for them."

My wife's grandfather Sergeant Major Walter "Rex" Raney, your archetypal American GI from a small farming community in western Colorado, personally told me a slightly different story. Rex fought with the Forty-Fifth "Thunderbird" Infantry Division across North Africa, Sicily, Salerno, Monte Cassino, Anzio, and Rome, before taking part in the Battle of the Bulge and the push over the Rhine River into Germany. While he was genuinely saddened by the loss of mules from enemy fire, with a wry grin and his typical dry wit, he told me, "Not to worry, Tim, their meat did not go to waste."

On April 29, the eve of Hitler's suicide, Rex and his comrades liberated the Dachau concentration camp outside Munich and came face-to-face with the horrors perpetrated by the führer's now crumbling Third Reich. Humans, however, were not the only ones in need of liberation from perverse Nazi camps. Horses, too, needed rescue.

As part of their twisted eugenics breeding program, the Nazis appropriated prized horses from across Europe, including the finest Arabian stallions, valuable racing Thoroughbreds, and the entire herd of the world-famous white Lipizzaners from the Spanish Riding School in Vienna, Austria. By 1943, the Nazis had collected nearly every Lipizzaner in the world at a fortified estate in Hostoun, Czechoslovakia.

In late April, during the closing days of the war, this facility was on the verge of being overrun by starving Soviet soldiers. Fanatical SS officers had also been instructed to slaughter the horses and burn the stables and research station before fleeing. The Lipizzaners faced the prospect of extinction.

Approaching Czechoslovakia from the west, Colonel Charles "Hank" Reed, a former US Cavalry officer now commanding an armored reconnaissance regiment, learned of the horse camp at Hostoun from a defecting German veterinarian who begged for help to save the treasured animals. Wanting to "do something beautiful" to counter the senseless inhumanity of war, Reed immediately sought permission from his superior, George Patton, to launch a raid. The general, not only a former equestrian but also a onetime cavalryman, authorized the daring mission dubbed "Operation Cowboy."

In an unbelievable rescue, the task force of American soldiers and volunteer German prisoners of war deceived the remaining SS guards into surrendering by staging a mock firefight. Reed was delighted to report the successful liberation of "about 300 Lipizzaners, the Piber [Austrian] Breeding Herd plus the Royal Lipizzaner Stud from Yugoslavia—well mixed together. Over one-hundred of the best Arabs in Europe, about two-hundred thoroughbred and trotting bred racehorses collected from all over Europe—finally about 600 Cossack breeding horses—Don and Urals." The cherished Lipizzaners evaded extinction.

At the insistence of General Patton, a handful of these horses found their way to the United States, where their offspring presently number almost a thousand. One of these Lipizzaner horses played Buffalo Bill's steed in the 1976 revisionist Western comedy *Buffalo Bill and the Indians, or Sitting Bull's History Lesson*, starring Paul Newman as the legendary cowboy and showman. According to the Lipizzan International Federation, there are currently more than 11,600 recognized horses across nineteen countries.

While I was writing this chapter, my mind kept projecting a scene from the award-winning 2001 television miniseries *Band of Brothers*, which dramatizes the authentic wartime events and personnel of "Easy" Company of the US 101st Airborne Division. Following the unconditional German surrender in May 1945, Private David Webster (portrayed by Eion Bailey) is riding in the back of a truck passing an endless line of dejected German prisoners when he is suddenly overcome by a surge of anger and resentment.

Webster, who left his English literature studies at Harvard University to enlist in the wake of Pearl Harbor, seemingly vents years of repressed trauma by screaming, among a highly articulate string of obscenities, "Say hello to Ford! Say hello to General fucking Motors! Look at you! You have horses! What were you thinking?!" In a way, it was precisely the heavy use of those horses by the Germans that helped Webster and his band of brothers win the war and save the world from Hitler and his sadistic Nazi regime.

Examining the outcome of the Second World War through this lens, the horse, along with American industry and twenty-seven million dead Soviet soldiers and civilians, delivered or safeguarded democracy, at least for the Western world. "But victory proved a poison chalice," concludes Overy. "The Soviet people did not win freedom or prosperity, but their sacrifices have made it possible for all the other warring states to enjoy them both."

CHAPTER 17

Equus Rising
Wild Horses, Therapeutic Healing, and Worldwide Sports

For most of us today, it is hard to believe that only a century ago, the horse was indispensable to human society and dominated all avenues of our existence. From a global peak of roughly 130 million to 150 million in 1920, there are currently an estimated 58 million horses—making them the ninth most abundant mammal on the planet, followed by the donkey at 40 million (and 15 million mules). Humans top the list, followed by cows, sheep, pigs, goats, domestic cats and dogs, and water buffaloes. Ten countries, led by China (7.4 million) and the United States (7.25 million), have horse populations over 1 million.* Rwanda, in Africa, and the British overseas territory of Saint Helena, an island in the South Atlantic, are the only countries on Earth that reported having no horses.

While the practical transport, military, and agricultural applications of the horse faded quickly following the Second World War, our longtime companion did not vanish from all sectors of society. As it has for millennia, the horse evolved with human civilization and redefined its role within our modern, machine-driven world.

From wild horse herds and equestrian sports to mounted military/police units and equine-based therapy programs, after 5,500 years of marriage, horses and humans continue to reinvent and renew the vows of our undying Centaurian Pact. While our modern relationship is

---

* The others are Mexico (6.3 million), Brazil (5.8 million), Argentina (3.7 million), Colombia (2.5 million), Mongolia (2.0 million), Ethiopia (1.7 million), Russia (1.3 million), and Kazakhstan (1.2 million).

fraught with changes, challenges, and controversies, one unbreakable bond remains constant: horses are not truly horses without humans, and humans are not truly humans without horses.

Our destinies and interactive histories have been so hitched, we would be naïve to think that we can unharness them now. Horses might be what grounds us in an increasingly cybernetic world consumed by artificial intelligence, virtual realities, microchips, and the constantly spinning cycle of disposable social media. The contentious discussion and debate surrounding the fate of global feral horse populations is a poignant example of this enduring paradigm.

At the onset of the Columbian Exchange, abundant bands of wild horses soon roamed the Pampas and Great Plains of the Americas and the outback of Australia. Horses, which had been absent from American ecosystems for nine thousand years, or, alternatively, were introduced into the quarantined ecology of Australia, were quickly viewed as an invasive species.

They cause environmental degradation in the form of soil erosion and the sedimentation of rivers and streams. Feral horses also compete directly with native animals and farmed domesticates for limited grazing and water resources in the comparatively dry regions of Australia and the western United States. As capitalist settlement, including farming, ranching, mining, and oil drilling, slowly crawled across these continents, nomadic horses, like Indigenous peoples, were viewed as an inconvenient obstruction to economic expansion.

By the dawn of the twentieth century, these equine pests, who also trampled and ate crops, fell directly into the crosshairs of farmers and ranchers. "The wild horse of Texas has become one of the greatest nuisances within the borders of the Lone Star State," reported the *El Paso Daily Herald* in 1897. "The Texas ranchman regards him as an emissary of the evil one, for he brings to his ranch despair and loss." A rancher in northern Arizona complained the following year, "It has reached the point where we cannot safely turn out a riding horse to graze" without it being swept up in the wild bands. "The mustang has become a nuisance," scolded the *National Tribune* in 1901. "He grew to be considered an outlaw and a thief, and then he was shot by the cowboys whenever possible." A 1908 *New York Tribune* story about these "varmints" reported, "Authorities in Washington are besieged with petitions from stockmen and farmers begging them to put a stop to the nuisance . . . orders have been received by the forest rangers to begin a systematic war of extermination." The meddlesome and detested creatures were wearing out their commercial value and nostalgic welcome.

While there are feral herds from the Marquesas Islands, Romania, and France to New Zealand, Namibia, and Argentina, the substantial and problematic populations in the United States, Australia, and Canada present mounting ethical debates, legal wranglings, and financial burdens. Conservational management of wild horses through gathering, culling, and fertility control is a contentious hot-button issue. A slice of this controversy also revolves around their consumption, bringing us back to the initial human relationship with horses: meat.

With the speedy American transition to a motorized society in the decades straddling the First World War, the demand for horses dried up. No one needed or wanted them anymore. Baseball covers, which remained horsehide until the switch to cowhide in 1974, did not account for all that many mustangs.* As a result of this reduced demand, wild populations soared to upwards of two million by 1920. A new economic market, however, quickly gobbled up these surplus horses: canned pet food. Unencumbered by any laws or policies, feral horses equaled free meat—and higher profit margins.

The first mass-produced canned horsemeat dog food was marketed by the Chappel Brothers Corporation of Rockford, Illinois, founded in 1923 by Philip Chappel and his siblings, Ernst and Earl.† The best cuts of the animal were exported, primarily to Europe, for human consumption. The remainder was transformed into Ken-L-Ration tinned dog food. "Chappels maintains the world's largest herd of healthy meat horses on more than 1,600,000 western acres," an early advertisement in *Time* magazine declared. "And that's why you should insist upon Ken-L-Ration. The dog food *assayed* in the crucible of modern science."

---

* This substitution was blamed for the 9 percent decrease in home runs that season, including criticism from legendary slugger Hank Aaron.

† Ironically, the name Philip means "lover of horses."

The massive Chappel factory along the rail lines in Rockford received steady shipments of wild horses, burros, and smaller batches of washed-up and unwanted workhorses and mules. Across the West, hired guns called "mustangers" rounded up feral horses for slaughter. Charles "Pete" Barnum, a Nevada horse wrangler dubbed by *Life* magazine the "King of the Wild Horse Catchers," was purported to have snagged fourteen thousand mustangs in sixteen years. With endorsements from the silent-movie star and canine celebrity Rin-Tin-Tin, by 1925 the processing plant was spitting out almost six million cans of Ken-L-Ration a year "to provide the meat your dog prefers."

While these feral herds were also foraged by companies processing chicken feed, fertilizers, and soaps, it was the assembly-line pet food industry that extracted the most horses from proximate herds. Dr. Ross' Dog and Cat Food, in Los Angeles, for example, drained the horses and burros from California, Arizona, and Nevada. The Schlesser brothers of Portland, Oregon, gathered from their state, as well as Wyoming and Nevada, for both European cuisine and their canned pet brand Man-Kind Dog and Cat Food, which was also claimed to be "Fit for human food. So good you can eat it yourself!" Mustangers employed by Chappel scavenged the herds across the Great Plains from Montana and Wyoming through Colorado and Texas.

When the most approachable herds had been liquidated, low-flying propeller planes (and eventually helicopters) were used to drive scared, stampeding horses from remote areas into accessible pre-sighted corrals or kill zones. *Time* magazine bluntly summarized this grisly money-spinning business as "round-up and ground-up." Not everyone, however, supported the commercial equine carnage.

In the fall of 1925, just as the Chappel processing facility reached full production capacity, a series of strange incidents set off alarm bells. Small fires were sparked around the plant and rail yard, the main electrical lines were cut purposely, and gasoline was splashed across the main horse intake gates. Although these threats were easily contained, clearly someone was trying to sabotage operations. To keep his horse-based profits flowing, Philip Chappel hired armed security to patrol his facilities.

Undeterred, the arsonist struck again in November, and this time the fire could not be contained. Large portions of the slaughterhouse, cannery, and corrals were incinerated. Although production was up and running again within a few weeks, the stakes had been raised perilously.

In early December a pair of guards conducting a midnight perimeter sweep spotted a mysterious figure crouched in the shadows over a box-shaped object. As they approached, the prowler fired a pistol in their direction. One of the guards returned fire, emptying his shotgun as the intruder vanished into the darkness. The saboteur had been attempting to light the fuse to an overstuffed suitcase held together with baling wire and bulging with seventy-five pounds of dynamite.

A few hours later, Frank Litts, a drifting cowboy and miner, was found unconscious, peppered with buckshot, and bleeding in a nearby field. The authorities had finally found their man. During his trial, Litts tried to flee the courtroom twice and adamantly rejected an insanity plea. He stated repeatedly that he was simply acting as a benevolent advocate for wild horses by endeavoring to blow up the factory. "The killing of horses is inhumane and contrary to Biblical ordinances," he protested. "I would rather see my body or my mother's body ground up and used for fertilizer than to have horses killed like they are here." His defense, while twistedly altruistic, was summarily dismissed. Litts was shipped off to a hospital for the criminally insane. His wild horse crusade, however, did not end here.

Shockingly, he escaped in late 1926 and was apprehended on his way back to Rockford in possession of 150 pounds of dynamite. After being reunited with the asylum, five years later he was shot during yet *another* failed breakout. By this time, more than two hundred American companies were crafting pet food from horses. When Litts died in custody in 1938, the feral herds had been so depleted that horsemeat canneries were either shelving their grinders or switching to other animal stocks. The few surviving western herds had been pushed into secluded pockets to eke out a harsh existence in austere surroundings. Even these remote sanctuaries did not shelter the fugitive horses for long. Settlement, ranchers, and the law were hot on their trail.

The passage of the Taylor Grazing Act of 1934—administered through the US Grazing Service within the Department of the

Interior—was nothing short of a death sentence for feral horses. The Grazing Service merged with the General Land Office to create the Bureau of Land Management (BLM) in 1946. These successive organizations sold grazing permits on government lands to cattle and sheep ranchers across western districts administered by local boards. "A wild horse consumes forage needed for domestic livestock," declared the director of the Grazing Service in 1939, "brings in no return, and serves no useful purpose." The agency then authorized "range clearance."

Watering holes frequented by horses and burros were marked and poisoned, and unsolicited animals were shot on sight. Smaller numbers were still sold for meat, although the market was in steep decline and not nearly as robust as it once was. Between 1946 and 1950 in Nevada alone, more than one hundred thousand feral horses were killed or removed from BLM lands. Into this slaughter stepped Velma Bronn Johnston, an unlikely candidate to become the national wild horse hero.

On an ordinary Nevada spring morning in 1950, Velma was driving her routine twenty-six-mile commute from the family ranch to the insurance company in Reno where she worked as a secretary. This particular morning, however, she found herself behind a livestock truck that had blood pouring from its back doors. Unsure if the driver was aware of the situation, she followed the transport into a stockyard. As she approached, she peered into the openings between the side panels of the cargo trailer.

She was immediately shaken and sickened by the chilling blood-soaked scene of numerous deceased, mutilated, suffering, and dying horses. Stepping from his seat, the teamster casually told her that they had been "run in by plane out there. No use crying your eyes out over a bunch of useless mustangs. They'll all be dead soon anyway." This gruesome encounter launched thirty-eight-year-old Velma Johnston on her lengthy and tumultuous campaign to ensure just treatment for America's feral equines.

Nicknamed "Wild Horse Annie" by her detractors and critics, after years of battling politicians, red tape, misogyny, scorn, and ridicule, she was instrumental in the ratification of the first laws defending these horses. Adopted in 1959, Public Law 86-234, dubbed the "Wild Horse

Annie Act," prohibited the use of poison and any mechanical land or aerial vehicles in hunting, capturing, killing, or corralling wild horses and burros for sale or slaughter.

Johnston was not satisfied, however, as the law did not afford complete protection, still allowing for indiscriminate killing and catching. As the leading voice of what was now a politically formative national grassroots campaign involving millions of people, she continued to champion American mustangs through federal legislation. Her tireless efforts over more than two decades eventually paid off.

The US Congress unanimously passed the comprehensive Wild Free-Roaming Horses and Burros Act in 1971. Under this statute, which has been amended four times (as recently as 2004), the BLM was entrusted with the absolute and unqualified protection of these animals on government lands: "Congress finds and declares that wild free-roaming horses and burros are living symbols of the historic and pioneer spirit of the West; that they contribute to the diversity of life forms within the Nation and enrich the lives of the American people and that these horses and burros are fast disappearing from the American scene. It is the policy of Congress that free wild-roaming horses and burros shall be protected from capture, branding, harassment, or death."

This decree reads like a prototypical Chevrolet or Ford pickup truck commercial: a noble and powerful wild horse with its free-flowing mane galloping and splashing through majestic American panoramas. In *Wild Horse Country*, author David Philipps declares: "They are freedom. They are independence. They are the ragtag misfits defying incredible odds. They are the lowborn outsiders whose nobility springs from the adversity of living a simple life. In short, they are American. Or at least they are what we tell ourselves we are, and what we aspire to be. If you think I'm laying it on a little thick, consider this: There are only two animals for which the United States Congress has ever specifically passed laws to protect from harm. The first was the bald eagle. The second was the wild horse." Natural predation and licensed hunting keep all other animal populations, including an estimated six million invasive and highly destructive feral hogs across thirty-five states, in check while providing nature-raised sustainable meat to millions of Americans.

This extraordinary untouchable status afforded to feral horses and burros is burdened with complications and controversy. Herd populations typically increase between 15 percent and 20 percent a year. Left unchecked, as mentioned, they cause environmental damage and monopolize already scarce resources, putting undue survival pressures on native wildlife and domestic livestock. These massive herds are simply not sustainable. Eventually, slow starvation, disease, and death stalk feral horses and donkeys.

To maintain ecological equilibrium and benefit the long-term health and welfare of the herd collective, the BLM is forced to "gather" select animals through ground or aerial roundups. Efforts are made to have the animals adopted, but placement rates are in steady decline, compounded by an increase in unwanted domestic horses.* Remember, any form of killing, including euthanasia or dying with dignity, is currently a crime punishable by up to a year in prison and a $2,000 fine.

After countless generations of free-roaming ancestry and existence, the majority of these gathered feral horses and burros are relocated to BLM holding facilities to breathe out the remainder of their lives in captivity. Holding the horses until they die of disease or old age eats up 60 percent to 64 percent of the bureau's total annual budget. A handful of fugitive domestic horses, however, manage to find their way back home.

Take Shane Adams of Utah and his beloved horse Mongo, who was swept up by a feral herd in 2014. Adams spent every weekend for the next three years scouring the West Desert before coming to terms with the fact that his friend was gone. After running with the mustangs for almost a decade, Mongo was captured during a 2022 BLM sweep. The brand on his shoulder matched the description reported by Adams years earlier. "There was no way. You have got to be kidding me," he remarked

---

* Although expenses can vary greatly depending on boarding and medical expenditures, the average recreational horse will run about $7,000 to $10,000 a year. Since 2009, an estimated 180,000 American horses have been abandoned. Private sale for consumption is also no longer an option since the closure of American slaughter plants in 2007 after the government shut down the mandatory horsemeat inspection program under pressure from lobbyists. Many of these unwanted animals are released into the wild, where they join feral herds, are killed by their owners, or are shipped to Mexico or Canada to be butchered and sold for human consumption.

upon receiving the news that Mongo had been found. "It didn't even seem real. To have him back is still not real. He is very special and has always been a part of my life." Although Mongo was now eighteen years old and four hundred pounds skinnier, he immediately settled back into domestic life, snacking on his favorite treat: Sour Patch Kids candy. "Every day it's like, 'Oh my gosh, there he is. He's here!'" Adams says. "It's a miracle feeling every day." Mongo is one of the lucky ones.

Currently, there are roughly sixty thousand horses and burros in twenty-eight BLM-managed holding facilities and forty-eight fenced grazing plots. Short-term holding costs $1,829 per horse per year, while long-term green pastures retirement homes are somewhat cheaper at $664 per animal per year. In 2020 the bureau spent $91 million on these animals, including $57 million on housing. The horses have cost the American taxpayer $450 million over the past five years, and more than $1 billion since 1975.

There are cost-effective and humane alternatives that are slowly being incorporated into BLM conservation strategies. Herd control has been accomplished by darting mares with PZP birth control at $30 a shot. Contraception for the average mare over her reproductive lifetime costs about $700.

This method has been effective with numerous herds, including those (totaling 450 animals) on Assateague, a barrier island off the coasts of Maryland and Virginia. Recent DNA studies suggest that these are some of the closest living descendants of Christopher Columbus's original horses. Farther south, the Outer Banks islands of North Carolina are also home to feral horses. The herds are regarded, and widely marketed, as tourist attractions, bringing visitors and revenue to these communities.

This is also true of the relatively accessible Little Book Cliffs Wild Horse Range in my adopted home of Grand Junction, Colorado, where I now live and teach. Roaming thirty-six thousand acres of high desert BLM land, these PZP-managed horses (ranging from 90 to 150 in number) are advertised in all tourist brochures and by visible road signs. They even greet visitors as they exit the local airport along the "Gateway to Grand Junction" in the form of six rearing and running wild horse statues to "reinforce the strength and individuality of both the wild horses

Spirit of the West: Little Book Cliffs wild horses running free near Grand Junction, Colorado. *(Sloane Milstein)*

and the people of this western community." These massive, engaging sculptures forged from eight and a half tons of steel were purposefully commissioned "to encourage tourism and visitors to the area." These horses are less remote and skittish than other feral herds, allowing for intimate vacationer photo opportunities.

When the Wild Free-Roaming Horses and Burros Act was passed in 1971, there were 17,300 horses and 8,000 burros on BLM lands in ten states. By 2014, these numbers had swelled to 34,000 horses and 6,800 burros. The peak population of 95,000 total equids was recorded in 2020. Global warming, drought, and dwindling water resources have heightened survival pressures on growing feral herds across the globe. Consequently, in 2021 and 2022 the United States, Australia, and Canada all conducted the largest roundups in history.

The total gathering across the American West in 2021 was 13,666 animals at a cost of $112 million. Roughly 63 percent of the horses were adopted, and an additional 1,622 mares were treated with fertility control. The 2022 capture totaled 20,193 feral equines (with another 1,160 darted with PZP) at a price tag of $138 million. Only 39 percent of this gather were adopted. In Colorado, for example, over 2,000 horses were removed in a series of roundups, marking a record gathering.

There are approximately 64,600 horses and 17,800 burros (82,400) currently roaming twenty-seven million acres of protected lands across ten western states: Arizona, California, Colorado, Idaho, Montana, Nevada, New Mexico, Oregon, Utah, and Wyoming. This is three times the BLM sustainability target number of 26,785 animals.

The debate instigated by Velma Johnston on that bloody spring morning in Nevada rages on to determine the fate of America's wild horses and donkeys. Lobbyists fight for their continued blanket protection, while others see euthanasia and PZP birth control as a more viable, cost-saving, and humane path forward. Unlike the United States, however, both Australia and Canada view culling as the most pragmatic method to regulate herds.

There are roughly 2,500 feral horses in Canada, primarily in isolated ranges of British Columbia, Alberta, and Saskatchewan. Since Canada views these animals as an introduced foreign species, there is no blanket federal protection. Jurisdiction remains at the provincial level. Although laws differ in the details, the horses are generally not allowed to be hunted but are harvested for herd management and, if adoption fails, sold for meat. We have now come full circle to the original and longest-serving human use for horses: food.

As mentioned in the opening chapters, the taboo against eating horseflesh across myriad cultures and faiths was driven largely by spiritual decree and economic expediencies. In many societies, however, horsemeat remained standard fare or was unceremoniously eaten during leaner times. During the French Revolution (1789–1799), for example, beef became an extravagant luxury. Horsemeat became such a common substitute that it quickly won approval as a socially acceptable and well-received staple of French cuisine. In 1990 the world consumed 530,000 tons of horsemeat. Since this time, global consumption has risen by 28 percent. Horsemeat is a common meal for more than a billion people primarily in Europe, China, Russia, Central Asia, and localized regions of the Americas, including French Canada, Mexico, and Argentina. These countries, along with Brazil and Australia, are also the leading producers.

Australia has the largest feral equine populations (and problems) in the world, increasing at an annual rate of 20 percent. According to government reports, there are currently four hundred thousand brumbies, more than five million burros, and one million feral camels roaming the land down under. These horses have evolved to withstand extreme temperatures easily topping 100°F (38°C), need less water, and have developed extremely strong hooves to deal with the harsh, abrasive terrain and constant trekking to find water and forage. Some herds have been recorded walking continuously for more than two days to increasingly scarce watering holes.

While these herds pose a serious threat to themselves by overtaxing their already strained resources, they also devastate ecosystems through erosion and the sedimentation of extremely precious, and disappearing, waterways. Cattle and sheep ranchers, vital components of the Australian economy, detest these scrawny pests, as they compete directly with their own livestock. This is not a new phenomenon. In the 1930s, government bounties were offered for "horse ears," encouraging friendly community competition. Two men shot more than four thousand in one year, while another hunter killed four hundred in one night.

Although government-sanctioned culling continues in Australia, it has also been met with fierce opposition and protests from environmental groups and animal rights activists. The proposed gather of ten thousand horses in 2021, for instance, ignited a firestorm of opposition, substantial government and ranching pushback against these environmentalist campaigners, and a reevaluation and renewal of the debate surrounding the growing problem of wild equines in Australia.

While the fate of feral *Equus ferus caballus* is a polemic and inflammatory issue, the opposite is true of *Equus ferus przewalskii*. This horse, also known as the takhi, Mongolian wild horse, or Dzungarian horse, has been resurrected from the brink of extinction. Based on DNA models, it appears to have diverged from the caballine branch sometime between seventy-two thousand and thirty-eight thousand years ago. Although Przewalski's horses have sixty-six chromosomes compared with sixty-four for the common horse, these two can still produce healthy, fertile offspring (unlike mules), suggesting that they may not be entirely distinct species.

The squat, stocky horses are named after the Russian colonel, naturalist, and explorer of Polish descent Nikolay Przhevalsky (Przewalski in Polish), who obtained a specimen in 1878 from the Mongolian-Chinese border. He would later mount an expedition to observe these horses in the Dzungarian/Tarim Basins. By this time, overhunting had taxed naturally modest and dwindling wild populations.

Around 1900, at the insistence of Carl Hagenbeck, a German merchant of wild animals and founder of the modern zoo, a handful of Przewalski's horses ended up in permanent exhibitions across Germany alongside exotic creatures—*and* human beings—from around the globe. Only thirty-one captive domesticates scattered throughout Europe survived the Second World War.

Although there were whispered sightings, Przewalski's horses had gone extinct in the wild by the 1960s. A stud book was created for the dozen remaining domesticates, to record and share genetic information. Thanks to captive breeding, by 1965, there were 130 animals across thirty-two zoos. Breeding was regulated to ensure the most diverse genetic combinations, and populations slowly began to stabilize. By 1979, seventy-five institutions throughout Europe and the Americas maintained 385 Przewalski's horses.

A coordinated international effort returned the first horses to their native habitat in the Chinese Dzungarian Desert/Tarim Basin of Xinjiang in 1985. Reintroductions to China and Mongolia continued over the next few decades. Przewalski's horses were also released into the contaminated sanctuary of the Chernobyl Exclusion Zone in Ukraine and Belarus in 1998. Despite poaching, the herds have thrived even as war currently rages around them. Russia, Kazakhstan, Germany, and Hungary have also established protected populations in national parks. Currently, the Askania-Nova reserve in Ukraine maintains the largest captive breeding program. The effort to save and reintroduce these animals is a testament to the unshakable impact that horses continue to have on the human experience and psyche.

Over the last few decades, horses have been coming back into vogue. "The numbers clearly show upward trends in all things equestrian over the past

Horsing around: Przewalski's horses engage in some frolicking horseplay at the Smithsonian Conservation Biology Institute's nature reserve in Front Royal, Virginia. *(Lawrence Layman/Smithsonian)*

fifty years," reports Emily Kilby, a writer for *Equus* magazine. "So far, the equine species has flourished in its nonutilitarian role." Horses are still surprisingly well represented in modern societies through human consumption, various sports and recreation, therapy, pets, and mounted military and police forces. Horses have also found a niche-farming clientele, as green, eco-friendly cultivators are reviving true organic horsepower.

The annual economic impact of the estimated 7.25 million horses in the United States (and 270,000 donkeys and 200,000 mules) is $122 billion, supporting 1.74 million full-time jobs. Racing accounts for roughly $50 billion to $60 billion of this total. "Those involved in the equine industry already know how important it is to the US economy," says Julie Broadway of the American Horse Council, "but also for the industry to demonstrate to the general public how much of a role the equine has in American households."*

---

* According to its website, the American Horse Council is a nonprofit trade organization that "works daily to advocate for the social, economic and legislative interests of the United States equine industry."

Roughly 43 percent of US horses are considered private, recreational, or pets. Another 34 percent are classified as breeding, showing, racing, or competition animals. Workhorses in farming, ranching, police, military, or education/therapy make up 16 percent. The remaining 7 percent are divided among Indigenous and Amish communities, wild horse herds, and BLM holding facilities.

The greater European Union contains some 7 million horses, with France and the UK each housing around 850,000. The annual UK financial impression is about £7 billion ($9.5 billion), with racing alone valued at half this total. Horses add US$15 billion to the Canadian and US$5 billion to the Australian economies, predominantly through thriving racing and ranching portfolios.* Horses are now generally most associated with sports such as racing, polo, rodeo, Olympic equestrian, and leisure-based horseback riding.

Racing dates to our domesticating teenage daredevil and the earliest riders on the Pontic-Caspian Steppe. All early horse cultures—from the Indo-Europeans, Assyrians, and Scythians to the Xiongnu, Han, and Huns—had informal or organized riding or chariot races, typically made more entertaining by friendly wagers. There are Greek writings recording Scythian equestrian games that included lassoing, standing on the back of a horse while it galloped, switching horses at full speed, and hunting rabbits with a spear.

The Greeks and Romans both built massive hippodromes and chariot-racing venues such as the Circus Maximus, the Circus of Nero, and the sites at Olympia and Constantinople. These stadiums displayed dazzling equine spectacles to spirited crowds that swelled, according to Pliny the Elder, to upwards of 250,000 passionate spectators.

Formal racing was brought to England by the Romans, who commissioned the first racetrack around 210 CE. A detailed account of official racing at Smithfield, London, during the twelfth century is provided by William Fitzstephen in his *Description of the City of London* (c. 1174). Between this time and the sixteenth century, racing became known as "the

---

* Canada and Australia are each home to roughly five hundred thousand to six hundred thousand domestic horses.

sport of kings," as the monarchy and aristocracy reveled in breeding and entering contestant horses. As an avid jouster and hippophile, King Henry VIII amassed a personal stable of more than two hundred prized horses. Racing expanded during the next few centuries among all classes. The five British Classics are represented by the St. Leger Stakes (1776), the Oaks (1779), the Derby (1780), the 2,000 Guineas Stakes (1809), and the 1,000 Guineas Stakes (1814).

As mentioned, all modern Thoroughbreds trace their lineage to just three stallions that were imported to England: Byerley Turk (c. 1686), Darley Arabian (c. 1704), and Godolphin Arabian (c. 1730). "Several hallowed bloodlines," observes Philipps, "are more inbred than a medieval monarchy." The legendary horses Man o' War, Seabiscuit, and War Admiral, for instance, were all descendants of the Godolphin Arabian, while roughly 95 percent of all Thoroughbreds trace their direct male line to the Darley Arabian, including the revered eighteenth-century racehorse Eclipse. These pedigrees are carefully logged into meticulous, now usually national, stud books. The first such catalog was the British *General Stud Book* of 1791, which received a thorough revision in 1793.

Although racing itself landed in the Americas with the initial equine immigrants, the first Thoroughbred—Bulle Rock (a son of the Darley Arabian)—was brought to the Virginia Colony in 1730. Horse racing was present at Jamestown as early as 1620 and was legalized by the governor of Virginia ten years later. The first formal racetrack in the American Colonies was established in 1665 at Newmarket on Long Island, New York, shortly after its surrender by the Dutch. The Belmont Stakes (1867), the Preakness Stakes (1873), and the Kentucky Derby (1875) now comprise the premier American Triple Crown races.

The annual international horse racing market is currently valued at $473 billion and expected to hit $794 billion by 2030. "You know horses are smarter than people," conceded humorist and cowboy movie star Will Rogers. "You never heard of a horse going broke betting on people."

Although racing was the initial equine sport, polo also has ancient roots. It appears to have its origins among Indo-European equestrian nomads of Central Asia to keep both warrior and horse combat ready. Polo became somewhat more standardized after it achieved Persian patronage under Darius I around 500 BCE, morphing into its current form, known as *Chovgan,* over the next thousand years.

Piggybacking on the rapid expansion of Islam, the game poured out of Persia across the Middle East and Eurasian Steppe, making its way to China in the 670s. By the Middle Ages, a form of polo was being played from Constantinople to Kyoto, Japan. It was imported from India to Britain in the 1860s by returning colonial officers and administrators. Polo was an event at five Olympics between 1900 and 1936. Although it is a sport of financial means, the Federation of International Polo currently has seventy-seven member countries, and the game is played professionally in sixteen nations.

Unlike racing and polo, American rodeo can be traced through two relatively recent lines: the flamboyant horse-trick performers of Buffalo Bill's Wild West show and the more authentic competitive expression of work-related cowboy skills. Composed of various individual events, there are governing rodeo organizations and associations at every level of competition in the United States and Canada, from "Little Britches" and juniors, through high school and university, to professional circuits. Of course, there are countless other equine competitions among global cultures, showcasing various horse-related talents, including skijoring, *buzkashi,* and international Olympic equestrian.*

According to the International Olympic Committee, "The horse made its first appearance at the ancient Olympic Games in 680 BC, when chariot racing was introduced—and was by far the most exciting and spectacular event on the programme." Equestrian sports were inaugurated into the modern Olympics in 1900 and have been a regular feature since 1912. The three Olympic disciplines are: dressage, eventing, and jumping. Until 1952, however, only commissioned military officers and "gentlemen" were allowed to compete. Women were permitted to participate in dressage in 1952, and subsequently in jumping (1956) and eventing (1964).

All Olympians—men, women, and horses—now compete on equal terms, and both individual and team medals are awarded in each disci-

---

* *Buzkashi* is a traditional mounted sport played across the Stans, in which mounted players attempt to place a headless sheep, goat, or calf carcass in the opposing goal or scoring circle. In skijoring, a person on skis is pulled by a horse and rider around a track or course.

pline. Olympic equestrian has become a source of national pride among numerous nations, with Germany dominating the pack since the 1960s.* Although horses are now generally associated with sport, they still serve indispensable and pragmatic functions in police and military forces.

Highly trained horses are employed by constabularies around the globe for patrols and crowd control. The use of these massive horses deescalates riotous or potentially dangerous situations through the psychological impact of both admiration and fear. Simply stated, people do not want to hurt horses; nor do they want to get hurt by horses. Mounted officers also benefit from a higher vantage point and broader operating space as crowds disperse and diffuse when they move. For most of us, even as adults, these police horses still inspire a healthy respect for their raw power, and a childlike sense of wonderment.

This dichotomy is precisely the reason they are still used. For example, the NYPD Mounted Unit still employs fifty-five elite patrol horses, the LAPD Mounted Platoon currently rosters thirty-two equine officers, while the Toronto Police Service has twenty-six, and the Metropolitan Police Mounted Branch operates one hundred and ten horses across London. "People are aware that horses have a dual nature and that, though they are routinely gentle, paradoxically they have the ability to evoke fear in certain situations. It is this quality of engendering fear that enables police horses to control unruly crowds and restore order," explains Elizabeth Atwood Lawrence, a veterinary anthropologist, in *Hoofbeats and Society: Studies of Human-Horse Interactions*. "That crowds fear and respond to horses in a uniquely effective way is a well-known fact. Police records document the successful work of mounted officers in this regard, and patrolmen vividly recount experiences of participation in activities in the line of duty during which crowds were broken up and riots put down." The tactical impact of a handful of mounted officers goes well beyond what can be accomplished by those on foot or bicycle.

Select mounted and pack-transport units are also still deployed by

---

* The gymnastics "pommel horse" originates with the Persian, Macedonian, and Roman militaries, which used wooden horses to teach mounting and dismounting.

modern militaries to patrol inaccessible frontiers and borderlands. These equine formations are trained at specialized schools such as the American facility at Fort Carson, Colorado. During operations in Kosovo and Afghanistan in the early 2000s, for example, NATO forces deployed equine pack units across rugged terrain that negated the use of vehicles. In both instances, horses and mules provided invaluable service resupplying forward operating bases and remote outposts.

The remarkable mounted escapades of US Special Forces fighting the Afghan Taliban on horseback in the immediate aftermath of 9/11 were Hollywoodized in 2018 by the action film *12 Strong*. The striking Horse Soldier Statue commemorating this daring misson stands in Liberty Park overlooking the National September 11 Memorial & Museum in New York City.

Horses are also increasingly being employed to treat and heal soldiers, veterans, and others suffering from post-traumatic stress disorder (PTSD) and other mental and physical disabilities. This innovative and growing field of equine therapy is also available for those with special needs, autism, at-risk populations, and children who have suffered trauma and abuse. The results have been extremely positive and promising.

Through our millennia-old bond, this is a simple case of horses healing humans. "There is something about the outside of a horse," recognized Winston Churchill, "that is good for the inside of a man." The idea of hippotherapy originates with Hippocrates, who advocated that horseback riding heals the soul. Currently, the two broad categories of treatment used at centers across the globe are equine-assisted psychotherapy and equine-assisted learning.

Similarly, Indigenous communities in the United States and Canada are reintroducing horses to promote healing through the preservation and transference of cultural traditions and horsemanship skills. "Crows have always been close to horses," said Robert Yellowtail (1889–1988), a Crow leader from Montana who earned a law degree and worked tirelessly to advance Indigenous rights, including the right to vote, won in 1924. "Without them, we're sunk. Today Crows still love horses. To separate us from horses, you might as well take one of our family away."

A Crow riding instructor explained the intersection between horses and communal and individual self-worth for Indigenous peoples: "The kids were losing their identities. Bicycles and Hondas just don't give them a sense of who they are. The kids are looking for something to make them Indian. Long hair, beads, and floppy hats just don't do it. Those things make you a hippie, not an Indian. An Indian is an individual. With the horse, he gets a sense of his own identity and individuality. Individuality was always part of the Plains Indian." There are roughly 120,000 horses residing on Indigenous territory across the United States.

Another therapeutic initiative is the US Wild Horse Inmate Program (WHIP), which began in 1985 as a cooperative recovery program. Mustangs gathered through BLM herd management are used to rehabilitate inmates, who recondition and train them for their new domesticated life. Many of these horses receive specific police, border patrol, or military education, preparation, and subsequent employment. Others find homes as therapy animals or family pets. The largest program at Cañon City, Colorado, home to thirteen prisons, has successfully trained and adopted out more than five thousand mustangs.

Corrections departments in various states have reported decreased recidivism rates among participants. "From horses, we may learn not only about the horse itself," wrote the pioneering horse evolutionist George Gaylord Simpson, "but also about animals in general, indeed about ourselves and about life as a whole." While WHIP is certainly a worthwhile endeavor for the rehabilitation of both horses and humans, taken as a multifaceted whole, the fate of wild horses remains an extremely contentious and hot-blooded issue.

The unique feral horses at Sable Island in the North Atlantic—roughly 110 miles off the coast of Nova Scotia, Canada—have been safeguarded permanently. Although it makes for a far more adventurous story, these five hundred or so animals are not the descendants of shipwrecked horses. Their forebears were seized during the Great Expulsion of the Acadians and purposefully introduced in 1760 in the wake of the British annexation of Canada during the Seven Years' War (1756–63).

Throughout more than 260 years of sheltered geographic and genetic seclusion, these horses have evolved through natural selection into a shaggy, squat (thirteen to fourteen hands and 650 to 800 pounds) horse with short, mountain goat–like limbs. This adaptation allows them to traverse the slippery rocks and steep thousand-foot sand dunes without breaking their legs.

Over the centuries, the abandoned horses were nabbed by passing sailors and local fishermen, and periodically slaughtered for pet food. By the 1950s, they were teetering on extinction. The Canadian government stepped up in 1960, declaring Sable horses fully protected. The island is predator free, and the horses are not interfered with in any way. "The Sable Island horses," writes Wendy Williams, "may be the modern world's *only* genuinely free-roaming, entirely unmanaged horse population."

Like their original ancestors who roamed the Americas millions of years ago, horses, and their role within our inseparable bond, continue to evolve along our eternal trails. While the final chapter of our enduring relationship with this majestic animal has yet to be written, it will no doubt be as intertwined with our own historic ride as it has always been.

Conclusion

The horse was an unlikely candidate to preside over human history. As its natural habitat receded, by 3500 BCE it had largely disappeared from its global landscapes. Domestication reined in the horse from the precipice of extinction, and, in the process, fused our destinies and redirected the historical flight path of our collective species. With the first comforting touch and calming whisper on the Pontic-Caspian Steppe, humans and horses forged an unbreakable bond and the most dominant animal dyad ever witnessed.

Within this unparalleled Centaurian Pact, horses redefined not only what it meant to be human but also our entire history by single-handedly igniting a revolution in transportation, trade, traction, and war. The overriding reign of the horse spanned an unprecedented 5,500 years. It is hard to believe that only a century ago, horses ruled over the global village they helped create.

The unique equine behavioral and physical attributes wrought by sixty million years of evolutionary natural selection produced an exploitable, multidimensional living machine. When humans harnessed this unrivaled utilitarian horsepower, the world was irrevocably changed. Horses were the complete, transformative package contained in a thousand-pound animal.

Possessing a rare combination of size, speed, strength, and stamina, the horse became the pinnacle weapon of war, a political leviathan, a prime economic mover, and an agricultural powerhouse. It was a universal multipurpose engine enhanced through intense selective breeding, various riding tackle, and specialized traction vehicles. As the apex of biotechnology, the horse drove profits and power, and was the prime mover of civilizations.

Horses were so common, so omnipresent, however, that no one took much notice of their paramount role at the forefront of history. While continuously leading the charge of change from center stage, they seemingly blended into the blurry background of supporting characters. They were, in reality, lead actors in our shared epic sagas, wielding immeasurable historical agency alongside their codependent human sidekicks.

While we recite and eulogize the names of the people and empires that rode to fame atop these conscripted mounts, the horses themselves remain anonymous, save a select handful. What would Alexander be without Bucephalus, the Duke of Wellington without Copenhagen, General Grant without Cincinnati, the Lone Ranger without Silver, and film and TV cowboy Roy Rogers without his faithful Trigger? Certainly, the historic influence and unshakable impact of the Indo-Europeans, Assyrians, Scythians, Xiongnu, Huns, Mongols, Comanche, and Lakota, among other equestrian cultures, would be erased from the historical record. Revered conquerors and reviled conquistadors would remain nameless.

It is generally only when something vanishes or its value wanes that we tend to sit back and reevaluate its importance. Stop for a moment and try to imagine our world, our heroes, and our history *without* horses. It is impossible. Our current space-time continuum and global order would be unrecognizable. The horse hauled and assembled the foundational building blocks of our modern world and made critical contributions to our cultural development. For more than five millennia, the horse was the invisible hand driving human history.

Horses changed the way we hunted, traded, traveled, farmed, fought, worshipped, and interacted with one another. They reconfigured the global human genome and the languages we speak. They were a potent medieval cudgel of feudal subjugation and manorial farming during the Middle Ages and were an overpowering weapon of colonization during the Columbian Exchange. Horses gave rise to nation-states, pulled into place modern international borders, and were instrumental in constructing the superpowers of Britain, the United States, and China. They made vital contributions to modern medicine, lifesaving vaccines, and developments in sanitation. They even inspired recreational drug use, sports, invention, entertainment, architecture, furniture, and fashion. Horses dominated every facet of our formative history. Across an unrivaled 5,500-year span, the horse carried the fate of human civilizations on its back. We still live among the galloping shadows of a world built by horses.

We casually walk through that same world never thinking about the incomparable power of the horse in creating its social, cultural, economic, and political contours. Perhaps now, you'll give these countless generations of unsung horses a passing thought and their rightful place in history when you speak or hear a Proto-Indo-European dialect, put on your pants, smell marijuana or manure, visit a museum or library, ride the subway, adjust your car seat, stare at a skyscraper, enter a place of worship, apply for your passport, get your DTaP shot, or peruse the international aisles of your local grocery store.

Horses are still an integral part of what it means to be human. Like everything else, our enduring history-making partnership and Centaurian Pact continue to quietly evolve. After all, humans still love horses.

With the domestication of the horse, the unrivaled transference of goods, information, knowledge, genetics, germs, and language became *the* hallmark of human civilization. The horse created a much larger interconnected world by making it exceedingly smaller. This manifold transmission, including my excessive selection of corn, wheat flour, beans, rice, potatoes, and Steven's dog food, has not really changed but merely quickened its steps from human feet, to horse hooves, to the engines of planes, trains, and automobiles. But for every successive generation, these inventions, including the horse, produced the same outcome. They increasingly put the larger world at our fingertips and the reins of destiny directly in our hands.

Acknowledgments

Upon the publication of my fifth book, *The Mosquito*, as customary, my dad and I sat down to brainstorm ideas for a follow-up. Although he is a physician of emergency medicine back home in Canada, he really should have been a historian. After several rounds of dialogue and dismissing an endless barrage of topics, we seemingly always circled back to the familiar list of usual suspects that delivered modern civilization to the hearth of humanity: the farming package, urbanization, vocational specialization, trade, war, imperialism, and disease. But what was the one defining factor, aside from humans, that assembled and secured these foundational building blocks?

Our pursuit to answer this question was rudely interrupted by the global coronavirus pandemic. My dad's essential medical skills were interminably needed in the ER. This viral pathogen and its collective repercussions, however, led directly to the last of a series of personal horse-related events that finally triggered me to write this book.

On a routine, everyday morning while driving to work, I saw a procession of kids walking to school. They were quickly outpaced and overtaken by my neighbors leisurely trotting past on horseback, who were summarily relegated to the rearview mirror of my car. As I headed toward the university, I began thinking about the timeline shift in transportation from our feet to horses to the automobile—millions of years of bipedalism and an astounding 5,500 years of horsepower compared to a brief century with the internal combustion engine. This was a timely reminder of the unrivaled impact of horses on generations of global civilizations, while satisfying many of those core features my dad and I kept revisiting.

A few weeks later, while walking my dog Steven, we stumbled upon a fresh pile of horse manure. As he curiously circled and sniffed this new scent, I imagined the unsanitary conditions, foul stench, and fecal stew oozing through sprawling horse-powered cities such as London, Toronto, Paris, or New York around the turn of the twentieth century. As I came to find out, by the 1890s, the crowded, rubbish-strewn streets were alive with steaming, fly-buzzing stacks of horse manure, urine, equine carrion, and a festering compost of disease-ridden filth. I crossed off a few more cornerstones on our list.

The final compelling nudge occurred in March 2020 during the initial COVID lockdown. As a rabid hockey fan, I was desperately missing my beloved NHL games. Flipping through the channels, while trying to avoid the chaotic and depressing news cycle, I decided to tune in to the pandemic treat ESPN8 ("The Ocho") broadcasting live from Leadville, Colorado. Like Steven and his initial taste of horse manure, I caught my first glimpse of something I had never seen before: skijoring.

I watched in surprised delight as a rider on a galloping horse pulled a skier around a snow-covered track. I glanced over at my equally astonished wife. "That's it!" she exclaimed, waving her arms. "You have to write your horse book!" As it turns out, skijoring is an ancient sport. During my research, I came across a dagger unearthed near Omsk, Siberia, dating from 1600 to 1500 BCE, with a bronze figurine adorning the pommel clearly depicting a bridled horse pulling a skier. For our ancestors, the utilitarian value of the horse was apparently endless. The horse was the unsurpassed driving force of civilizations and *that* defining factor we had been seeking.

This series of fortunate events led me to delve deeper into the sweeping impact of the horse on human history. The power of the human-horse dyad went far beyond anything I had previously imagined. I was, however, acutely aware of the staggering volume of literature devoted to the horse. I view this vast catalog from a wide swath of academic disciplines as a testament to the status of the horse across our historical trails. To help navigate these disparate fields, numerous colleagues and new acquaintances were consulted, and proffered their invaluable expertise and advice. To these horse lovers and to those academics on whose writings and research this story is partially built, I am indebted and extend my metaphorical hand of appreciation in thanks.

I would like to individually recognize: Dr. John Seebach for everything archaeology. Dr. Adam Rosenbaum for our deconstruction of Herodotus and the Beatles. Dr. Colin Carman for your anapestic insights into the literary landscapes of the horse, including Byron. Dr. Warren MacEvoy, Dr. Sloane Milstein, and the volunteers of the Friends of the Mustangs for their sensible care and prudent conservation of our local Little Book Cliffs Wild Horse Range. Branden Edwards for graciously allowing me to visit with and ride your beautiful horses. Amy Beisel and Emily Lozon at the International Museum of the Horse and the Library and Archives at the Kentucky Horse Park. The staff at the Benson Ford Research Center. A warm thanks to the resourceful librarians at Colorado Mesa University for procuring my endless catalog of requests. I also wish to acknowledge Colorado Mesa University for providing funding to offset the otherwise prohibitive cost of procuring photographs.

A special thanks to Sir Hew Strachan, my doctoral supervisor at the University of Oxford, who taught me to see beyond the words on the pages and interact with history as an evolving creature. Dr. Erika Jackson, Dr. Justin Gollob, Dr. Tim Casey, Bruno and Katie Lamarre, Dr. Sean Maloney, Dr. David Murray, Dr. P. Whitney Lackenbauer, Dr. Tim Cook, Terry Copp, Dr. Jeremy Tost, Dr. Alan Anderson, Jeff Obermeyer, the reclusive Dr. Hoko-Shodee, Dr. Susan Becker, Dr. Pamela Krch, Dr. Gregory Liedtke, Dr. Douglas O'Roark, Reese Kegans, Wendy West, Cathy Rickley, Dr. Jake Jones, and all my past and present players on the Colorado Mesa University Hockey Team (I am Canadian, after all). You have all positively impacted my career and, far more importantly, my life.

To my agent, Rick Broadhead, I am grateful to be the beneficiary of your expertise and friendship. My editors John Parsley, Nick Garrison, and Ella Kurki at Penguin Random House, thank you for your keen eyes, instrumental feedback, and invaluable guidance during the revision and editing stages. To my publicist, Emily Canders, you are so appreciated that you are the only person who is allowed to interrupt our family suppertime.

My parents-in-law, Larry and Vernann Raney, thanks for your continued support and friendship. Jaxson, I love you forever and in every galaxy far, far away. Mom and Dad, thank you for teaching me the ways of the Force and filling my formative years with love, travel, and curiosity. With my apologies to Alexander the Great, Wayne Gretzky, and Yoda, you are also tops on my list of heroes. I love every member of my Longhouse and miss our lakefront home in Canada.

Finally, to my wife, Becky, thank you for holding down the fort during my work- and hockey-related absences and my seeming absence while at home, writing. As always, you have heeded the sage counsel of "Patience" espoused by the esteemed philosopher Axl Rose and mastered it.

Thank you all,
Tim

Selected Bibliography

Adelman, Miriam, and Kirrilly Thompson, eds. *Equestrian Cultures in Global and Local Contexts*. Cham, Switz.: Springer International, 2017.
Ambrus, Victor G. *Horses in Battle*. London: Oxford University Press, 1975.
Amirsadeghi, Hossein, ed. *Drinkers of the Wind: The Arabian Horse—History, Mystery and Magic*. London: Thames & Hudson, 1998.
Amitai, Reuvan, and Michal Biran, eds. *Nomads as Agents of Cultural Change: The Mongols and Their Eurasian Predecessors*. Honolulu: University of Hawai'i Press, 2014.
Anderson, Virginia DeJohn. *Creatures of Empire: How Domestic Animals Transformed Early America*. Oxford: Oxford University Press, 2004.
Angelakis, Andreas N., Daniele Zaccaria, Jens Krasilnikoff, Miguel Salgot, Mohamed Bazza, Paolo Roccaro, Blanca Jiménez, et al. "Irrigation of World Agricultural Lands: Evolution Through the Millennia." *Water* 12, no. 5 (May 2020): 1–50.
Anthony, David W. *The Horse, the Wheel, and Language: How Bronze-Age Riders from the Eurasian Steppes Shaped the Modern World*. Princeton, NJ: Princeton University Press, 2007.
———. "The Sintashta Genesis: The Roles of Climate Change, Warfare, and Long-Distance Trade." In *Social Complexity in Prehistoric Eurasia: Monuments, Metals, and Mobility*, edited by Bryan K. Hanks and Katheryn M. Linduff, 47–73. Cambridge: Cambridge University Press, 2010.
Anthony, David W., and Dorcas R. Brown. "The Secondary Products Revolution, Horse-Riding, and Mounted Warfare." *Journal of World Prehistory* 24, nos. 2/3 (September 2011): 131–60.
Anthony, David W., Dorcas R. Brown, Aleksandr A. Khokhlov, Pavel E. Kuznetsov, and Oleg D. Mochalov, eds. *A Bronze Age Landscape in the Russian Steppes: The Samara Valley Project*. Los Angeles: UCLA Cotsen Institute of Archaeology Press, 2016.
Anthony, David W., and Jennifer Y. Chi, eds. *The Lost World of Old Europe: The Danube Valley, 5000–3500 BC*. Princeton, NJ: Princeton University Press, 2009.
Anthony, David, Dmitri Y. Telegin, and Dorcas Brown. "The Origin of Horseback Riding." *Scientific American* 265, no. 6 (December 1991): 94–101.
Armistead, Gene C. *Horses and Mules in the Civil War: A Complete History with a Roster of More Than 700 War Horses*. Jefferson, NC: McFarland, 2013.
Arrian. *The Campaigns of Alexander*. Edited by J. R. Hamilton. Translated by Aubrey de Sélincourt. New York: Penguin Classics, 1978.

Askwith, Richard. *Unbreakable: The Woman Who Defied the Nazis in the World's Most Dangerous Horse Race.* New York: Pegasus Books, 2019.

Astill, Grenville, and John Langdon, eds. *Medieval Farming and Technology: The Impact of Agricultural Change in Northwest Europe.* New York: E. J. Brill, 1997.

Ayton, Andrew. *Knights and Warhorses: Military Service and the English Aristocracy Under Edward III.* Woodbridge, UK: Boydell Press, 1994.

Ayton, Andrew, and J. L. Price, eds. *The Medieval Military Revolution: State, Society, and Military Change in Medieval and Early Modern Europe.* London: I. B. Tauris, 1995.

Bacon, Benjamin W. *Sinews of War: How Technology, Industry, and Transportation Won the Civil War.* Novato, CA: Presidio, 1997.

Bak, Richard. *Henry and Edsel: The Creation of the Ford Empire.* Hoboken, NJ: John Wiley & Sons, 2003.

Bakker, Jan Albert, Janusz Kruk, Albert E. Lanting, and Sarunas Milisauskas. "The Earliest Evidence of Wheeled Vehicles in Europe and the Near East." *Antiquity* 73, no. 282 (December 1999): 778–90.

Bankoff, Greg, and Sandra Swart, with Peter Boomgaard, William Clarence-Smith, Bernice de Jong Boers, and Dhiravat na Pombejra. *Breeds of Empire: The "Invention" of the Horse in Southeast Asia and Southern Africa, 1500–1950.* Copenhagen, Den.: NIAS Press, 2007.

Barber, Elizabeth Wayland. *The Mummies of Ürümchi.* New York: W. W. Norton, 1999.

Barclay, Harold B. *The Role of the Horse in Man's Culture.* London: J. A. Allen, 1980.

Barfield, Thomas J. "The Hsiung-nu Imperial Confederacy: Organization and Foreign Policy." *Journal of Asian Studies* 41, no. 5 (November 1981): 45–61.

———. *The Perilous Frontier: Nomadic Empires and China.* Oxford, UK: Basil Blackwell, 1989.

Barker, Graeme. *The Agricultural Revolution in Prehistory: Why Did Foragers Become Farmers?* Oxford: Oxford University Press, 2006.

Barras, Colin. "Story of Most Murderous People of All Time Revealed in Ancient DNA." *New Scientist,* March 2019, 29–33.

Bartosiewicz, Laszlo, and Erika Gal, eds. *Care or Neglect? Evidence of Animal Disease in Archaeology.* Oxford, UK: Oxbow, 2018.

Beal, Richard H. "Hittite Military Organization." In *Civilizations of the Ancient Near East, vol. 1,* edited by Jack M. Sasson, 545–54. Peabody, MA: Hendrickson, 2000.

Beard, Mary. *SPQR: A History of Ancient Rome.* New York: Liveright, 2015.

Beattie, Gladys Mackey. *The Canadian Horse: A Pictorial History.* Lennoxville, Can.: Sun Books, 1981.

Beck, Ulrike, Mayke Wagner, Xiao Li, Desmond Durkin-Meisterernst, and Pavel E. Tarasov. "The Invention of Trousers and Its Likely Affiliation with Horseback Riding and Mobility: A Case Study of Late 2nd Millennium BC Finds from Turfan in Eastern Central Asia." *Quaternary International* 348, no. 20 (October 2014): 224–35.

Beckwith, Christopher I. *Empires of the Silk Road: A History of Central Eurasia from the Bronze Age to the Present.* Princeton, NJ: Princeton University Press, 2009.

———. *The Scythian Empire: Central Eurasia and the Birth of the Classical Age from Persia to China.* Princeton, NJ: Princeton University Press, 2023.

Beevor, Antony. *Stalingrad: The Fateful Siege: 1942–1943.* New York: Viking, 1998.

Bellwood, Peter. *First Farmers: The Origins of Agricultural Societies.* Oxford, UK: Blackwell, 2005.

Bemmann, Jan, and Michael Schmauder, eds. *Complexity of Interaction Along the Eurasian Steppe Zone in the First Millennium CE*. Bonn, Ger.: Rheinische Friedrich-Wilhelms-Universität Bonn, 2015.

Bendrey, Robin. "New Methods for the Identification of Evidence for Bitting on Horse Remains from Archaeological Sites." *Journal of Archaeological Science* 34, no. 7 (July 2007): 1036–50.

Bendrey, Robin, Nick Thorpe, Alan Outram, and Louise H. van Wijngaarden-Bakker. "The Origins of Domestic Horses in North-West Europe: New Direct Dates on the Horses of Newgrange, Ireland." *Proceedings of the Prehistoric Society* 79 (December 2013): 91–103.

Betts, Alison V. G., and Peter Wei Ming Jia. "A Re-analysis of the Qiemu'erqieke (Shamirshak) Cemeteries, Xinjiang, China." *Journal of Indo-European Studies* 38, nos. 3/4 (Winter/Fall 2010): 275–317.

Bidwell, Percy Wells, and John I. Falconer. *History of Agriculture in the Northern United States, 1620–1860*. Washington, DC: Carnegie Institution of Washington, 1925.

Bird, Norman. *The Distribution of Indo-European Root Morphemes: A Checklist for Philologists*. Wiesbaden, Ger.: Harrassowitz, 1982.

Birrell, Anne. *Chinese Mythology: An Introduction*. Baltimore: Johns Hopkins University Press, 1993.

Bolles, Albert S. *Industrial History of the United States. 1881*. Reprint, New York: Augustus M. Kelley, 1966.

Boot, Max. *War Made New: Technology, Warfare, and the Course of History, 1500 to Today*. New York: Gotham Books, 2006.

Bose, Partha. *Alexander the Great's Art of Strategy: The Timeless Leadership Lessons of History's Greatest Empire Builder*. New York: Gotham Books, 2003.

Bou, Jean. *Light Horse: A History of Australia's Mounted Arm*. Cambridge: Cambridge University Press, 2010.

Bouman, Inge. *The Reintroduction of Przewalski Horses in the Hustain Nuruu Mountain Forest Steppe Reserve in Mongolia: An Integrated Conservation Development Project*. Leiden, Neth.: University of Leiden, 1998.

Bowen, Edward L. *The Jockey Club's Illustrated History of Thoroughbred Racing in America*. New York: Bulfinch Press, 1994.

Bower, Bruce. "First Pants Worn by Horse Riders 3,000 Years Ago." Science News online. Last modified May 30, 2014. https://www.sciencenews.org/article/first-pants-worn-horse-riders-3000-years-ago.

———. "How Asia's First Nomadic Empire Broke the Rules of Imperial Expansion." Science News online. Last modified July 2, 2023. https://www.sciencenews.org/article/asia-xiongnu-nomadic-empire-expansion.

Boyd, Lee, and Katherine A. Houpt, eds. *Przewalski's Horse: The History and Biology of an Endangered Species*. Albany: State University of New York Press, 1994.

Boyle, Katie, Colin Renfrew, and Marsha Levine, eds. *Ancient Interactions: East and West in Eurasia*. Cambridge, UK: McDonald Institute for Archaeological Research, 2002.

Bracegirdle, Hilary. *The National Horseracing Museum: A Concise History of British Horse Racing*. Derby, UK: Heritage House Group, 1999.

Bradbury, Jim. *The Battle of Hastings*. Stroud, UK: Sutton, 1998.

Brassley, Paul, and Richard Soffe. *Agriculture: A Very Short Introduction*. Oxford: Oxford University Press, 2016.

Braudel, Fernand. *Civilization and Capitalism, 15th–18th Century.* Vol. 1, *The Structures of Everyday Life: The Limits of the Possible.* New York: Harper & Row, 1979.

———. *Civilization and Capitalism, 15th–18th Century,* Vol. 2, *The Wheels of Commerce.* New York: Harper & Row, 1982.

———. *Civilization and Capitalism, 15th–18th Century.* Vol. 3, *The Perspective of the World.* New York: Harper & Row, 1984.

Brereton, J. M. *The Horse in War.* New York: Arco, 1976.

Briant, Pierre. *The First European: A History of Alexander in the Age of Empire.* Translated by Nicholas Elliott. Cambridge, MA: Harvard University Press, 2017.

Brinkley, Douglas. *Wheels for the World: Henry Ford, His Company, and a Century of Progress, 1903–2003.* New York: Penguin, 2004.

Bryce, Trevor. *The World of the Neo-Hittite Kingdoms: A Political and Military History.* Oxford: Oxford University Press, 2012.

Budiansky, Stephen. *The Nature of Horses: Exploring Equine Evolution, Intelligence, and Behavior.* New York: Free Press, 1997.

Büntgen, Ulf, and Nicola Di Cosmo. "Climatic and Environmental Aspects of the Mongol Withdrawal from Hungary in 1242 CE." *Scientific Reports* 6, art. 25606 (May 26, 2016). https://doi.org/10.1038/srep25606.

Bureau of Agricultural Economics. *Statistical Bulletin No. 5: Horses, Mules, and Motor Vehicles: Year Ended March 31, 1924, with Comparable Data for Earlier Years* (United States Department of Agriculture). Washington, DC: US Government Printing Office, 1925.

Busby, Debbie, and Catrin Rutland. *The Horse: A Natural History.* Princeton, NJ: Princeton University Press, 2019.

Butler, Simon. *The War Horses: The Tragic Fate of a Million Horses Sacrificed in the First World War.* Wellington, UK: Halsgrove, 2011.

Campbell, Brian, and Lawrence A. Tritle, eds. *The Oxford Handbook of Warfare in the Classical World.* Oxford: Oxford University Press, 2013.

Campbell, Bruce M. S. *English Seigniorial Agriculture, 1250–1450.* Cambridge: Cambridge University Press, 2000.

Campbell, Gordon Lindsay, ed. *The Oxford Handbook of Animals in Classical Thought and Life.* Oxford: Oxford University Press, 2014.

Cantor, Norman F. *Alexander the Great: Journey to the End of the Earth.* New York: HarperCollins, 2005.

Cantrell, Deborah O'Daniel. *The Horsemen of Israel: Horses and Chariotry in Monarchic Israel (Ninth–Eighth Centuries B.C.E.).* Winona Lake, IN: Eisenbrauns, 2011.

Carter, William Harding. *The Horses of the World: The Development of Man's Companion in War Camp, on Farm, in the Marts of Trade, and in the Field of Sports.* Washington, DC: National Geographic Society, 1923.

Cartledge, Paul. *Alexander the Great: The Hunt for a New Past.* New York: Overlook Press, 2004.

Cassidy, Rebecca, and Molly Mullin, eds. *Where the Wild Things Are Now: Domestication Reconsidered.* New York: Berg, 2007.

Chamberlin, J. Edward. *Horse: How the Horse Has Shaped Civilizations.* New York: BlueBridge, 2006.

Chambers, James. *The Devil's Horsemen: The Mongol Invasion of Europe.* New York: Atheneum, 1979.

Chang, Kwang-chih. *Shang Civilization.* New Haven, CT: Yale University Press, 1980.

Chen, Hao, ed. *Competing Narratives Between Nomadic People and Their Sedentary Neighbours: Papers of the 7th International Conference on the Medieval History of the Eurasian Steppe, Nov. 9–12, 2018, Shanghai University, China*. Szeged, Hung.: University of Szeged / Innovariant, 2019.

Chevalier de Monroy. *Foreign Agriculture, or, An Essay on the Comparative Advantages of Oxen for Tillage, in Competition with Horses*. Edited by John Talbot Dillon. London: G. Nicol, 1796.

Christian, David. "Silk Roads or Steppe Roads? The Silk Roads in World History." *Journal of World History* 11, no. 1 (Spring 2000): 1–26.

Clabby, John. *The Natural History of the Horse*. New York: Taplinger, 1976.

Clark, La Verne Harrell. *They Sang for Horses: The Impact of the Horse on Navajo and Apache Folklore*. Phoenix: University of Arizona Press, 1966.

Clark, Lloyd. *Blitzkrieg: Myth, Reality, and Hitler's Lightning War—France, 1940*. New York: Atlantic Monthly Press, 2016.

Clausewitz, Carl Von. *On War*. Edited by Michael Howard and Peter Paret. Princeton, NJ: Princeton University Press, 1989.

Clavin, Tom. *Reckless: The Racehorse Who Became a Marine Corps Hero*. New York: NAL Caliber, 2014.

Cline, Eric H. *1177 B.C.: The Year Civilization Collapsed*. Princeton, NJ: Princeton University Press, 2014.

Clutton-Brock, Juliet. *Horse Power: A History of the Horse and the Donkey in Human Societies*. Cambridge, MA: Harvard University Press, 1992.

———. *A Natural History of Domesticated Mammals*. Cambridge: Cambridge University Press, 1999.

———, ed. *The Walking Larder: Patterns of Domestication, Pastoralism, and Predation*. London: Unwin Hyman, 1989.

Cole, H. H., and Magnar Ronning, eds. *Animal Agriculture: The Biology of Domestic Animals and Their Use by Man*. San Francisco: W. H. Freeman, 1974.

Conkin, Paul K. *A Revolution Down on the Farm: The Transformation of American Agriculture Since 1929*. Lexington: University Press of Kentucky, 2008.

Cook, Noble David. *Born to Die: Disease and New World Conquest, 1492–1650*. Cambridge: Cambridge University Press, 1998.

Cooper, Jilly. *Animals in War: Valiant Horses, Courageous Dogs, and Other Unsung Animal Heroes*. Guilford, CT: Lyons Press, 2002.

Corum, James S. *The Roots of Blitzkrieg: Hans von Seeckt and German Military Reform*. Lawrence: University of Kansas Press, 1992.

Corvi, Steven J. "Men of Mercy: The Evolution of the Royal Army Veterinary Corps and the Soldier-Horse Bond During the Great War." *Journal of the Society for Army Historical Research* 76, no. 308 (Winter 1998): 272–84.

Cotterell, Arthur. *Chariot: From Chariot to Tank, The Astounding Rise and Fall of the World's First War Machine*. New York: Overlook Press, 2005.

Creel, H. G. "The Role of the Horse in Chinese History." *American Historical Review* 70, no. 3 (April 1965): 647–72.

Crook, Paul. *Darwinism, War and History: The Debate over the Biology of War from the "Origin of Species" to the First World War*. Cambridge: Cambridge University Press, 1994.

Crosby, Alfred W. *The Columbian Exchange: Biological and Cultural Consequences of 1492*. Westport, CT: Greenwood Press, 1973.

———. *Ecological Imperialism: The Biological Expansion of Europe, 900–1900*. 1986. Reprint. Cambridge: Cambridge University Press, 2009.

Cunfer, Geoff. *On the Great Plains: Agriculture and Environment*. College Station: Texas A&M University Press, 2005.

Cunha, Luis Sá. "The Horse in Ancient Chinese History, Symbolism and Myth: The Horse in Ancient China." *Review of Culture: Instituto Cultural de Macau* 4, no. 3 (2009): 3–19.

Cunliffe, Barry. *By Steppe, Desert, and Ocean: The Birth of Eurasia*. Oxford: Oxford University Press, 2015.

———. *Europe Between the Oceans: Themes and Variations: 9000 BC–AD 1000*. New Haven, CT: Yale University Press, 2008.

———. *Scythians: Nomad Warriors of the Steppe*. Oxford: Oxford University Press, 2019.

Curry, Andrew. "Horse Nations." *Science* 379, no. 6639 (March 2023): 1288–93.

———. "Milk Fueled Bronze Age Expansion of 'Eastern Cowboys' into Europe." *Science* online. Last modified September 15, 2021. doi:10.1126/science.acx9119.

———. "Patagonian People Were Riding Horses Long Before Europeans Arrived." *Science* online. Last modified December 8, 2023. doi: 10.1126/science.z541din.

Curtis, John, Nigel Tallis, and Astrid Johansen. *Horse and History: From Arabia to Ascot*. Burlington, VT: Lund Humphries, 2012.

Dabashi, Hamid. *Persophilia: Persian Culture on the Global Scene*. Cambridge, MA: Harvard University Press, 2015.

Dalley, Stephanie, and John Peter Oleson. "Sennacherib, Archimedes, and the Water Screw: The Context of Invention in the Ancient World." *Technology and Culture* 44, no. 1 (January 2003): 1–26.

Daston, Lorraine, and Gregg Mitman, eds. *Thinking: New Perspectives on Anthropomorphism with Animals*. New York: Columbia University Press, 2005.

Davies, Stephen. "The Great Horse-Manure Crisis of 1894: The Problem Solved Itself." Foundation for Economic Education online. Last modified September 1, 2004. https://fee.org/articles/the-great-horse-manure-crisis-of-1894.

Davis, Dona Lee, and Anita Maurstad, eds. *The Meaning of Horses: Biosocial Encounters*. Abingdon, UK: Routledge, 2016.

Davis, Paul K. *100 Decisive Battles: From Ancient Times to the Present*. Oxford: Oxford University Press, 2001.

Dawson, Anthony. *Real War Horses: The Experience of the British Cavalry, 1814–1914*. Barnsley, UK: Pen & Sword Military, 2020.

Day, Herbert L. *When Horses Were Supreme: The Age of the Working Horse*. North Humberside, UK: Hutton, 1985.

Dean, Joanna, Darcy Ingram, and Christabelle Sethna, eds. *Animal Metropolis: Histories of Human-Animal Relations in Urban Canada*. Calgary, Can.: University of Calgary Press, 2017.

De Barros Damgaard, Peter, Rui Martiniano, Jack Kamm, J. Víctor Moreno-Mayar, Guus Kroonen, Michaël Peyrot, Gojko Barjamovic, et al. "The First Horse Herders and the Impact of Early Bronze Age Steppe Expansions into Asia." *Science* 360, no. 6396 (June 29, 2018). doi:10.1126/science.aar7711.

Deichmann, Ute. *Biologists Under Hitler*. Translated by Thomas Dunlap. Cambridge, MA: Harvard University Press, 1996.

Denhardt, Robert Moorman. *The Horse of the Americas*. Norman: University of Oklahoma Press, 1947.

Denison, Colonel George T. *A History of Cavalry from the Earliest Times, with Lessons for the Future*. 2nd ed. London: Macmillan, 1913.

Dent, Anthony. *The Horse, Through Fifty Centuries of Civilization.* New York: Holt, Rinehart and Winston, 1974.
Derry, Margaret E. *Horses in Society: A Story of Animal Breeding and Marketing, 1800–1920.* Toronto: University of Toronto Press, 2006.
Diamond, Jared. *Guns, Germs, and Steel: The Fates of Human Societies.* New York: W. W. Norton, 1999.
———. "The Worst Mistake in the History of the Human Race." *Discover,* May 1987. 64–66.
Di Cosmo, Nicola. *Ancient China and Its Enemies: The Rise of Nomadic Power in East Asian History.* Cambridge: Cambridge University Press, 2002.
DiMarco, Louis A. *War Horse: A History of the Military Horse and Rider.* Yardley, PA: Westholme, 2012.
DiNardo, R. L. *Mechanized Juggernaut or Military Anachronism? Horses and the German Army of WWII.* New York: Greenwood Press, 1991.
DiNardo, R. L., and Austin Bay. "Horse-Drawn Transport in the German Army." *Journal of Contemporary History* 23, no. 1 (January 1988): 129–42.
Dorondo, David R. *Riders of the Apocalypse: German Cavalry and Modern Warfare, 1870–1945.* Annapolis, MD: Naval Institute Press, 2012.
Douglas, David C. *William the Conqueror: The Norman Impact upon England.* Berkeley: University of California Press, 1964.
Downs, James F. "The Origin and Spread of Riding in the Near East and Central Asia." *American Anthropologist* 63, no. 6 (December 1961): 1193–203.
Downs, James F., Miwa Karoku, Suezaki Masumi, and Motoyoshi Shigekatsu. *Horses and Humanity in Japan.* Tokyo: Japan Association for International Horse Racing, 1999.
Drews, Robert. *The Coming of the Greeks: Indo-European Conquests in the Aegean and the Near East.* Princeton, NJ: Princeton University Press, 1988.
———. *Early Riders: The Beginnings of Mounted Warfare in Asia and Europe.* New York: Routledge, 2004.
———. *The End of the Bronze Age: Changes in Warfare and the Catastrophe ca. 1200 B.C.* Princeton, NJ: Princeton University Press, 1993.
———. *Militarism and the Indo-Europeanizing of Europe.* New York: Routledge, 2017.
DuPuy, William. *The History of the Horse.* New York: Holt, Rinehart and Winston, 1965.
Edwards, Elwyn Hartley. *Horses: Their Role in the History of Man.* London: Willow Books, 1987.
———. *Wild Horses: The World's Last Surviving Herds.* Irvington, NY: Hylas, 2003.
Edwards, Peter. *Horse and Man in Early Modern England.* London: Continuum, 2007.
Ehringer, Gavin. *Leaving the Wild: The Unnatural History of Dogs, Cats, Cows, and Horses.* New York: Pegasus, 2017.
Ellis, John. *Cavalry: The History of Mounted Warfare.* New York: G. P. Putnam's Sons, 1978.
Essin, Emmett M. *Shavetails and Bell Sharps: The History of the U.S. Army Mule.* Lincoln: University of Nebraska Press, 1997.
Everitt, Anthony. *Alexander the Great: His Life and His Mysterious Death.* New York: Random House, 2019.
Ewers, John C. *The Horse in Blackfoot Indian Culture: With Comparative Material from Other Western Tribes.* Washington, DC: US Government Printing Office, 1955.

Fagan, Brian. *The Intimate Bond: How Animals Shaped Human History*. New York: Bloomsbury Press, 2015.

Fages, Antoine, Kristian Hanghøj, Naveed Khan, Charleen Gaunitz, Andaine Seguin-Orlando, Michela Leonardi, Christian McCrory Constantz, et al. "Tracking Five Millennia of Horse Management with Extensive Ancient Genome Time Series." *Cell* 177, no. 6 (May 30, 2019): 1419–35.

Favereau, Marie. *The Horde: How the Mongols Changed the World*. Cambridge, MA: Belknap Press of Harvard University Press, 2021.

———. "Introduction: The Islamisation of the Steppe." *Revue des mondes musulmans et de la Méditerranée*, no. 143 (2018). https://doi.org/10.4000/remmm.106.

Federico, Giovanni. *Feeding the World: An Economic History of Agriculture, 1800–2000*. Princeton, NJ: Princeton University Press, 2005.

Felber, Bill. *The Horse in War*. Philadelphia: Chelsea House, 2002.

Felton, Mark. *Ghost Riders: When US and German Soldiers Fought Together to Save the World's Most Beautiful Horses in the Last Days of World War II*. New York: Da Capo Press, 2018.

Fincke, Jeanette C. "The Babylonian Texts of Nineveh: Report on the British Museum's Ashurbanipal Library Project*." *Archiv für Orientforschung* 50 (2003): 111–49.

Fitzgerald, Deborah. *Every Farm a Factory: The Industrial Ideal in American Agriculture*. New Haven, CT: Yale University Press, 2003.

Fitzhugh, William W., Morris Rossabi, and William Honeychurch, eds. *Genghis Khan and the Mongol Empire*. Fyshwick, Aus.: Odyssey Books, 2013.

Flower, William Henry. *The Horse: A Study in Natural History*. London: Kegan Paul, Trench, Trübner, 1891.

Flynn, Jane. *Soldiers and Their Horses: Sense, Sentimentality and the Soldier-Horse Relationship in the Great War*. New York: Routledge, 2020.

Forrest, Susanna. *The Age of the Horse: An Equine Journey Through Human History*. New York: Atlantic Monthly Press, 2017.

Fouracre, Paul. *The Age of Charles Martel*. New York: Longman, 2000.

Fox, Charles Philip. *Working Horses: Looking Back 100 Years to America's Horse-Drawn Days*. Whitewater, WI: Heart Prairie Press, 1990.

Frahm, Eckart. *Assyria: The Rise and Fall of the World's First Empire*. New York: Basic Books, 2023.

Francis, Richard. *Domesticated: Evolution in a Man-Made World*. New York: W. W. Norton, 2015.

Frankopan, Peter. *The Silk Roads: A New History of the World*. New York: Vintage Books, 2017.

Freeman, Philip. *Alexander the Great*. New York: Simon & Schuster Paperbacks, 2011.

Gabriel, Richard A. *The Culture of War: Invention and Early Development*. New York: Greenwood Press, 1990.

———. *The Great Armies of Antiquity*. Westport, CT: Praeger, 2002.

Gaebel, Robert E. *Cavalry Operations in the Ancient Greek World*. Norman: University of Oklahoma Press, 2002.

Gardiner, Juliet. *The Animals' War: Animals in Wartime from the First World War to the Present Day*. London: Piatkus Books, 2006.

Gassmann, Jürg. "Combat Training for Horse and Rider in the Early Middle Ages." *Acta Periodica Duellatorum* 6, no. 1 (2018): 63–98.

———. "East Meets West: Mounted Encounters in Early and High Mediaeval Europe." *Acta Periodica Duellatorum* 5, no. 1 (2017): 75–107.

Gawronski, Radoslaw Andrzej. *Roman Horsemen Against Germanic Tribes: The Rhineland Frontier Cavalry Fighting Styles, 31 BC–AD 256*. Warsaw: Institute of Archaeology of the Cardinal Stefan Wyszyński University in Warsaw, 2018.
Gearin, Conor. "Mongol Hordes Gave Up on Conquering Europe Due to Wet Weather." *New Scientist* online. Last modified May 26, 2016. https://www.newscientist.com/article/2090335-mongol-hordes-gave-up-on-conquering-europe-due-to-wet-weather/.
Gerhold, Dorian, ed. *Road Transport in the Horse-Drawn Era*. Aldershot, UK: Scolar Press, 1996.
Gernet, Jacques. *Daily Life in China, on the Eve of the Mongol Invasion, 1250–1276*. Translated by H. M. Wright. Stanford, CA: Stanford University Press, 1962.
Goldsworthy, Adrian. *Cannae: Hannibal's Greatest Victory*. New York: Basic Books, 2019.
———. *Pax Romana: War, Peace and Conquest in the Roman World*. New Haven, CT: Yale University Press, 2016.
———. *Philip and Alexander: Kings and Conquerors*. New York: Basic Books, 2020.
———. *The Punic Wars*. London: Cassell, 2001.
———. *The Roman Army at War: 100 BC–AD 200*. Oxford: Clarendon Press, 1996.
Gommans, Jos. *The Indian Frontier: Horse and Warband in the Making of Empires*. New York: Routledge, 2018.
Gonzaga, Paulo Gaviä. *A History of the Horse*. Vol. 1. *The Iberian Horse from Ice Age to Antiquity*. London: J. A. Allen, 2004.
Goodwin, D., and N. Davidson. "The Role of the Horse in Europe." *Equine Veterinary Journal* 31, suppl. S28 (April 1999): 5.
Gordon, W. J. *The Horse-World of London*. London: Religious Tract Society, 1893.
Green, Peter. *Alexander of Macedon, 356–323 B.C.: A Historical Biography*. Berkeley: University of California Press, 1991.
Greene, Ann Norton. *Horses at Work: Harnessing Power in Industrial America*. Cambridge, MA: Harvard University Press, 2008.
Griggs, Mary Beth. "The World's Oldest Pants Were Developed for Riding Horses." *Smithsonian* online. Last modified June 3, 2014. https://www.smithsonianmag.com/smart-news/worlds-oldest-pants-were-developed-riding-horses-180951638/.
Grigoriev, Stanislav A. *Ancient Indo-Europeans*. Chelyabinsk, Russ.: RIFEI, 2002.
Grousset, René. *The Empire of the Steppes: A History of Central Asia*. Translated by Naomi Walford. New Brunswick, NJ: Rutgers University Press, 1970.
Guest, Kristen, and Monica Mattfeld, eds. *Equestrian Cultures: Horses, Human Society, and the Discourse of Modernity*. Chicago: University of Chicago Press, 2019.
Hacker, Barton C. "Horse, Wheel, and Saddle: Recent Works on Two Ancient Military Revolutions." *International Bibliography of Military History* 32, no. 2 (January 2012): 175–91.
———. "Mounted Archery and Firearms: Late Medieval Muslim Military Technology Reconsidered." *Vulcan* 3, no. 1 (May 2015): 42–65.
Hacker, J. David, and Michael R. Haines. "American Indian Mortality in the Late Nineteenth Century: The Impact of Federal Assimilation Policies on a Vulnerable Population." *Annales de démographie historique* 110, no. 2 (2005): 17–29.
Haines, Francis. *Horses in America*. New York: Thomas Y. Crowell, 1971.
———. "Where Did the Plains Indians Get Their Horses?" *American Anthropologist* 40, no. 1 (January–March 1938): 112–17.

———. "The Northward Spread of Horses Among the Plains Indians." *American Anthropologist* 40, no. 3 (July–September 1938): 429–37.
Hämäläinen, Pekka. *The Comanche Empire*. New Haven, CT: Yale University Press, 2009.
———. *Lakota America: A New History of Indigenous Power*. New Haven, CT: Yale University Press, 2019.
———. "Reconstructing the Great Plains: The Long Struggle for Sovereignty and Dominance in the Heart of the Continent." *Journal of the Civil War Era* 6, no. 4 (December 2016): 481–509.
———. "The Rise and Fall of Plains Indian Horse Cultures." *Journal of American History* 90, no. 3 (December 2003): 833–62.
Hamblin, William J. *Warfare in the Ancient Near East to 1600 BC: Holy Warriors at the Dawn of History*. New York: Routledge, 2006.
Hammond, N. G. L. *The Genius of Alexander the Great*. Chapel Hill: University of North Carolina Press, 1997.
Hansen, Skylar. *Roaming Free: Wild Horses of the American West*. Flagstaff, AZ: Northland Press, 1983.
Harari, Yuval Noah. *Sapiens: A Brief History of Humankind*. New York: Harper, 2015.
Harper, Kyle. *The Fate of Rome: Climate, Disease, and the End of an Empire*. Princeton, NJ: Princeton University Press, 2017.
Harris, Albert W. *The Blood of the Arab: The World's Greatest War Horse*. Chicago: Arabian Horse Club of America, 1941.
Harris, David R., ed. *The Origins of Agriculture in Western Central Asia: An Environmental-Archaeological Study*. Philadelphia: University of Pennsylvania Museum of Archeology and Anthropology, 2010.
———. *The Origins and Spread of Agriculture and Pastoralism in Eurasia*. Washington, DC: Smithsonian Institution Press, 1996.
Harrison, Lorraine. *Horse: From Noble Steeds to Beasts of Burden*. New York: Watson-Guptill, 2000.
Hausberger, Martine, Hélène Roche, Séverine Henry, and E. Kathalijne Visser. "A Review of the Human-Horse Relationship." *Applied Animal and Behaviour Science* 109, no. 1 (January 2008): 1–24.
Hawting, G. R. *The First Dynasty of Islam: The Umayyad Caliphate, AD 661–750*. 2nd ed. New York: Routledge, 2000.
Headrick, Daniel R. *Power over Peoples: Technology, Environments, and Western Imperialism, 1400 to the Present*. Princeton, NJ: Princeton University Press, 2010.
———. *Technology: A World History*. New York: Oxford University Press, 2009.
Henderson, W. O. *The Rise of German Industrial Power, 1834–1914*. Berkeley: University of California Press, 1975.
Herodotus. *The Histories*. Edited by John M. Marincola. Translated by Aubrey de Sélincourt. New York: Penguin Classics, 2003.
Hildinger, Erik. *Warriors of the Steppe: A Military History of Central Asia, 500 B.C. to 1700 A.D.* New York: Sarpedon, 1997.
Hindley, Geoffrey. *A Brief History of the Crusades: Islam and Christianity in the Struggle for World Supremacy*. London: Constable & Robinson, 2004.
Holmes, Richard. *Redcoat: The British Soldier in the Age of Horse and Musket*. New York: W. W. Norton, 2001.
Holt, Frank L. *Into the Land of Bones: Alexander the Great in Afghanistan*. Berkeley: University of California Press, 2012.

Honeychurch, William. "From Steppe Roads to Silk Roads: Inner Asian Nomads and Early Interregional Exchange." In *Nomads as Agents of Cultural Change: The Mongols and Their Eurasian Predecessors*, edited by Reuvan Amitai and Michal Biran, 50–87. Honolulu: University of Hawaiʻi Press, 2014.

———. *Inner Asia and the Spatial Politics of Empire: Archaeology, Mobility, and Culture Contact*. New York: Springer, 2015.

Hood, John. "The Horse at War: The Experience of the Horse in Human Warfare, with Particular Reference to the Adoption of the Stirrup in the Middle Ages." *ISAA Review: Journal of Independent Scholars Association of Australia* 15, no. 1 (April 2016): 29–41.

Hooper, Frederick. *The Military Horse: The Equestrian Warrior Through the Ages*. London: Marshall Cavendish, 1976.

Horn, Christian, and Kristian Kristiansen, eds. *Warfare in Bronze Age Society*. Cambridge: Cambridge University Press, 2018.

Horse and Mule Association of America. *Horses-Mules, Power-Profit*. Chicago, 1934.

———. *Horse and Mule Power in American Agriculture: Advantages of Farm Grown Power*. Chicago, 1939.

Horse Capture, George P., and Emil Her Many Horses, eds. *A Song for the Horse Nation: Horses in Native American Cultures*. Washington, DC: National Museum of the American Indian, Smithsonian Institution, 2006.

Horwitz, Tony. *A Voyage Long and Strange: Rediscovering the New World*. New York: Henry Holt, 2008.

Howard, Robert West. *The Horse in America*. Chicago: Follett, 1965.

Hoyland, Robert G., and Philip F. Kennedy, eds. *Islamic Reflections, Arabic Musings: Studies in Honour of Professor Alan Jones*. Oxford, UK: Gibb Memorial Trust, 2004.

Hucker, Charles O. *China's Imperial Past: An Introduction to Chinese History and Culture*. Stanford, CA: Stanford University Press, 1975.

Humphreys, Eileen. *The Royal Road: A Popular History of Iran*. London: Scorpion, 1991.

Hunt, Frazier, and Robert Hunt. *Horses and Heroes: The Story of the Horse in America for 450 Years*. New York: Scribner, 1949.

Hunt, Patrick N. *Hannibal*. New York: Simon & Schuster, 2017.

Hurt, R. Douglas. *American Agriculture: A Brief History*. Rev. West Lafayette, IN: Purdue University Press, 2002.

Hutton, Robin. *War Animals: The Unsung Heroes of World War II*. Washington, DC: Regnery History, 2018.

Hyland, Ann. *Equus: The Horse in the Roman World*. New Haven, CT: Yale University Press, 1990.

———. *The Horse in the Ancient World*. Stroud, UK: Sutton, 2003.

———. *The Horse in the Middle Ages*. Stroud, UK: Sutton, 1999.

———. *The Medieval Warhorse from Byzantium to the Crusades*. Strode, UK: Sutton, 1994.

———. *The War Horse in the Modern Era: Breeder to Battlefield, 1600 to 1865*. Stockton-on-Tees, UK: Black Tent, 2009.

———. *The Warhorse, 1250–1600*. Stroud, UK: Sutton, 1998.

Hyllested, Adam, Benedicte Nielsen Whitehead, Thomas Olander, and Birgit Anette Olsen, eds. *Language and Prehistory of the Indo-European Peoples: A Cross-Disciplinary Perspective*. Copenhagen, Den.: Museum Tusculanum Press, 2017.

Iarocci, Andrew. "Engines of War: Horsepower in the Canadian Expeditionary Force, 1914–18." *Journal of the Society for Army Historical Research* 87, no. 349 (Spring 2009): 59–83.

Imbler, Sabrina. "The Horse You Rode In On May Have Been Made in Southern Russia." *New York Times* online, October 20, 2021. https://www.nytimes.com/2021/10/20/science/horse-domestication-russia.html?searchResultPosition=1.
Isett, Christopher, and Stephen Miller. *The Social History of Agriculture: From the Origins to the Current Crisis*. Lanham, MD: Rowman & Littlefield, 2017.
Jackson, Peter. *The Mongols and the West, 1221–1410*. London: Routledge, 2014.
Jagchid, Sechin, and Van Jay Symons. *Peace, War, and Trade Along the Great Wall: Nomadic-Chinese Interaction Through Two Millenia*. Bloomington: Indiana University Press, 1989.
Jansen, Thomas, Peter Forster, Marsha A. Levine, Hardy Oelke, Matthew Hurles, Colin Renfrew, Jurgen Weber, and Klaus Olek. "Mitochondrial DNA and the Origins of the Domestic Horse." *Proceedings of the National Academy of Sciences of the United States of America* 99, no. 16 (August 6, 2002): 10905–10.
Jarymowycz, Roman. *Cavalry from Hoof to Track*. Westport, CT: Praeger Security International, 2008.
Johns, Catherine. *Horses: History-Myth-Art*. Cambridge, MA: Harvard University Press, 2006.
Johnson, Ben. "The Great Horse Manure Crisis of 1894." Historic UK, accessed September 27, 2022. https://www.historic-uk.com/HistoryUK/HistoryofBritain/Great-Horse-Manure-Crisis-of-1894/#:~:text=by%20Ben%20Johnson,the%20streets%20of%20London%20alone.
Johnson, Lonnie R. *Central Europe: Enemies, Neighbors, Friends*. Oxford: Oxford University Press, 2002.
Johnson, Paul C. *Farm Animals in the Making of America*. Des Moines: Wallace-Homestead, 1975.
Johnson, Paul Louis, ed. *Horses of the German Army in World War II*. Atglen, PA: Schiffer Military History, 2006.
Jones, Martin, ed. *Traces of Ancestry: Studies in Honour of Colin Renfrew*. Cambridge, UK: McDonald Institute for Archaeological Research, 2004.
Jones, Robert Leslie. "The Old French-Canadian Horse: Its History in Canada and the United States." *Canadian Historical Review* 28, no. 2 (June 1947): 125–55.
Jones, Spencer. "The Influence of Horse Supply upon Field Artillery in the American Civil War." *Journal of Military History* 74, no. 2 (April 2010): 357–77.
———. "Scouting for Soldiers: Reconnaissance and the British Cavalry, 1899–1914." *War in History* 18, no. 4 (December 2011): 495–513.
Joseph, Brian D. "The Indo-Europeanization of Europe: An Introduction to the Issues." *Proceedings of the American Philosophical Society* 162, no. 1 (March 2018): 15–23.
Jurmain, Suzanne. *Once Upon a Horse: A History of Horses—And How They Shaped Our History*. New York: Lothrop, Lee & Shepard Books, 1989.
Kahn, Paul, ed. *The Secret History of the Mongols: The Origin of Chingis Khan*. San Francisco: North Point Press, 1984.
Karolevitz, Robert F. *Old-Time Agriculture in the Ads: Being a Compendium of Magazine and Newspaper Sales Literature Reminiscent of the Days When Farming Was a Way of Life and Horsepower Came in Horses*. Aberdeen, SD.: North Plains Press, 1970.
Karunanithy, David. *The Macedonian War Machine: Neglected Aspects of the Armies of Philip, Alexander and the Successors (359–281 BC)*. Barnsley, UK: Pen & Sword Military, 2013.

Keegan, John. *A History of Warfare*. New York: Alfred A. Knopf, 1993.
———. *The Mask of Command: Alexander the Great, Wellington, Ulysses S. Grant, Hitler, and the Nature of Leadership*. New York: Penguin Books, 1988.
Keeley, Lawrence H. *War Before Civilization: The Myth of the Peaceful Savage*. Oxford: Oxford University Press, 1996.
Kean, Sam. *The Icepick Surgeon: Murder, Fraud, Sabotage, Piracy, and Other Dastardly Deeds Perpetrated in the Name of Science*. New York: Little, Brown, 2021.
Kelekna, Pita. *The Horse in Human History*. Cambridge: Cambridge University Press, 2009.
———. "The Politico-Economic Impact of the Horse on Old World Cultures: An Overview." *Sino-Platonic Papers* 190 (June 2009): 1–24.
Kelly, William, and Nora Hickson Kelly. *The Horses of the Royal Canadian Mounted Police: A Pictorial History*. Toronto: Doubleday Canada, 1984.
Kennedy, Hugh. *The Great Arab Conquests: How the Spread of Islam Changed the World We Live In*. Philadelphia: Da Capo Press, 2007.
Kentucky Horse Park. *All the Queen's Horses: The Role of the Horse in British History*. Lexington, KY: Kentucky Horse Park / Harmony House, 2003.
———. *Imperial China: The Art of the Horse in Chinese History*. Lexington, KY: Kentucky Horse Park / Harmony House, 2000.
Kenyon, David. *Horsemen in No Man's Land: British Cavalry and Trench Warfare, 1914–1918*. Barnsley, UK: Pen & Sword, 2011.
Keyser, Christine, Vincent Zvénigorosky, Angéla Gonzalez, Jean-Luc Fausser, Florence Jagorel, Patrice Gérard, Turbat Tsagaan, et al. "Genetic Evidence Suggests a Sense of Family, Parity and Conquest in the Xiongnu Iron Age Nomads of Mongolia." *Human Genetics* 140, no. 2 (February 2021): 349–59.
Khazanov, Anatoly M. "The Eurasian Steppe Nomads in World History." In *Nomad Aristocrats in a World of Empires*, edited by Jurgen Paul, 187–207. Wiesbaden, Ger.: Dr. Ludwig Reichert Verlag, 2013.
———. "The Scythians and Their Neighbors." In *Nomads as Agents of Cultural Change: The Mongols and Their Eurasian Predecessors*, edited by Reuvan Amitai and Michal Biran, 32–49. Honolulu: University of Hawai'i Press, 2014.
Kilby, Emily R. "The Demographics of the U.S. Equine Population." In *The State of the Animals*, edited by Deborah J. Salem and Andrew N. Rowan, 175–205. Washington, DC: Humane Society Press, 2007.
Kimball, Cheryl. *The Complete Horse: An Entertaining History of Horses*. Saint Paul, MN: Voyageur Press, 2006.
Kiple, Kenneth F., and Stephen V. Beck, eds. *Biological Consequences of the European Expansion, 1450–1800*. Aldershot, UK: Ashgate, 1997.
Kohl, Philip L. *The Making of Bronze Age Eurasia*. Cambridge: Cambridge University Press, 2007.
Kohlstedt, Kurt. "The Big Crapple: NYC Transit Pollution from Horse Manure to Horseless Carriages." 99% Invisible online. Last modified November 6, 2017. https://99percentinvisible.org/article/cities-paved-dung-urban-design-great-horse-manure-crisis-1894/.
Kolbert, Elizabeth. "Hosed: Is There a Quick Fix for the Climate?" *The New Yorker*, November 16, 2009, 75–77.
Kristiansen, Kristian. *Archeology and the Genetic Revolution in European Prehistory*. Cambridge: Cambridge University Press, 2022.

Kuan, Jeffrey Kah-Jin. *Neo-Assyrian Historical Inscriptions and Syria-Palestine: Israelite /Judean-Tyrian-Damascene Political and Commercial Relations in the Ninth–Eighth Centuries BCE.* Eugene, OR: Wipf & Stock, 2016.

Kuhrt, Amélie. *The Persian Empire: A Corpus of Sources from the Achaemenid Period.* London: Routledge, 2007.

Kust, Matthew J. *Man and Horse in History.* Alexandria, VA: Plutarch Press, 1983.

Laffaye, Horace A. *Polo in the United States: A History.* Jefferson, NC: McFarland, 2011.

Landry, Donna. *Noble Brutes: How Eastern Horses Transformed English Culture.* Baltimore: Johns Hopkins University Press, 2009.

Langdon, John. *Horses, Oxen and Technological Innovation: The Use of Draught Animals in English Farming from 1066–1500.* Cambridge: Cambridge University Press, 2002.

Lanning, Michael Lee. *The Battle 100: The Stories Behind History's Most Influential Battles.* Naperville, IL: Sourcebooks, 2005.

Larson, Greger, Dolores R. Piperno, Robin G. Allaby, Michael D. Purugganan, Leif Andersson, Manuel Arroyo-Kalin, Loukas Barton, et al. "Current Perspectives and the Future of Domestication Studies." *Proceedings of the National Academy of Sciences of the United States of America* 111, no. 17 (April 29, 2014): 6139–46.

Law, Robin. *The Horse in West African History: The Role of the Horse in the Societies of Pre-Colonial West Africa.* Oxford: Oxford University Press, 1980.

Lawrence, Elizabeth Atwood. *Hoofbeats and Society: Studies of Human-Horse Interactions.* Bloomington: Indiana University Press, 1985.

Lee, Wayne E. *Waging War: Conflict, Culture, and Innovation in World History.* Oxford: Oxford University Press, 2016.

Letts, Elizabeth. *The Perfect Horse: The Daring U.S. Mission to Rescue the Priceless Stallions Kidnapped by the Nazis.* New York: Ballantine Books, 2016.

Levin, H. A. *A History of Horses Told by Horses: Horse Sense for Humans.* Garden City, NY: Morgan James, 2009.

Levin, Jonathan V. *Where Have All the Horses Gone?: How Advancing Technology Swept American Horses from the Road, the Farm, the Range, and the Battlefield.* Jefferson, NC: McFarland, 2017.

Levine, M. A. "Investigating the Origins of Horse Domestication." *Equine Veterinary Journal* 31, no. S28 (April 1999): 6–14.

Levine, Marsha A. "Botai and the Origins of Horse Domestication." *Journal of Anthropological Archaeology* 18, no. 1 (March 1999): 29–78.

Levine, Marsha, Colin Renfrew, and Katie Boyle, eds. *Prehistoric Steppe Adaptation and the Horse.* Cambridge, UK: McDonald Institute for Archaeological Research, 2003.

Levine, Marsha, Yuri Rassamakin, Aleksandr Kislenko, and Nataliya Tatarintseva. *Late Prehistoric Exploitation of the Eurasian Steppe.* Cambridge, UK: McDonald Institute for Archaeological Research, 1999.

Levitt, Steven D., and Stephen J. Dubner. *Super Freakonomics: Global Cooling, Patriotic Prostitutes, and Why Suicide Bombers Should Buy Life Insurance.* New York: William Morrow, 2010.

Lewis, David Levering. *God's Crucible: Islam and the Making of Europe, 570 to 1215.* New York: W. W. Norton, 2008.

Li, Chunxiang, Hongjie Li, Yinqiu Cui, Chengzhi Xie, Dawei Cai, Wenying Li, Victor H. Mair, et al. "Evidence That a West-East Admixed Population Lived in the Tarim Basin as Early as the Early Bronze Age." *BMC Biology* 8 (February 2010): art. 15. doi.org/10.1186/1741-7007-8-15.

Li, Chunxiang, Chao Ning, Erika Hagelberg, Hongjie Li, Yongbin Zhao, Wenying Li, Idelisi Abuduresule, et al. "Analysis of Ancient Human Mitochondrial DNA from the Xiaohe Cemetery: Insights into Prehistoric Population Movements in the Tarim Basin, China." *BMC Genetics* 16 (July 2015): art. 78. doi.org/10.1186/s12863-015-0237-5.

Librado, Pablo, Naveed Khan, Antoine Fages, Mariya A. Kusliy, Tomasz Suchan, Laure Tonasso-Calvière, Stéphanie Schiavinato, et al. "The Origins and Spread of Domestic Horses from the Western Eurasian Steppes." *Nature* 598, no. 7882 (October 2021): 634–40.

Liedtke, Gregory. *Enduring the Whirlwind: The German Army and the Russo-German War, 1941–1943.* Solihull, UK: Helion, 2016.

Littauer, M. A., and J. H. Crouwel. *Wheeled Vehicles and Ridden Animals in the Ancient Near East.* Leiden, Neth.: E. J. Brill, 1979.

Liu, Li, and Xingcan Chen. *The Archaeology of China: From the Late Paleolithic to the Early Bronze Age.* Cambridge: Cambridge University Press, 2012.

Liu, Xinru. *The Silk Road in World History.* Oxford: Oxford University Press, 2010.

Lobell, Jarrett A., and Eric A. Powell. "The Story of the Horse: How Its Unique Role in Human Culture Transformed History." *Archaeology* 68, no. 4 (July/August 2015): 28–34.

Lubow, Robert E. *The War Animals.* Garden City, NY: Doubleday, 1977.

Ludwig, Arne, Melanie Pruvost, Monika Reissmann, Norbert Benecke, Gudrun A. Brockmann, Pedro Castaños, Michael Cieslak, et al. "Coat Color Variation at the Beginning of Horse Domestication." *Science* 324, no. 5926 (April 24, 2009): 485.

Lundgren, Erick J., Daniel Ramp, Juliet C. Stromberg, Jianguo Wu, Nathan C. Nieto, Martin Sluk, Karla T. Moeller, et al. "Equids Engineer Desert Water Availability." *Science* 372, no. 6541 (April 30, 2021): 491–95.

Lyons, Justin D. *Alexander the Great and Hernán Cortés: Ambiguous Legacies of Leadership.* Lanham, MD: Lexington Books, 2015.

MacClintock, Dorcas. *A Natural History of Zebras.* New York: Scribner, 1976.

Macdonald, Janet. *Horses in the British Army, 1750–1950.* Barnsley, UK: Pen & Sword Military, 2017.

MacEwan, Grant. *Heavy Horses: Highlights of Their History.* Saskatoon, Can.: Western Producer Prairie Books, 1986.

MacFadden, Bruce J. *Horse Fossils: Systematics, Paleobiology, and Evolution of the Family Equidae.* Cambridge: Cambridge University Press, 1992.

Maenchen-Helfen, J. Otto. *The World of the Huns: Studies in Their History and Culture.* Berkeley: University of California Press, 1973.

Main, Douglas. "Wild Horses and Donkeys Dig Wells in the Desert, Providing Water for Wildlife." *National Geographic*, April 29, 2021.

Mair, Victor H. "Ancient Mummies of the Tarim Basin: Discovering Early Inhabitants of Eastern Central Asia." *Expedition* 58, no. 2 (Fall 2016): 25–29.

Mallapaty, Smriti. "DNA Reveals Surprise Ancestry of Mysterious Chinese Mummies." *Nature* 599, no. 7883 (November 2021): 19–20.

Mallory, J. P. "Bronze Age Languages of the Tarim Basin." *Expedition* 52, no. 3, November 2010, 44–53.

———. *In Search of the Indo-Europeans: Language, Archaeology and Myth.* New York: Thames & Hudson, 1991.

Mallory, J. P., and D. Q. Adams, eds. *Encyclopedia of Indo-European Culture.* Chicago: Fitzroy Dearborn, 1997.

———. *The Oxford Introduction to Proto-Indo-European and the Proto-Indo-European World*. Oxford: Oxford University Press, 2006.

Mallory, J. P., and Victor H. Mair. *The Tarim Mummies: Ancient China and the Mystery of the Earliest Peoples from the West*. New York: Thames & Hudson, 2000.

Man, John. *Empire of Horses: The First Nomadic Civilization and the Making of China*. New York: Pegasus Books, 2020.

Mancall, Peter C., ed. *Envisioning America: English Plans for the Colonization of North America, 1580–1640—A Brief History with Documents*. Boston: Bedford–St. Martin's Press, 2017.

Mann, Charles C. *1491: New Revelations of the Americas Before Columbus*. New York: Vintage Books, 2006.

———. *1493: Uncovering the New World Columbus Created*. New York: Alfred A. Knopf, 2011.

Martin, H. Desmond. "The Mongol Army." *Journal of the Royal Asiatic Society* 75, no. 1/2 (January 1943): 46–85.

Martin, Thomas R., and Christopher W. Blackwell. *Alexander the Great: The Story of an Ancient Life*. Cambridge: Cambridge University Press, 2012.

May, Timothy. *The Mongol Art of War: Chinggis Khan and the Mongol Military System*. Yardley, PA: Westholme, 2007.

Mayor, Adrienne. *The Amazons: Lives and Legends of Warrior Women Across the Ancient World*. Princeton, NJ: Princeton University Press, 2014.

Mazoyer, Marcel, and Laurence Roudart. *A History of World Agriculture: From the Neolithic Age to the Current Crisis*. Translated by James H. Membrez. New York: Monthly Review Press, 2006.

McCall, Jeremiah B. *The Cavalry of the Roman Republic: Cavalry Combat and Elite Reputations in the Middle and Late Republic*. New York: Routledge, 2002.

McLynn, Frank. *Genghis Khan: His Conquests, His Empire, His Legacy*. Boston: Da Capo Press, 2016.

McNeill, William H. *Europe's Steppe Frontier, 1500–1800*. Chicago: University of Chicago Press, 1964.

———. *Plagues and Peoples*. New York: Anchor Books, 1976.

———. *The Pursuit of Power: Technology, Armed Force, and Society Since A.D. 1000*. Chicago: University of Chicago Press, 1982.

———. *The Rise of the West: A History of the Human Community*. Chicago: University of Chicago Press, 1963.

McShane, Clay, and Joel A. Tarr. *The Horse in the City: Living Machines in the Nineteenth Century*. Baltimore: Johns Hopkins University Press, 2007.

Metcalfe, Tom. "Bronze Age Tarim Mummies Aren't Who Scientists Thought They Were." Live Science. Last modified October 27, 2021. https://www.livescience.com/tarim-mummies-origins-uncovered.

Meyer, Robinson. "The Weekly Planet: What Extremely Muscular Horses Teach Us About Climate Change." *The Atlantic* online. Last modified December 8, 2020. https://www.theatlantic.com/science/archive/2020/12/new-visual-history-american-energy/617329/.

Michaud, Roland, and Sabrina Michaud. *Horsemen of Afghanistan*. London: Thames & Hudson, 1988.

Miller, John Anderson. *Fares, Please!: A Popular History of Trolleys, Horse-Cars, Street-Cars, Buses, Elevateds, and Subways*. New York: Dover, 1961.

Mills, Daniel, and Sue McDonnell, eds. *The Domestic Horse: The Origins, Development and Management of Its Behaviour*. Cambridge: Cambridge University Press, 2005.

Mitcham, Samuel W., Jr. *Hitler's Legions: The German Army Order of Battle, World War II*. New York: Stein and Day, 1985.

Mitchell, Peter. *The Donkey in Human History: An Archaeological Perspective*. Oxford: Oxford University Press, 2018.

———. *Horse Nations: The Worldwide Impact of the Horse on Indigenous Societies Post-1492*. Oxford: Oxford University Press, 2015.

Moberly, F. J. *The Campaign in Mesopotamia, 1914–1918*. Vol. 4. London: His Majesty's Stationery Office (HMSO), 1927.

Mohr, Erna. *The Asiatic Wild Horse*. London: J. A. Allen, 1971.

Moquin, Wayne, with Charles van Doren, eds. *Great Documents in American Indian History*. New York: Da Capo Press, 1995.

Morillo, Stephen, ed. *The Battle of Hastings: Sources and Interpretations*. Rochester, NY: Boydell Press, 1996.

Morton, Nicholas. *The Mongol Storm: Making and Breaking Empires in the Medieval Near East*. New York: Basic Books, 2022.

Musgrove, David, and Michael Lewis. *The Story of the Bayeux Tapestry: Unravelling the Norman Conquest*. London: Thames & Hudson, 2021.

Neparáczki, Endre, Zoltán Maróti, Tibor Kalmar, Kitti Maár, István Nagy, Dóra Latinovics, Ágnes Kustár, et al. "Y-chromosome Haplogroups from Hun, Avar and Conquering Hungarian Period Nomadic People of the Carpathian Basin." *Scientific Reports* 9, no. 1 (November 2019): art. 16569. https://doi.org/10.1038/s41598-019-53105-5.

Nicolle, David. *Poitiers AD 732: Charles Martel Turns the Islamic Tide*. Oxford, UK: Osprey, 2008.

Nikiforuk, Andrew. "The Big Shift Last Time: From Horse Dung to Car Smog." *The Tyee*. Last modified March 6, 2013. https://thetyee.ca/News/2013/03/06/Horse-Dung-Big-Shift/.

Noble, Duncan. *Dawn of the Horse Warriors: Chariot and Cavalry Warfare, 3000–600 BC*. Barnsley, UK: Pen & Sword Military, 2015.

Novozhenov, Viktor A., Aibek Zh. Sydykov, and Elina K. Altynbekova. "Indo-European Communications: The Model of 'Nomadic Homeland.'" *International Journal of Psychosocial Rehabilitation* 24, no. 8 (2020): 13922–43. https://www.psychosocial.com/archives/volume%2024/Issue%208/28654.

O'Connell, Robert L. *The Ghosts of Cannae: Hannibal and the Darkest Hour of the Roman Republic*. New York: Random House, 2011.

Olalde, Iñigo, Selina Brace, Morten E. Allentoft, Ian Armit, Kristian Kristiansen, Thomas Booth, Nadin Rohland, et al. "The Beaker Phenomenon and the Genomic Transformation of Northwest Europe." *Nature* 555, no. 7695 (March 8, 2018): 190–96.

Olsen, Birgit Anette, Thomas Olander, and Kristian Kristiansen, eds. *Tracing the Indo-Europeans: New Evidence from Archaeology and Historical Linguistics*. Oxford, UK: Oxbow Books, 2019.

Olsen, Sandra L., ed. *Horses Through Time*. Boulder, CO: Roberts Rinehart / Carnegie Museum of Natural History, 1996.

Olsen, Sandra L., Susan Grant, Alice M. Choyke, and László Bartosiewicz, eds. *Horses and Humans: The Evolution of Human-Equine Relationships*. Oxford, UK: BAR, 2006.

Outram, Alan K., Natalie A. Stear, Robin Bendrey, Sandra Olsen, Alexei Kasparov, Victor Zaibert, Nick Thorpe, and Richard P. Evershed. "The Earliest Horse Harnessing and Milking." *Science* 323, no. 5919 (March 6, 2009): 1332–35.

Overy, Richard. *Why the Allies Won*. London: Pimlico, 1995.
Paludan, Ann. *Chronicle of the Chinese Emperors: The Reign-by-Reign Record of the Rulers of Imperial China*. London: Thames & Hudson, 1998.
Papayanis, Nicholas. *Horse-Drawn Cabs and Omnibuses in Paris: The Idea of Circulation and the Business of Public Transit*. Baton Rouge: Louisiana State University Press, 1996.
Parker, Jack. "Turning Manure into Gold." *EMBO Reports* 3, no. 12 (December 2002): 1114–16.
Parpola, Asko. *The Roots of Hinduism: The Early Aryans and the Indus Civilization*. Oxford: Oxford University Press, 2015.
Parry, V. J., and M. E. Yapp, eds. *War, Technology and Society in the Middle East*. London: Oxford University Press, 1975.
Patent, Dorothy Hinshaw. *The Horse and the Plains Indians: A Powerful Partnership*. New York: Clarion Books, 2012.
Peachey, Stuart. *Horses and Oxen on the Farm, 1580–1660*. Bristol, UK: Stuart Press, 2003.
Peers, Chris. *Genghis Khan and the Mongol War Machine*. Barnsley, UK: Pen & Sword Military, 2015.
Pennisi, Elizabeth. "Horses Domesticated Multiple Times." *Science* 291, no. 5503 (January 19, 2001): 412.
Pereltsvaig, Asya, and Martin W. Lewis. *The Indo-European Controversy: Facts and Fallacies in Historical Linguistics*. Cambridge: Cambridge University Press, 2015.
Philbrick, Nathaniel. *The Last Stand: Custer, Sitting Bull, and the Battle of the Little Big Horn*. New York: Viking, 2010.
Philipps, David. *Wild Horse Country: The History, Myth, and Future of the Mustang*. New York: W. W. Norton, 2017.
Phillips, E. D. "The Scythian Domination in Western Asia: Its Record in History, Scripture and Archaeology." *World Archaeology* 4, no. 2 (October 1972): 129–38.
Phillips, Gervase. "Writing Horses into American Civil War History." *War in History* 20, no. 2 (April 2013): 160–81.
Piggott, Stuart. *Ancient Europe: From the Beginnings of Agriculture to Classical Antiquity—A Survey*. Edinburgh: Edinburgh University Press, 1965.
———. *The Earliest Wheeled Transport: From the Atlantic Coast to the Caspian Sea*. Ithaca, NY: Cornell University Press, 1983.
———. *Wagon, Chariot and Carriage: Symbol and Status in the History of Transport*. New York: Thames & Hudson, 1992.
Plets, Gertjan. "Exceptions to Authoritarianism? Variegated Sovereignty and Ethno-Nationalism in a Siberian Resource Frontier." *Post-Soviet Affairs* 35, no. 4 (May 2019): 308–22.
Plummer, Alexander, and Richard H. Power, comps. *The Army Horse in Accident and Disease, for the Instruction of Farriers and Horseshoers at the School of Application for Cavalry and Field Artillery, Fort Riley, Kansas*. Washington, DC: US Government Printing Office, 1903.
Plutarch. *The Age of Alexander*. Edited by Timothy E. Duff. New York: Penguin Classics, 2012.
Pollan, Michael. *The Omnivore's Dilemma: A Natural History of Four Meals*. New York: Penguin Press, 2006.
Polo, Marco. *The Travels of Marco Polo*. Edited by Milton Rugoff. New York: Signet Classics, 2004.
Porter, Pamela. *Medieval Warfare in Manuscripts*. London: The British Library, 2000.

Potts, Daniel T. "Technological Transfer and Innovations in Ancient Eurasia." In *The Globalization of Knowledge in History*, edited by Jurgen Renn, 105–23. Berlin: Max Planck Research Library, 2012.
Preece, Rod. *Brute Souls, Happy Beasts, and Evolution: The Historical Status of Animals*. Vancouver, Can.: University of British Columbia Press, 2005.
Price, Steve. *America's Wild Horses: The History of the Western Mustang*. New York: Skyhorse Press, 2017.
Price, T. Douglas, ed. *Europe's First Farmers*. Cambridge: Cambridge University Press, 2000.
Prothero, Donald R., and Robert M. Schoch. *Horns, Tusks, and Flippers: The Evolution of Hoofed Mammals*. Baltimore: Johns Hopkins University Press, 2002.
Puett, Michael. "China in Early Eurasian History: A Brief Review of Recent Scholarship on the Issue." In *The Bronze Age and Early Iron Age Peoples of Eastern Central Asia*, vol. 1, edited by Victor H. Mair, 699–715. Washington, DC: Institute for the Study of Man, 1998.
Purdy, C. S. *The Equine Legacy: How Horses, Mules, and Donkeys Shaped America*. Vista, CA: Mozaic Press, 2016.
Pydyn, Andrzej. *Exchange and Cultural Interactions: A Study of Long-Distance Trade and Cross-Cultural Contacts in the Late Bronze Age and Early Iron Age in Central and Eastern Europe*. Oxford, UK: Archaeopress, 1999.
Pyhrr, Stuart W., Donald J. LaRocca, and Dirk H. Breiding. *The Armored Horse in Europe, 1480–1620*. New York: Metropolitan Museum of Art / Yale University Press, 2005.
Raber, Karen, and Treva J. Tucker, eds. *The Culture of the Horse: Status, Discipline, and Identity in the Early Modern World*. New York: Palgrave Macmillan, 2005.
Ramsdell, Charles W. "General Robert E. Lee's Horse Supply, 1862–1865." *American Historical Review* 35, no. 4 (July 1930): 758–77.
Raulff, Ulrich. *Farewell to the Horse: A Cultural History*. Translated by Ruth Ahmedzai Kemp. New York: Liveright, 2018.
Raulwing, Peter. *Horses, Chariots and Indo-Europeans: Foundations and Methods of Chariotry Research from the Viewpoint of Comparative Indo-European Linguistics*. Budapest, Hung.: Archaeolingua Alapítvány, 2000.
Reck, Franklin M. *Horses to Horsepower: A Study of the Effect of the Motor Truck on American Living*. Detroit: Automobile Manufacturers Association, 1947.
Redford, Donald B. *Egypt, Canaan, and Israel in Ancient Times*. Princeton, NJ: Princeton University Press, 1992.
Reich, David. "Ancient DNA Suggests Steppe Migrations Spread Indo-European Languages." *Proceedings of the American Philosophical Society* 162, no. 1 (March 2018): 39–55.
———. *Who We Are and How We Got Here: Ancient DNA and the New Science of the Human Past*. New York: Pantheon Books, 2018.
Reynolds, Henry. *The Other Side of the Frontier: Aboriginal Resistance to the European Invasion of Australia*. Rev. ed. Sydney, Aus.: University of New South Wales Press, 2006.
Reynoldson, L. A. *Influence of the Tractor on Use of Horses. Farmers' Bulletin 1093 (United States Department of Agriculture)*. Washington, DC: Office of Farm Management, 1920.
Rice, Tamara Talbot. *The Scythians*. New York: Praeger, 1957.
Richter, Christian Klaus. *Cavalry of the Wehrmacht, 1941–1945*. Atglen, PA: Schiffer, 1995.

Ridgeway, William. *The Origin and Influence of the Thoroughbred Horse.* Cambridge: Cambridge University Press, 1905.
Rink, Bjarke. *The Centaur Legacy: How Equine Speed and Human Intelligence Shaped the Course of History.* n.p.: Long Riders' Guild Press, 2004.
Roberts, Alice. *Tamed: Ten Species That Changed the World.* London: Windmill Books, 2018.
Robinson, I. *The Waltham Book of Human-Animal Interaction: Benefits and Responsibilities of Pet Ownership.* New York: Pergamon Press, 1995.
Robles, Heather A. "Spanish Additions to the Cowboy Lexicon from 1850 to the Present." *Deseret Language and Linguistic Society Symposium* 25, no. 1 (1999): art. 4.
Rockhill, William Woodville, ed. *The Journey of William of Rubruck to the Eastern Parts of the World, 1253–55, as Narrated by Himself, with Two Accounts of the Earlier Journey of John of Pian de Carpine.* London: Hakluyt Society, 1900.
Roe, Frank Gilbert. *The Indian and the Horse.* Norman: University of Oklahoma Press, 1968.
Rogers, Guy MacLean. *Alexander: The Ambiguity of Greatness.* New York: Random House, 2005.
Rolle, Renate. *The World of the Scythians.* Translated by Gayna Walls. London: B. T. Batsford, 1989.
Romm, James S. *Ghost on the Throne: The Death of Alexander the Great and the Bloody Fight for His Empire.* New York: Vintage Books, 2012.
Ropa, Anastasija, and Timothy Dawson, eds. *The Horse in Premodern European Culture.* Berlin: de Gruyter, 2019.
Rosenwein, Barbara. *A Short History of the Middle Ages.* Toronto: University of Toronto Press, 2014.
Rossdale, P. D., T. R. C. Greet, P. A. Harris, R. E. Green, and S. Hall, eds. *Guardians of the Horse: Past, Present and Future.* London: British Equine Veterinary Association, 1999.
Rowan, Andrew N., ed. *Animals and People Sharing the World.* Hanover, NH: University Press of New England, 1988.
Rowson, Alex. *The Young Alexander: The Making of Alexander the Great.* London: William Collins, 2022.
Roy, Kaushik. *A Global History of Pre-modern Warfare: Before the Rise of the West, 10,000 BCE–1500 CE.* New York: Routledge, 2021.
Rudenko, Sergei I. *Frozen Tombs of Siberia: The Pazyryk Burials of the Iron Age Horsemen.* Translated by M. W. Thompson. Berkeley: University of California Press, 1970.
Salensky, W. *Prjevalsky's Horse (Equus Prjewalskii Pol.).* Translated by Captain M. Horace Hayes and O. Charnock Bradley. London: Hurst and Blackett, 1907.
Sandler, Martin W. *Galloping Across the USA: Horses in American Life.* Oxford: Oxford University Press, 2003.
Saxon, A. H. *Enter Foot and Horse: A History of Hippodrama in England and France.* New Haven, CT: Yale University Press, 1968.
Scott, James C. *Against the Grain: A Deep History of the Earliest States.* New Haven, CT: Yale University Press, 2017.
Seth-Smith, Michael, ed. *The Horse in Art and History.* New York: Mayflower Books, 1978.
Shaughnessy, Edward L. "Historical Perspectives on the Introduction of the Chariot to China." *Harvard Journal of Asiatic Studies* 48, no. 1 (January 1988): 189–237.

Shaw, Jonathan. "Telling Humanity's Story Through DNA: Geneticist David Reich Rewrites the Ancient Human Past." *Harvard* online, July/August 2022. https://www.harvardmagazine.com/2022/06/feature-ancient-dna.

Shelach-Lavi, Gideon. "Steppe Land Interactions and Their Effects on Chinese Cultures During the Second and Early First Millennia BCE." In *Nomads as Agents of Cultural Change: The Mongols and Their Eurasian Predecessors*, edited by Reuvan Amitai and Michal Biran. Honolulu: University of Hawai'i Press, 2014: 10–31.

Sidnell, Philip. *Warhorse: Cavalry in Ancient Warfare*. London: Hambledon Continuum, 2006.

Simmons, Sue, ed. *The Military Horse: A Story of Equestrian Warriors*. London: Marshall Cavendish Books, 1984.

Simpson, George Gaylord. *Horses: The Story of the Horse Family in the Modern World and Through Sixty Million Years of History*. Oxford: Oxford University Press, 1951.

Sinclair, Thomas R., and Carol Janas Sinclair. *Bread, Beer and the Seeds of Change: Agriculture's Imprint on World History*. Wallingford, UK: CABI, 2010.

Singleton, John. "Britain's Military Use of Horses, 1914–1918." *Past & Present* 139, no. 1 (May 1993): 178–203.

Sinor, Denis. *Inner Asia and Its Contacts with Medieval Europe*. London: Variorum, 1977.

Slicher van Bath, B. H. *The Agrarian History of Western Europe, AD 500–1850*. London: Edward Arnold, 1963.

Smith, Brad, and Carol Ann Browne. "The Day the Horse Lost Its Job." *Microsoft: Today in Technology* (blog), December 2017. https://blogs.microsoft.com/today-in-tech/day-horse-lost-job/.

Smith, Charles Hamilton. *Equus: Comprising the Natural History of the Horse, Ass, Onager, Quagga and Zebra*. Edinburgh: W. H. Lizars, 1841.

Smith, Donald J. *Horses at Work*. Wellingborough, UK: Patrick Stephens, 1985.

Smith, Gene. *Mounted Warriors: From Alexander the Great and Cromwell to Stuart, Sheridan, and Custer*. Hoboken, NJ: John Wiley & Sons, 2009.

Smith, Paul J. *Taxing Heaven's Storehouse: Horses, Bureaucrats, and the Destruction of the Sichuan Tea Industry, 1074–1224*. Cambridge, MA: Harvard University, Council on East Asian Studies, 1991.

Soplop, Julia. *Equus Rising: How the Horse Shaped U.S. History*. Pittsboro, NC: Hill Press, 2020.

Sorabji, Richard. *Animal Minds and Human Morals: The Origins of the Western Debate*. Ithaca, NY: Cornell University Press, 1993.

Spencer, Diana. *Roman Landscape: Culture and Identity*. Cambridge: Cambridge University Press, 2010.

Standage, Tom. *An Edible History of Humanity*. New York: Walker, 2009.

———. *A Brief History of Motion: From the Wheel, to the Car, to What Comes Next*. New York: Bloomsbury, 2021.

Stanton, Doug. *Horse Soldiers: The Extraordinary Story of a Band of U.S. Soldiers Who Rode to Victory in Afghanistan*. New York: Scribner, 2009.

Starr, Stephen Z. *The Union Cavalry in the Civil War*. Vol. 1. *From Fort Sumter to Gettysburg, 1861–1863*. Baton Rouge: Louisiana State University Press, 1979.

———. *The Union Cavalry in the Civil War*. Vol. 2. *The War in the East from Gettysburg to Appomattox, 1863–1865*. Baton Rouge: Louisiana State University Press, 1981.

———. *The Union Cavalry in the Civil War*. Vol. 3, *The War in the West, 1861–1865*. Baton Rouge: Louisiana State University Press, 1985.

Stillman, Deanne. *Mustang: The Saga of the Wild Horse in the American West*. New York: Houghton Mifflin Harcourt, 2008.
Stock, Chester, and Hildegarde Howard. *The Ascent of Equus: A Story of the Origin and Development of the Horse*. 2nd ed. Los Angeles: Los Angeles County Museum Science Series No. 8 Paleontology Publication No. 5, 1963.
Sun Tzu. *The Art of War*. Boston: Shambhala, 2002.
Sverdrup, Carl Fredrik. *The Mongol Conquests: The Military Operations of Genghis Khan and Sübe'etei*. Warwick, UK: Helion, 2017.
Swart, Sandra. "Horses in the South African War, c. 1899–1902." *Society and Animals* 18 (2010): 348–66.
———. *Riding High: Horses, Humans and History in South Africa*. Johannesburg, SA: Wits University Press, 2010.
———. "'The World the Horses Made': A South African Case Study of Writing Animals into Social History." *International Review of Social History* 55, no. 2 (August 2010): 241–63.
Tauger, Mark B. *Agriculture in World History*. New York: Routledge, 2011.
Taylor, Louis. *Bits: Their History, Use and Misuse*. New York: Harper & Row, 1966.
Taylor, William Timothy Treal, Jamsranjav Bayarsaikhan, Tumurbaatar Tuvshinjargal, Scott Bender, Monica Tromp, Julia Clark, K. Bryce Lowry, et al. "Origins of Equine Dentistry." *Proceedings of the National Academy of Sciences of the United States of America* 115, no. 29 (July 2018): 1–9.
Taylor, William Timothy Treal, and Christina Isabelle Barrón-Ortiz. "Rethinking the Evidence for Early Horse Domestication at Botai." *Nature: Scientific Reports* 11 (April 2021): art. 7440. doi.org/10.1038/s41598-021-86832-9.
Thirsk, Joan. *Horses in Early Modern England: For Service, for Pleasure, for Power*. Reading, UK: University of Reading, 1978.
Thompson, F. M. L., ed. *Horses in European Economic History: A Preliminary Canter*. Reading, UK: British Agricultural History Society, 1983.
———. *Victorian England: The Horse-Drawn Society: An Inaugural Lecture*. London: Bedford College, 1970.
Thompson, Tosin. "Ancient DNA Points to Origins of Modern Domestic Horses." *Nature* 598, no. 7882 (October 28, 2021): 550.
Thucydides. *The Peloponnesian War*. Edited by T. E. Wick. New York: Random House, 1982.
Tokyo Metropolitan Teien Art Museum. *The Noble Horse: Man and Horse in Western Art History*. Tokyo Metropolitan Foundation for History and Culture, 1998.
Tozer, Basil. *The Horse in History*. London: Methuen, 1908.
Trautmann, Martin, Alin Frînculeasa, Bianca Preda-Bălănică, Marta Petruneac, Marin Focșaneanu, Stefan Alexandrov, Nadezhda Atanassova, et al. "First Bioanthropological Evidence for Yamnaya Horsemanship." *Science Advances* 9, no. 9 (March 3, 2023). https://www.science.org/doi/10.1126/sciadv.ade2451.
Trench, Charles Chenevix. *A History of Horsemanship: The Story of Man's Ways and Means of Riding Horses from Ancient Times to the Present*. Garden City, NY: Doubleday, 1970.
Trifonov, Viktor, Denis Petrov, and Larisa Savelieva. "Party Like a Sumerian: Reinterpreting the 'Sceptres' from the Maikop Kurgan." *Antiquity* 96, no. 385 (February 2022): 67–84.
Trimm, Charlie. *Fighting for the King and the Gods: A Survey of Warfare in the Ancient Near East*. Atlanta: SBL Press, 2017.

Trippett, Frank. *The First Horsemen*. New York: Time-Life Books, 1974.
Turchin, Peter. *War and Peace and War: The Rise and Fall of Empires*. New York: Plume, 2007.
United States Army Historical Division, Headquarters Europe. *Horses in the German Army (1941–1945)*. Edited by General Major Burkhart Mueller-Hillebrand. Washington, DC: Foreign Military Studies Branch, 1952.
———. *Horses in the Russian Campaign*. Washington, DC: Foreign Military Studies Branch, 1946.
United States Department of Agriculture, Bureau of Animal Industry. *Special Report on Diseases of the Horse*. Washington, DC: US Government Printing Office, 1903.
Utley, Robert M., and Wilcomb E. Washburn. *Indian Wars*. New York: Mariner Books, 2002.
Van Creveld, Martin. *Supplying War: Logistics from Wallenstein to Patton*. 2nd ed. Cambridge: Cambridge University Press, 2004.
———. *The Transformation of War:* New York: Free Press, 1991.
Van De Mieroop, Marc. *A History of the Ancient Near East, ca. 3000–23 BC*. 3rd ed. Hoboken, NJ: Wiley-Blackwell, 2015.
Van Emden, Richard. *Tommy's Ark: Soldiers and Their Animals in the Great War*. London: Bloomsbury, 2010.
Vandervort, Bruce. *Indian Wars of Mexico, Canada, and the United States, 1812–1900*. New York: Routledge, 2006.
Vernon, Arthur. *The History and Romance of the Horse*. Boston: Waverly House, 1939.
Vilà, Carles, Jennifer A. Leonard, Anders Götherström, Stefan Marklund, Kaj Sandberg, Kerstin Lidén, Robert K. Wayne, et al. "Widespread Origins of Domestic Horse Lineages." *Science* 291, no. 5503 (January 19, 2001): 474–77.
Volti, Rudi. *Society and Technological Change*. 8th ed. New York: Macmillan Learning, 2017.
Wallace, Jenelle P., Brooke E. Crowley, and Joshua H. Miller. "Investigating Equid Mobility in Miocene Florida, USA, Using Strontium Isotope Rations." *Palaeogeography, Palaeoclimatology, Palaeoecology* 516 (February 15, 2019): 232–43.
Wallace, John H. *The Horse of America in His Derivation, History, and Development*. New York: published by the author, 1897.
Wallner, Barbara, Nicola Palmieri, Claus Vogl, Doris Rigler, Elif Bozlak, Thomas Druml, Vidhya Jagannathan, et al. "Y Chromosome Uncovers the Recent Oriental Origin of Modern Stallions." *Current Biology* 27, no. 13 (July 10, 2017): 2029–35.
Watson, J. N. P. *Through Fifteen Reigns: A Complete History of the Household Cavalry*. Staplehurst, UK: Spellmount, 1997.
Warmuth, Vera, Anders Eriksson, Mim Ann Bower, Graeme Barker, Elizabeth Barrett, Bryan Kent Hanks, Shuicheng Li, et al. "Reconstructing the Origin and Spread of Horse Domestication in the Eurasian Steppe." *Proceedings of the National Academy of Sciences of the United States of America* 109, no. 21 (May 2012): 8202–6.
Weatherford, Jack. *Genghis Khan and the Making of the Modern World*. New York: Broadway Books, 2004.
Weil, Kari. *Precarious Partners: Horses and Their Humans in Nineteenth-Century France*. Chicago: University of Chicago Press, 2020.
West, Stephanie. "Introducing the Scythians: Herodotus on Koumiss (4.2)." *Museum Helveticum* 56, no. 2 (1999): 76–86.

White, Lynn, Jr. *Medieval Technology and Social Change*. Oxford: Oxford University Press, 1962.
Wilkin, Shevan, Alicia Ventresca Miller, Ricardo Fernandes, Robert Spengler, William T.-T. Taylor, Dorcas R. Brown, David Reich, et al. "Dairying Enabled Early Bronze Age Yamnaya Steppe Expansions." *Nature* 598, no. 7882 (October 28, 2021): 629–33.
Willekes, Carolyn. "Equine Aspects of Alexander the Great's Macedonian Cavalry." In *Greece, Macedon and Persia: Studies in Social, Political and Military History in Honour of Waldemar Heckel*, edited by Timothy Howe, E. Edward Garvin, and Graham Wrightson, 47–58. Oxford, UK: Oxbow Books, 2015.
———. *The Horse in the Ancient World: From Bucephalus to the Hippodrome*. London: I. B. Taurus, 2016.
William of Rubruck. *The Mission of Friar William of Rubruck: His Journey to the Court of the Great Khan Möngke, 1253–1255*. Edited by Peter Jackson and David Morgan. Translated by Peter Jackson. Indianapolis: Hackett, 2009.
Williams, Wendy. *The Horse: The Epic History of Our Noble Companion*. New York: Scientific American / Farrar, Straus and Giroux, 2015.
Willoughby, David P. *The Empire of Equus: The Horse, Past, Present and Future*. New York: A. S. Barnes, 1974.
Winegard, Timothy C. *The First World Oil War*. Toronto: University of Toronto Press, 2016.
———. *Indigenous Peoples of the British Dominions and the First World War*. Cambridge: Cambridge University Press, 2011.
———. *The Mosquito: A Human History of Our Deadliest Predator*. New York: Dutton, 2019.
Winton, Harold R., and David R. Mets, eds. *The Challenge of Change: Military Institutions and New Realities, 1918–1941*. Lincoln: University of Nebraska Press, 2000.
Wissler, Clark. "The Influence of the Horse in the Development of Plains Culture." *American Anthropologist* 16, no. 1 (January–March 1914): 1–25.
Wofford, Jim. "Hoofprints Through History." *Practical Horseman*, November 2018, 12–15.
Wood, Harriet Harvey. *The Battle of Hastings: The Fall of Anglo-Saxon England*. New York: Overlook Press, 2009.
Wormser, Richard. *The Yellowlegs: The Story of the United States Cavalry*. Barnsley, UK: Frontline Books, 2018.
Worthington, Ian. *By the Spear: Philip II, Alexander the Great, and the Rise and Fall of the Macedonian Empire*. Oxford: Oxford University Press, 2014.
Wright, Gordon, ed. *The Cavalry Manual of Horsemanship and Horsemastership: Education of the Rider—The Official Manual of the United States Cavalry School at Fort Riley (Kansas)*. Garden City, NY: Doubleday, 1962.
Wright, Quincey, and Louis Leonard Wright. *A Study of War*. Chicago: University of Chicago Press, 1983.
Wu, Hsiao-yun. *Chariots in Early China: Origins, Cultural Interaction, and Identity*. Oxford: Archaeopress, 2013.
Wu, Shuanglei, Yongping Wei, Brian Head, Yan Zhao, and Scott Hanna. "The Development of Ancient Chinese Agricultural and Water Technology from 8000 BC to 1911 AD." *Palgrave Communications* 5 (July 2019): art. 77. doi.org/10.1057/s41599-019-0282-1.
Wynn, Stephen, and Tanya Wynn. *Animals in the Great War*. Barnsley, UK: Pen & Sword Military, 2019.

Xenophon. *The Art of Horsemanship*. Translated by Morris H. Morgan. London: J. A. Allen & Co., 1962.

Yang, Liang Emlyn, Hans-Rudolf Bork, Xiuqi Fang, and Steffen Mischke, eds. *Socio-Environmental Dynamics Along the Historical Silk Road*. Cham, Switz.: Springer, 2019.

Zhang, Fan, Chao Ning, Ashley Scott, Qiaomei Fu, Rasmus Bjørn, Wenying Li, Dong Wei, et al. "The Genomic Origins of the Bronze Age Tarim Basin Mummies." *Nature* 599 (2021): 256–61.

Zukosky, Michael L. *Przewalski's Horses in Eurasia: Pluralism in International Reintroduction Biology*. Lanham, MD: Lexington Books, 2016.

Notes

These notes, and the larger bibliography, are designed to provide access to the literature and research but are by no means exhaustive. To cite every source would produce a reference list and endnotes that would far exceed the length of the book. This is neither practical nor feasible. One equine historian claimed that more than fifty thousand books have been penned on horse-related topics since Kikkuli, "the master horse trainer of the land of Mitanni," first chiseled his knowledge in tablets around 1400 BCE. I speculate that the total number of horse-specific publications is considerably higher. This seemingly infinite catalog across a wide academic breadth is a testament to the status of the horse, its inseparable relationship to humans, and its historical power.

The chapter notes below are intended to offer further readings and suggestions to those who are seeking more detailed explanations and to recognize the authors who provided the main, but certainly not exclusive, secondary source material for each chapter. Others not referenced here are cited directly in the text. Every attempt was made to bring statistics, estimates, and dates in line with the most recent scholarship.

CHAPTERS 1, 2, & 3

Detailed studies of equine evolution, migration, and behavior include Wendy Williams, *The Horse: The Epic History of Our Noble Companion*; Stephan Budiansky, *The Nature of Horses: Exploring Equine Evolution, Intelligence, and Behavior*; Donald R. Prothero and Robert M. Schoch, *Horns, Tusks, and Flippers: The Evolution of Hoofed Mammals*; David Philipps, *Wild Horse Country: The History, Myth, and Future of the Mustang*; *The Domestic Horse: The Origins, Development and Management of Its Behavior*, edited by D. S. Mills and S. M. McDonnell; and Debbie Busby and Catrin Rutland, *The Horse: A Natural History*. The Bone Wars are covered in a cross-disciplinary collection of primary and secondary sources, including a narrative cloak-and-dagger chapter in Sam Kean, *The Icepick Surgeon: Murder, Fraud, Sabotage, Piracy, and Other Dastardly Deeds Perpetrated in the Name of Science*.

Juliet Clutton-Brock, *Horse Power: A History of the Horse and the Donkey in Human Societies*; *Horses Through Time*, edited by Sandra L. Olsen; *The Horse in Human History*, by Pita Kelekna; Harold B. Barclay, *The Role of the Horse in Man's Culture*; Matthew J. Kust, *Man and Horse in History*; and Susanna Forrest, *The Age of the Horse: An Equine Journey Through Human History*, offer expansive thematic overviews. While not the focus of this book, donkeys are assiduously represented in Peter Mitchell, *The Donkey in Human History: An Archaeological Perspective*.

CHAPTER 4

The Agricultural Revolution is well covered by Graeme Barker, *The Agricultural Revolution in Prehistory: Why Did Foragers Become Farmers?*; James C. Scott, *Against the Grain: A Deep History of the Earliest States*; *First Farmers: The Origins of Agricultural Societies*, by Peter Bellwood; Mark B. Tauger, *Agriculture in World History*; Thomas R. Sinclair and Carol Janas Sinclair, *Bread, Beer and the Seeds of Change: Agriculture's Imprint on World History*; and Tom Standage, *An Edible History of Humanity*. Daniel R. Headrick's industrious long-view approach in *Technology: A World History* was invaluable to this chapter and numerous others, as was *A History of World Agriculture: From the Neolithic Age to the Current Crisis*, by Marcel Mazoyer and Laurence Roudart.

Specific to animal domestication are *The Walking Larder: Patterns of Domestication, Pastoralism, and Predation*, edited by Juliet Clutton-Brock, and her work *A Natural History of Domesticated Mammals*; Richard Francis, *Domesticated: Evolution in a Man-Made World*; Alice Roberts, *Tamed: Ten Species That Changed the World*; Gavin Ehringer, *Leaving the Wild: The Unnatural History of Dogs, Cats, Cows, and Horses*; and *The Intimate Bond: How Animals Shaped Human History*, by Brian Fagan. Both Yuval Noah Harari, *Sapiens: A Brief History of Humankind*, and Jared Diamond, *Guns, Germs, and Steel: The Fates of Human Societies*, are widely read, cited, and acclaimed for a reason—they are both brilliant and engaging feats of historical anthropology.

To cover the vast and open-ended historiography of horse domestication and early riding, it is far easier to mention the notable researchers rather than each scholarly entry on the brimming shelves: David W. Anthony, Dorcas Brown, Colin Renfrew, Robert Drews, William Timothy Treal Taylor, Marsha Levine, Mary Littauer, Pavel E. Kuznetsov, Barbara Wallner, Shevan Wilkin, Norbert Benecke, Alan K. Outram, Sandra L. Olsen, and Ludovic Orlando, among numerous others.

CHAPTER 5

Like domestication, the spread of Indo-Europeans and their horses, language, and genes is also well documented. The leading voices belong to J. P. Mallory, Victor H. Mair, D. Q. Adams, David W. Anthony, Sergei Rudenko, Andrew Sherratt, Hui Zhou, Chunxiang Li, Peter Raulwing, Chao Ning, Robert Drews, and Michael Frachetti, among a lengthy list of others. Recent genetic research and DNA modeling conducted by David Reich, Wolfgang Haak, Kristian Kristiansen, Iosif Lazaridis, Eske Willerslev, and Volker Heyd have reinforced the linguistic evidence for a steppe migration and hostile takeover of Europe, India, and other parts of the Eurasian Steppe by Indo-Europeans.

The multidiscipline volumes *Tracing the Indo-Europeans: New Evidence from Archeology and Historical Linguistics*, edited by Birgit Anette Olsen, Thomas Olander, and Kristian Kristiansen, and *Language and Prehistory of the Indo-European Peoples: A Cross-Disciplinary Perspective*, edited by Adam Hyllested, Benedicte Nielsen Whitehead, Thomas Olander, and Birgit Anette Olsen, are both valuable editions, as is *The Indo-European Controversy: Facts and Fallacies in Historical Linguistics*, by Asya Pereltsvaig and Martin W. Lewis. The diverse and extraordinary directory of contributors to the compilations *Prehistoric Steppe Adaptation and the Horse*, edited by Marsha Levine, Colin Renfrew, and Katie Boyle; *Horses and Humans: The Evolution of Human-Equine Relationships*, edited by Sandra L. Olsen, Susan Grant, Alice M. Choyke, and László Bartosiewicz; and *Ancient Interactions: East and West in Eurasia*, edited by Katie Boyle, Colin Renfrew, and Marsha Levine, are too numerous to record but are acknowledged for their valuable research.

As a single stand-alone entry, David W. Anthony's authoritative *The Horse, the Wheel, and Language: How Bronze-Age Riders from the Eurasian Steppes Shaped the Modern World* is a scholarly gem. *Militarism and the Indo-Europeanizing of Europe*, by Robert Drews, is also a highly readable account. Two sweeping chronological publications by Barry Cunliffe—*By Steppe, Desert, and Ocean: The Birth of Eurasia* and *Europe Between the Oceans: Themes and Variations, 9000 BC–AD 1000*—were valuable guides not only for this chapter but also for several others. The same can be said for the compilation *All the Queen's Horses: The Role of the Horse in British History*, published by the Kentucky Horse Park.

CHAPTERS 6 & 7

Early chariot and cavalry warfare are detailed in a variety of fields, from military history and technology to anthropology and archeology. Eric Cline's *1177 B.C.: The Year Civilization Collapsed* is an acclaimed yet still underappreciated assessment. Robert Drews's works *Early Riders: The Beginnings of Mounted Warfare in Asia and Europe* and *The End of the Bronze Age: Changes in Warfare and the Catastrophe ca. 1200 B.C.* are both excellent studies. *The Culture of War: Invention and Early Development*, by Richard Gabriel, and *A History of Warfare*, by John Keegan, provide a hard military edge to these and other chapters.

The following titles are all munificent vaults of information: *Wagon, Chariot and Carriage: Symbol and Status in the History of Transport*, by Stuart Piggott; William Hamblin, *Warfare in the Ancient Near East to 1600 BC: Holy Warriors at the Dawn of History*; *Chariot: From Chariot to Tank, The Astounding Rise and Fall of the World's First War Machine*, by Arthur Cotterell; *Dawn of the Horse Warriors: Chariot and Cavalry Warfare, 3000–600 BC*, by Duncan Noble; Philip Kohl's *The Making of Bronze Age Eurasia*; and "Silk Roads or Steppe Roads? The Silk Roads in World History," by David Christian.

The Scythian Empire: Central Eurasia and the Birth of the Classical Age from Persia to China, by Christopher L. Beckwith, is a skilled, specialized investigation, as are Barry Cunliffe's *Scythians: Nomad Warriors of the Steppe*, and *The Amazons: Lives and Legends of Warrior Women Across the Ancient World*, by Adrienne Mayor.

CHAPTER 8

I was elated to receive my copy of Adrian Goldsworthy's recent book *Philip and Alexander: Kings and Conquerors*. It did not disappoint. It is an indefatigable work of both historiography and analysis, as is the definitive work *Cavalry Operations in the Ancient Greek World*, by Robert E. Gaebel. As a deep military assessment, *The Macedonian War Machine: Neglected Aspects of the Armies of Philip, Alexander and the Successors (359–281 BC)*, by David Karunanithy, offers substantial reward. Alex Rowson provides a detailed investigation of the early years, upbringing, and education in *The Young Alexander: The Making of Alexander the Great*. Other thorough and insightful reflections on Philip and Alexander are offered by Ian Worthington, *By the Spear: Philip II, Alexander the Great, and the Rise and Fall of the Macedonian Empire*; Philip Freeman, *Alexander the Great*; Anthony Everitt, *Alexander the Great: His Life and His Mysterious Death*; James S. Romm, *Ghost on the Throne: The Death of Alexander the Great and the Bloody Fight for His Empire*; and Pierre Briant, *The First European: A History of Alexander in the Age of Empire*.

Cavalry-specific studies spanning numerous chapters include Louis A. DiMarco, *War Horse: A History of the Military Horse and Rider*; Philip Sidnell, *Warhorse: Cavalry in Ancient Warfare*; Ann Hyland, *The Horse in the Ancient World*; *The Oxford Handbook of*

Warfare in the Classical World, edited by Brian Campbell and Lawrence A. Tritle; Carolyn Willekes, *The Horse in the Ancient World: From Bucephalus to the Hippodrome*; and Roman Jarymowycz, *Cavalry from Hoof to Track*.

The Roman period is thoroughly covered by Goldsworthy in *The Roman Army at War: 100 BC–AD 200* and *Pax Romana: War, Peace and Conquest in the Roman World*; by Ann Hyland in *Equus: The Horse in the Roman World*; by Jeremiah B. McCall in *The Cavalry of the Roman Republic: Cavalry Combat and Elite Reputations in the Middle and Late Republic*; and across the works of Mary Beard.

CHAPTER 9

Four comprehensive works on the Xiongnu and the fluid Chinese frontier are Nicola Di Cosmo, *Ancient China and Its Enemies: The Rise of Nomadic Power in East Asian History*; Christopher I. Beckwith, *Empires of the Silk Road: A History of Central Eurasia from the Bronze Age to the Present*; Thomas J. Barfield, *The Perilous Frontier: Nomadic Empires and China*; and *Empire of Horses: The First Nomadic Civilization and the Making of China*, by John Man.

Other valuable sources include William Honeychurch, *Inner Asia and the Spatial Politics of Empire: Archaeology, Mobility, and Culture Contact*; Xinru Liu, *The Silk Road in World History*; *Peace, War, and Trade Along the Great Wall: Nomadic-Chinese Interaction Through Two Millenia*, by Sechin Jagchid and Van Jay Symons; *Complexity of Interaction Along the Eurasian Steppe Zone in the First Millennium CE*, edited by Jan Bemmann and Michael Schmauder; Jos Gommans, *The Indian Frontier: Horse and Warband in the Making of Empires*; Erik Hildinger, *Warriors of the Steppe: A Military History of Central Asia, 500 B.C. to 1700 A.D.*; *Nomads as Agents of Cultural Change: The Mongols and Their Eurasian Predecessors*, edited by Reuven Amitai and Michal Biran; *Nomad Aristocrats in a World of Empires*, edited by Jurgen Paul; the still relevant "The Role of the Horse in Chinese History," by H. G. Creel; and the Kentucky Horse Park collection *Imperial China: The Art of the Horse in Chinese History*.

CHAPTERS 10 & 11

For Charles Martel and the Battle of Tours, see G. R. Hawting, *The First Dynasty of Islam: The Umayyad Caliphate, AD 661–750*; Hugh Kennedy, *The Great Arab Conquests: How the Spread of Islam Changed the World We Live In*; David Levering Lewis, *God's Crucible: Islam and the Making of Europe, 570 to 1215*; *The Horse in Premodern European Culture*, edited by Anastasija Ropa and Timothy Dawson; Donna Landry, *Noble Brutes: How Eastern Horses Transformed English Culture*; "East Meets West: Mounted Encounters in Early and High Mediaeval Europe" and "Combat Training for Horse and Rider in the Early Middle Ages," by Jürg Gassmann; David Nicolle, *Poitiers AD 732: Charles Martel Turns the Islamic Tide*; Paul Fouracre, *The Age of Charles Martel*; and Wayne E. Lee, *Waging War: Conflict, Culture, and Innovation in World History*.

William the Conqueror and the Battle of Hastings are represented by Jim Bradbury, *The Battle of Hastings*; David C. Douglas, *William the Conqueror: The Norman Impact upon England*; Harriet Harvey Wood, *The Battle of Hastings: The Fall of Anglo-Saxon England*; and David Musgrove and Michael Lewis, *The Story of the Bayeux Tapestry: Unravelling the Norman Conquest*. Two general period books come from Ann Hyland, *The Horse in the Middle Ages* and *The Medieval Warhorse from Byzantium to the Crusades*.

Equine farming advancements and the medieval Agricultural Revolution are covered by Bruce M. S. Campbell, *English Seigniorial Agriculture, 1250–1450*; *Medieval Farming and Technology: The Impact of Agricultural Change in Northwest Europe*, edited by Grenville Astill and John Langdon; *Horse and Man in Early Modern England*, by Peter Edwards; and Christopher Isett and Stephen Miller, *The Social History of Agriculture: From the Origins to the Current Crisis*. John Langdon's *Horses, Oxen and Technological Innovation: The Use of Draught Animals in English Farming from 1066–1500* is a masterpiece of analytical and statistical research.

CHAPTER 12

The best depictions of Chinggis Khan and the Mongol era can be found in: Peter Frankopan, *The Silk Roads: A New History of the World*; Jack Weatherford, *Genghis Khan and the Making of the Modern World*; Frank McLynn, *Genghis Khan: His Conquests, His Empire, His Legacy*; Marie Favereau, *The Horde: How the Mongols Changed the World*; James Chambers, *The Devil's Horsemen: The Mongol Invasion of Europe*; Peter Jackson, *The Mongols and the West, 1221–1410*; *Genghis Khan and the Mongol Empire*, edited by William W. Fitzhugh, Morris Rossabi, and William Honeychurch; and Nicholas Morton, *The Mongol Storm: Making and Breaking Empires in the Medieval Near East*. From a military posture, the top tier is occupied by Timothy May, *The Mongol Art of War: Chinggis Khan and the Mongol Military System*; Carl Fredrik Sverdrup, *The Mongol Conquests: The Military Operations of Genghis Khan and Sübe'etei*; and Chris Peers, *Genghis Khan and the Mongol War Machine*.

CHAPTERS 13 & 14

For the Columbian Exchange, see Alfred W. Crosby, *Ecological Imperialism: The Biological Expansion of Europe, 900–1900*, and *The Columbian Exchange: Biological and Cultural Consequences of 1492*; William H. McNeill, *Plagues and Peoples*; Lawrence H. Keeley, *War Before Civilization: The Myth of the Peaceful Savage*; Noble David Cook, *Born to Die: Disease and New World Conquest, 1492–1650*; Daniel R. Headrick, *Power over Peoples: Technology, Environments, and Western Imperialism, 1400 to the Present*; Virginia DeJohn Anderson, *Creatures of Empire: How Domestic Animals Transformed Early America*; Greg Bankoff and Sandra Swart, *Breeds of Empire: The "Invention" of the Horse in Southeast Asia and Southern Africa, 1500–1950*; Robin Law, *The Horse in West African History: The Role of the Horse in the Societies of Pre-Colonial West Africa*; Sandra Swart, *Riding High: Horses, Humans and History in South Africa*.

Peter Mitchell, *Horse Nations: The Worldwide Impact of the Horse on Indigenous Societies Post-1492*, is a masterpiece of scholarship, as are the luminously brilliant publications of Pekka Hämäläinen: *The Comanche Empire*; *Lakota America: A New History of Indigenous Power*; and "The Rise and Fall of Plains Indian Horse Cultures." Other enlightening narratives include *A Song for the Horse Nation: Horses in Native American Cultures*, edited by George P. Horse Capture and Emil Her Many Horses; David Philipps, *Wild Horse Country*; Steve Price, *America's Wild Horses: The History of the Western Mustang*; Deanne Stillman, *Mustang: The Saga of the Wild Horse in the American West*; Andrew Currie, "Horse Nations"; and Ann Hyland, *The Warhorse, 1250–1600* and *The War Horse in the Modern Era: Breeder to Battlefield, 1600 to 1865*. Older writings from Clark Wissler, Francis Haines, Frank Gilbert Row, Robert Denhardt, and John C. Ewers remain valuable works, containing troves of primary testimonies.

CHAPTER 15

For advancements in mechanical farming (including the previous chapters), see Geoff Cunfer, *On the Great Plains: Agriculture and Environment*; R. Douglas Hurt, *American*

Agriculture: A Brief History; Paul K. Conkin, *A Revolution Down on the Farm: The Transformation of American Agriculture Since 1929*; Deborah Fitzgerald, *Every Farm a Factory: The Industrial Ideal in American Agriculture*; and Giovanni Federico, *Feeding the World: An Economic History of Agriculture, 1800–2000*. Three exceptional histories of urban horses are Clay McShane and Joel A. Tarr, *The Horse in the City: Living Machines in the Nineteenth Century*; Ann Norton Greene, *Horses at Work: Harnessing Power in Industrial America*; and Jonathan V. Levin, *Where Have All the Horses Gone?: How Advancing Technology Swept American Horses from the Road, the Farm, the Range, and the Battlefield*.

Other valuable entries are W. J. Gordon, *The Horse-World of London*; *Horses in European Economic History: A Preliminary Canter*, edited by F. M. L. Thompson, as well as Thompson's *Victorian England: The Horse-Drawn Society*; Nicholas Papayanis, *Horse-Drawn Cabs and Omnibuses in Paris: The Idea of Circulation and the Business of Public Transit*; Kari Weil, *Precarious Partners: Horses and Their Humans in Nineteenth-Century France*; Elizabeth Atwood Lawrence, *Hoofbeats and Society: Studies of Human-Horse Interactions*; *Animal Metropolis: Histories of Human-Animal Relations in Urban Canada*, edited by Joanna Dean, Darcy Ingram, and Christabelle Sethna; John Anderson Miller, *Fares, Please!: A Popular History of Trolleys, Horse-Cars, Street-Cars, Buses, Elevateds, and Subways*; Ulrich Raulff, *Farewell to the Horse: A Cultural History*; Richard Bak, *Henry and Edsel: The Creation of the Ford Empire*; and Douglas Brinkley, *Wheels for the World: Henry Ford, His Company, and a Century of Progress, 1903–2003*. The Great Manure Crisis is covered across a wide range of sources.

CHAPTER 16

Literature on horses during the Great War covers a lot of ground, providing a balance of history, commentary, and primary source material: *Horsemen in No Man's Land: British Cavalry and Trench Warfare, 1914–1918*, by David Kenyon; *Real War Horses: The Experience of the British Cavalry, 1814–1914*, by Anthony Dawson; Jilly Cooper, *Animals in War: Valiant Horses, Courageous Dogs, and Other Unsung Animal Heroes*; Juliet Gardiner, *The Animals' War: Animals in Wartime from the First World War to the Present Day*; *Animals in the Great War*, by Stephen Wynn and Tanya Wynn; Janet Macdonald, *Horses in the British Army, 1750–1950*; Richard Van Emden, *Tommy's Ark: Soldiers and Their Animals in the Great War*; and Simon Butler, *The War Horses: The Tragic Fate of a Million Horses Sacrificed in the First World War*—among shelves of others. More-focused academic contributions are provided by John Singleton, Steven Corvi, and Andrew Iarocci.

Mechanization during the interwar period is covered by James S. Corum, *The Roots of Blitzkrieg: Hans von Seeckt and German Military Reform*; Martin van Creveld, *The Transformation of War* and his *Supplying War: Logistics from Wallenstein to Patton*; *The Challenge of Change: Military Institutions and New Realities, 1918–1941*, edited by Harold R. Winton and David R. Mets; and *Riders of the Apocalypse: German Cavalry and Modern Warfare, 1870–1945* by David R. Dorondo.

Unmasking the myth of blitzkrieg and highlighting the German dependency on horses is the meticulously researched *Mechanized Juggernaut or Military Anachronism?: Horses and the German Army of WWII*, by R. L. DiNardo; his "Horse-Drawn Transport in the German Army," coauthored with Austin Bay; and Lloyd Clark, *Blitzkrieg: Myth, Reality, and Hitler's Lightning War—France, 1940*. The two primary source publications collected by the US Army Historical Division, *Horses in the German Army (1941–1945)* and *Horses in the Russian Campaign*, are mines of valuable firsthand information. Lastly, one of my favorite books, Richard Overy's *Why the Allies Won*, was as valuable as ever.

CHAPTER 17

On the current state of wild horses in the United States, Australia, and Canada (as well as Przewalski's horses), newspapers, journal articles, magazines, and government press releases were utilized as well as the works of Philipps, Price, and Stillman. For the Przewalski's horse, see *The Reintroduction of Przewalski Horses in the Hustain Nuruu Mountain Forest Steppe Reserve in Mongolia; An Integrated Conservation Development Project*, by Inge Bouman; *Przewalski's Horse: The History and Biology of an Endangered Species*, edited by Lee Boyd and Katherine A. Houpt; and Michael L. Zukosky, *Przewalski's Horses in Eurasia: Pluralism in International Reintroduction Biology*.

Index

Note: Page numbers in *italics* refer to images, figures, and tables.

Abbasid Empire, 249, 251–252
Abbot, Downing & Co., 380
Abdul Rahman al-Ghafiqi, 241, 246–248
Aboriginal Australians, 354–357
Acadians, 457–458
Achaemenid Empire, 179
Acoma Pueblo, 333
Acosta, José de, 304
Adams, Shane, 445–446
Aelian, 205
Against the Grain (Scott), 72
The Age of Charles Martel (Fouracre), 242
The Agrarian History of Western Europe (Slicher van Bath), 269
agricultural practices. *See also* Agricultural Revolutions; feudalism
    American Manifest Destiny and, *329*, *330*, 347–349, 364
    of Aztecs, 314
    of Great Plains Indigenous people, 334, 336–337
    during WWII, 429, 431, 431n
Agricultural Revolutions
    background, 5, 59–60
    calvary differences and, 146–147
    domestication and, 61–69, 63n, 77
    imperialism and, 120, 132
    industrial agriculture, 397–399
    land management, 268–269
    in medieval Europe, 261–263, 266–271, 273–274
    modern roles for horses, 451
    in Old Europe, 102–103
    post–Black Death, 296–297
    reality of, 76
    transition to tractors, 396–399, *397–398*
Ahmose I, 135–136
Ahnert, Gerald, 351
airag, 78
Akeley, Carl, 74
Akhal-Teke, 163
Akhenaten, 139–140
Akkadian Empire, 131–132
Alaric (king), 230–231
Alexander the Great, 174–206
    academia and cultural impact of, 201–203
    ascension to throne, 191
    calvary training, 188–189
    death of, 199
    education of, 174, 176–177
    historical influences on, 177, 178–186, 200
    horse breeding and procurement by, 189–190
    infrastructure projects, 201
    legacy of, 199–206
    military achievements, 186–187, 191–199, *192*
    military supply innovations, 190
    military tactics, 171–172, 178, 191–196, 258–259
    on Nisean herds, 148
    warhorse of, 40–41, 174–176, 178, *192*, 193, 194, 196

al-Ghafiqi, Abdul Rahman, 241, 246–248
Allenby, Edmund, 123–127, *125*, 402
All Quiet on the Western Front
  (Remarque), 404, 405
Ibn al-Wardi, 295
Amazons, 164–166, *172*
The Amazons (Mayor), 164
Amenope, 140
American Camel Company, 350
American Civil War Unbridled
  Veterans, 412
American Expeditionary Force (AEF),
  405, 408
American Horse Council, 451, 451n
American Humane Society, 373
American Progress (Gast), 329–330, *330*
American Society for the Prevention of
  Cruelty to Animals (ASPCA), 373
American Street Railway Association,
  391–392
Ammianus, 232
Anabasis (Xenophon), 174
Anacreon, 168
Anatolia, 139, 142, 145, 148, 156, 179, 182
anatomical adaptations, 32–48
  back of horses, 43–44
  brain and social behavior, 44–48
  for digesting grass, 36–39
  for eating grass, 32–36, 45–47
  feet and legs, 41–43
  size of horses, 40–42
Angkor Empire, 293
Anglo-Saxon culture, 233–234, 259–260
animal anti-cruelty bills, 373
Animals in War (Cooper), 423
Animals in War Memorial, 412, *413*
Antelope band, 364
Anthony, David, 1–2, 80, 83, 84–85, 87,
  91, 102, 110, 115, 119
antivenoms, 414
Apache, *xii*, 323, 334, 335–337, 359,
  363–364, 385
apocalypse horsemen, 127, 235, *237*
Appaloosas, 340
Arabian horses, 130, 239, 246, 249–251,
  253–254, 436, 453
Arapaho, *xii*, 335, 336, 341, 354, 362, 364
Arauco War, 322
Arawak, 307, 310

Arikara, *xii*, 336, 341, 362
Aristophanes, 168
Aristotle, 174, 176–177, 200
Arizona, 385, 439–440
Armageddon, 127
Arora, Namit, 105
Arrian, 189, 189n, 193, 199
Arthashastra, 198
Arthurian legends, 233–234
Ashur, 144
Ashurbanipal, 152–154, *153*, 155
Ashurnasirpal II, 143, 145, 151
Asinus, 29
Assateague, 446
Assiniboine, *xii*, 340–341, 354, 362
ass species, xviii
Assyria (Frahm), 144
Assyrian Empire, *xiv*
  academia and inventions, 151–152,
    153–154, 155, 155n
  chariot-to-cavalry transition, 127, 142,
    145–148, *146*
  downfall of, 154–156, 172–173
  –Egypt relations, 133, 142, 144
  horse plundering by, 147–149
  horse training regimen, 148–150
  infrastructure projects, 151–152, 182
  military campaigns of, 143–145,
    150–151, 152–153, 169–170
  weapons of, 139, 150–151, 179
Assyrians, 142
Atahuallpa (emperor), 318
Attila and the Huns, 225, 229–230,
  231–233, 241
Australia
  Columbian Exchange, 354–357
  current horse population, 453, 453n
  feral horses in, xx, 356, 439, 447, 448–449
  wartime armament production, 429
  WWI equine shipments, 411
Australopithecus, 28
automobiles. *See* cars
Avars, 235, 245, 245n
Azara, Félix de, 322
Aztec Empire, 311–313, 314–317, *317*, 318

Babel, Isaac, 388
Bacon, Roger, 298, 387–388
Bak, Richard, 393

INDEX

Baltimore Omnibus Company, 378–379
Ban Chao, 226–227
Band of Brothers (television miniseries), 436–437
Barbarika Nomina (Hellanicus), 157
barbarism, 166–170, 209, 231–232, 288, 288n
Barfield, Thomas, 225, 283
Barnum, Charles "Pete," 441
baseball, 344–345, 440
The Battle of Hastings (Bradbury), 256
Bay, Austin, 433
Bayeux Tapestry, 257, *258*
Beatus of Liébana, 235, *237*
Beaver Wars, 341
Beckwith, Christopher, 119, 129, 157, 164, 206, 228–229
Bede, 248
Behring, Emil von, 413
Benavides, Alonso de, 334
Bentham, Jeremy, 373
Benz, Carl, 390, 399
Berbers (Moors), 240–241
Bergmann, Karl, 40
Bering Bridge, 29
Betts, C. W., 13
bison. *See* buffalo
Black, David, 243
Black Beauty (Sewell), 373, 401
Black Death, 101–102, 275, 294–297
Blackfoot, *xii*, 336, 340–341, 354, 360–361, 362, 389
blitzkrieg ("lightning war"), 286
BLM (Bureau of Land Management), 443–447, 448, 452, 457
Blue Cross Fund, 373
Boer War Horse Memorial, 412
Bolivia, 313. *See also* Incan Empire
Bone Wars, 12–18
Book of the Marvels of the World (Rustichello), 300
Boston, 370, 378, 385, 391
Botai culture, 77–80, 82, 83, 85, 100
Bourke, John Gregory, 385
Bradbury, Jim, 256
Brave New World (Huxley), 117, 390
Breasted, James, 125
Brick Top (horse), 414
A Brief Account of the Destruction of the Indies (Las Casas), 309

A Brief History of Motion (Standage), 375
Brinkley, Douglas, 396
British Museum of Natural History, 19, 24
Broadway, Julie, 451
bronze, 114–115, 152
Brown, Dorcas, 80, 83
brumbies (Australian feral horses), xx, 356, 439, 447, 448–449
Bucephalus (horse), 40–41, 174–176, 178, *192*, *193*, *194*, *196*
Buchanan, James, 351
Budapest
  Mongols attack on, 290
  subway (public transportation), 391
Buddhism, 197
Budiansky, Stephen, 19, 39, 44, 60
Budweiser, 380
buffalo (bison), 326, 329–331, *332*, 340–347, *343*, 358–360, 362–363
Buffalo Bill and the Indians, or Sitting Bull's History Lesson (film), 436. *See also* Cody, William
Buffalo Bull's Back Fat, 358
Buffalo Running (film), 326
Bukhara, 240, 279–280
Bulle Rock (horse), 453
Büntgen, Ulf, 291
Bureau of Land Management (BLM), 443–447, 448, 452, 457
Butterfield, John, 351
Butterfield Overland Mail Company, 351–353
buzkashi, 454, 454n, 464
Byerley Turk (horse), 251, 453
Byron, George Gordon (lord), 152, *419*
By Steppe, Desert, and Ocean (Cunliffe), 81

Caddo, *xiii*, 336, 338
Cajamarca, Battle of, 318
Calamity Jane, 365
camels, 69, 149, 149–150n, 350–351, 449
Campbell, Bruce, 262–263, 267, 269, 273
Canada
  current horse population, 452, 453n
  feral horses in, 440, 447, 448, 457–458
  rodeos and, 454
  WWI and, 402, 408
  WWII and, 428, 429, *429*, 434

Canadian Criminal Code, 373
cannabis, 115–116, 163, 228
Capra, Frank, 420
Carmen de Hastingae Proelio, 256, 257–258
Carpini (friar), 281, 284, 286–287, 288
cars, 3–4, 128–129, 390–396, *393*, 416
Castañeda de Nájera, Pedro de, 333
Castillo, Bernal Díaz del, 314, 315, 316
Catalaunian Fields, 232–233, 241
cavalry. *See also* Alexander the Great; Assyrian Empire; Attila and the Huns; Columbian Exchange; Great Plains; Mongols; Scythians; World War I; World War II; Xiongnu
   of Africa, 309
   chariot transition to, 127–129, 142, 147–148
   of Chinese Empire, 22–25, 87–88, 207–212, *208*, 214, 216–219, 235
   feudalism and, 242–245, *244*, 254–259
   of Greece, 183–184, 184n, 185
   heavy cavalry, 235–237, 239, 241–249, 254–259
   horse behavioral patterns suited for, 46–48
   modern special ops forces, 455–456
   of Persian Empire, 179, 181
   of Roman Empire, 205, 206, 230–231
The Cavalry of the Roman Republic (McCall), 178
Cavalry Operations in the Ancient Greek World (Gaebel), 186
cave art, 53–55, *53*, *55*
Centaurian Pact, 2–3, 48, 81–82, 438–439, 459–461
The Centaur Legacy (Rink), 2
centaurs, 112, 310, 355
Central Overland California, 352
Chamberlin, J. Edward, 326
Chanakya, 197–198
Chandragupta Maurya, 197–198
Chao Cuo, 216–217
Chappel Brothers Corporation, 440–442
Chardon, Francis, 362
"The Charge of the Light Brigade" (Tennyson), 7

chariot warfare, 123–142
   apocalypse horsemen and, 127
   cavalry transition from, 127–129, 142, 147–148
   history of, 144
   horse training regimen, 130
   imperialism and, 131–142
   invention of, 129
   largest battle, 140–141, *141*
   legacy of, 123–131, 142
   Megiddo Plains battles, 123–127, 137
Charlemagne (king), 50, 254
Chengpu (battlefield), 124
Chernykh, Evgenij, 102
Cheyenne, *xii*, 336, 341, 354, 358, 362, 364
Chicago World's Fair, 368, *369*
Chief Joseph, 340, 365
Chief Many Horses, 361
Chinese Empire and civilization, *xv*
   academia and innovations, 116, 139, 235–236, 262–265, 298
   background, 98, 106–109, 120
   on barbarism, 167, 209–211
   Black Death and, 294–295
   borders of, 225–226, 234–235
   cavalry of, 87–88, 207–212, *208*, 214, 216–219, 220–225, 235
   current horse population, 438
   feral horses in, 450
   Han dynasty, 214–225
   horse sacrifice by, 163–164
   infrastructure projects, *xv*, 210, 211, 212–214, 220
   Islamic Empire and, 239–240
   Mongols and, 281–282, 289, 293
   polo and, 454
   Qin dynasty, 212–214
   Roman Empire and, 226–230, 234
   War of the Heavenly Horses, 223–225
   Warring interregnum, 214
   Warring States period, 212
   Xiongnu and, 87–88, 173, 206, 211, 213–216, 219–221, 225
Chinggis Khan, 279–284, 289–290, 294
Christian, David, 106
Chronica Majora (Paris), 285
Chronicle of 754 (*Mozarabic Chronicle*), 246, 247
Chronicle of Saint-Denis, 248

INDEX

Churchill, Winston, 428, 456
Cimmerians, 146, 154–155, 156, 169–170, 170n
cities. *See* urban horses
Clark, Grahame, 8
Clark, William, 339, 340
Clausewitz, Carl von, 191
Clutton-Brock, Juliet, 46–47, 60, 69, 71, 73, 112, 133, 171, 205
Cochise, 363–364, 385
Cody, William "Buffalo Bill," 14, 326, *327*, 352, 365–366, *366*, 368, 454
Colebourn, Harry, 411–412, *412*
Collier, John, 365
Coloradas, Mangas, 363–364
Columbian Exchange, 302–323, 329–347, 353–361
  American Manifest Destiny and, 329–331, 346–347, 361–367
  background, *xiii*, 6, 6n, 58, 300–301, 305
  British imperialism and, 354–357
  cattle ranches, 320–321, 334
  equine and livestock transport, 302–304
  Great Plains Indigenous horse culture, *332*, 337–344, 353–354, 357–361
  Spanish imperialism and, 305–321, 331–337, *332* (*See also* Aztec Empire; Incan Empire)
The Columbian Exchange (Crosby), 6n, 304, 320
Columbus, Christopher
  background, 300–301
  horse transported by, *xiii*, 302–304
  impact of, 6, 6n, 58, 300–301, 305–311 (*See also* Columbian Exchange)
Comanche, *xii–xiii*, 82, 110, 323, 336–339, 342, 354, 359, 360, 362–364
The Comanche Empire (Hämäläinen), 338
Commentary on the Apocalypse (Beatus), 235, *237*
Concord coach, 380
Consuming Habits (Sherratt), 116
Continuations of the Chronicle of Fredegar (anonymous), 247–248
Cook, James, 357
Cook, Noble David, 310
Cooke, Bill, 204, 211, 224, 236, 263–264

Cooper, Astley, 47–48
Cooper, Jilly, 423
Cope, Edward Drinker, 14, 16
Coronado, Francisco Vásquez de, 317, 333
Cortés, Hernán, 311–312, 314–317, 318, 321
Cosmo, Nicola Di, 291
Cossack breeding horses, 436
Crassus, Marcus Licinius, 205
Cree, *xii–xiii*, 336, 340–341
Creel, Herrlee, 209, 217
Crosby, Alfred W., 6n, 75, 304, 306, 320, 328
Crow, *xii*, 323, 336, 341, 354, 361, 362, 364, 456–457
Crusades, 202–203, 262, 291–292
The Culture of War (Gabriel), 131
Cuneo, Michele de, 302
Cunliffe, Barry, 81, 109, 119–120, 142, 156, 157, 216, 226, 232
Curtis, Edward S., 328
Custer, George Armstrong, 16
Cyaxares, 170
Cyrus the Great, 165, 177, 179–180, 200

Daimler, Gottlieb, 390, 399
dairy, 78–80, 100–101, *375*, 395–396, 414
Darius, 181–184
Darius I, 454
Darius III, 187, 192–193, *192*
Dark Ages, 142
Darley Arabian (horse), 251, 453
Darwin, Charles, 16–21, 18n, 23–24, 96, 307
Davis, Jefferson, 350
Davis, Paul, 184–185, 248–249, 260
Dawes Severalty Act (1887), 364, 365
De Medicina Equorum (Ruffo), 243, 270
de Mézières, Athanase, 337
de Posada, Alonso, 335, 336
Description of the City of London (Fitzstephen), 452–453
de Soto, Hernando, 318, 333
"The Destruction of Sennacherib" (Byron), 152
Diamond, Jared, 66, 119, 120, 306, 319, 323
Díaz del Castillo, Bernal, 314, 315, 316

INDEX

Digesta Artis Mulomedicinae (Vegetius), 205
Dilger, Anton, 409
Dilger, Carl, 409
DiMarco, Louis, 130, 172–173, 200, 230
DiNardo, R. L., 420, 422, 432, 433
diphtheria, 412–413n, 412–414, *415*
diseases
  influenza (equine), 384–387
  influenza (human), 409
  vaccines from equine research, 412–413n, 412–414, *415*
  zoonotic "spillover," 67, 67n, 75–76, 75n
Dobie, Frank, 346
dog food, 440–442
dogs, 69–70, 354, 414
Domesday Book, 259, 267
domestication, defined, 70, 73
domestication of animals
  as Agricultural Revolution by-product, 68–69
  in Great Plains, 349–350
  horseback riding, 81–83
  origination of horse domestication, 76–81, 83–84, 85–86
  secondary products resulting from, 75, 78–80, 96
  selective breeding, 84–85
  social transformation created by, 86–88
  spillover diseases and, 75–76, 75n
  stages of, 69–73
  suitable traits for, 73–75
  tame animals, 63, 69–72
domestication of plants
  animal domestication as by-product of, 68–69
  background, 59–60, 61–62, 63n
  hunter-gatherer transition to, 62–63, 63n
  social transformation created by, 63–68
donkeys, xvii, 69, 113–114
Dorondo, David, 417, 428–429
Douglas, David, 256, 259
Drews, Robert, 59, 191, 244
Dzungarian horses (Przewalski's horses), xviii, xx, 449–450, *451*

Early Riders (Drews), 59
Ecclesiastical History of the English People (Bede), 248
An Edible History of Humanity (Standage), 63–64
Egypt, xiv, 132–138, 139–141, *141*, 144, 193
Egypt, Canaan, and Israel in Ancient Times (Redford), 135
Elamites, 143–144
elephants, 55, 70, 192, 195–197, 228
Empire of Horses (Man), 210
Empires of the Silk Road (Beckwith), 119
England and British Empire
  Anglo-Saxon culture, 233–234, 259–260
  borders of, 260
  famines and plagues, 274–275, 297
  manure crises in, 370
  medieval period, 261–263, 266–273
  modern sport horses, 452–453
  Normans and, xiv, 237, 254–260, *258*
  racing events, 452–453
  WWI, 123–125, 127, *128*, 400–401, 402, 405–409, *407*
  WWII, 416, 428–429, *429*, 434
English Seigniorial Agriculture, 1250–1450 (Campbell), 273
Eohippus (dawn horse), 17, 24–25, *25*
Ephedra, 116–117
Equidae family, xviii–xix, *xix*
equids, xviii–xix
equine-assisted therapies, 456
equine influenza, 384–387
Equus (genus), xvii–xviii, *xix*, 26, 28–29, 28n, 57
Esarhaddon, 150
Espejo, Antonio de, 310
Eudes (duke), 241, 246
Eurasian Steppe, xiv–xv, xviiin, 1n, 59–60, 80, 158. *See also* Pontic-Caspian Steppe; Steppe/Silk Roads
Europe. *See* feudalism; medieval Europe; "Old" Europe; World War I; World War II; *specific countries and empires*
evolution of equines
  anatomy and (*See* anatomical adaptations)
  fossil hunting, 11–18, 21–23

global evolution and dispersion, 23–30, *31*
habitat loss, 55–60
human hunting for horsemeat, 49–57
modern horse genetics, 250–251
natural selection and, 18–21, 29–30, 59–60
religion and, 13, 17–18, 20–21
social organization and, 46–47

Fagan, Brian, 6, 146
Farewell to the Horse (Raulff), 8, 321
farm horses. *See* agricultural practices
Favereau, Marie, 287, 300
Federation of International Polo, 454
feral (wild) horses, xx, 320, 344–346, 356, 439–450, *447*, 457–458
Ferdinand (king), 300, 320, 321
Fergana, 223–224
feudalism
　background, 236–237, 238–240, 242, 255
　cavalry and, 242–245, *244*, 254–259
　defined, 242
　demise of, 296–297
　in Great Plains, 334–335
　Normans and, 259–260, 261–262
firehorses, 381, 385, 394, *395*, 414
First World War. *See* World War I
Fischer, Martin, 43
Fitzstephen, William, 452–453
Flathead, *xii*, 340
Forbes, Jack, 335
Ford, Henry, and Ford Motors, 128, 390–391, 393, 396–397
Forster-Cooper, Clive, 24
fossil hunting, 11–18
Fouracre, Paul, 242
Four Horsemen of the Apocalypse, 127, 235, *237*
Frachetti, Michael, 109
Frahm, Eckart, 144, 151
France, 392, 405–406, 408, 440, 452
Frankopan, Peter, 107, 238, 294, 296
Franks, 236–237, 239, 241–249, 254
French Society for the Protection of Animals, 373
Frozen Tombs of Siberia (Rudenko), 162

Gabriel, Richard, 131, 134–136, 155
Gaebel, Robert, 186, 194, 200
Gaiseric, 233, 234
gait photographs, 324–325, *325*
Galton, Francis, 69
Game of Thrones (series), 118
Gandhi, Mahatma, 7
Gan Ying, 227
Gast, John, 329–330, *329*
Gaul, 231, 232–233
Genghis Khan and the Making of the Modern World (Weatherford), 284
George III (king), 48
Gercke, Rudolf, 416
Germany, 392, 450, 455. *See also* World War I; World War II
Geronimo, 363–364, 365
Gesta Guillelmi ducis Normannorum et regis Anglorum (William of Poitiers), 257
Glock, Percival, 406
Godolphin Arabian (horse), 251, 453
God's Crucible (Lewis), 241
Godwinson, Harold, 254, 256–257
Goebbels, Joseph, 420–422
Goldsworthy, Adrian, 175, 202
Gómara, Francisco López de, 316
Good Roads Act (1916), 395
Gordon, W. J., 381
Göring, Hermann, 423, 432
Granicus (battle), *xiv*, 192–193
Grant, Ulysses S., 15, 344, 347, 384
Great American Interchange, 28–29
The Great Arab Conquests (Kennedy), 240
Great Britain. *See* England and British Empire
Great Epizootic (1872), 384–387
Great Famine (1315–1317), 274–275
Great Plains, 324–367
　agricultural production, 347–349, *349*
　camels and, 349–350
　domestication programs for native animals, 349–350
　feral herds, 344–346
　horse dispersion, *xii–xiii*
　Indigenous horse culture, 82, 337–344, *343*, 353–354, 357–361
　infrastructure projects, 351–353

Great Plains (cont.)
  Manifest Destiny and, 329–331, *329*, 346–349, 364
  "noble savage" image of, 326–327
  photography and films of horses in, 324–329, *325*, *327–328*, 367
  Spanish imperialism and, 331–337, *332*
Great Wall of China, xv, 210, 211, 213–214, 220
Great War. *See* World War I
Greece
  on barbarism, 166–170
  cannabis use in, 115–116
  cavalry of, 149, 183–184, 184n, 185
  Greco-Persian Wars, 178–180, 183–184, 184n
  military campaigns of, 155
  sports, 189, 452–453
Greene, Ann Norton, 348, 376, 394
Gregory III (pope), 50
Gregory of Tours, 241
Grey, Zane, 326
Gulliver's Travels (Swift), 76
Guns, Germs, and Steel (Diamond), 66, 119, 306
Guthrie, Dale, 54
Gyde, Arnold, 400–401

Hagenbeck, Carl, 450
Hämäläinen, Pekka, 338, 353, 354, 357–358, 359–360, 363
Hancock, Reginald, 406
Han dynasty, 214–225
Harari, Yuval Noah, 57, 62–63
Harington, Charles, 403
Harold (king), 254, 256–257
Hastings, Battle of, xiv, 237, 254–260, *258*
Hatshepsut, 136–137
Headrick, Daniel, 64, 253, 263, 268, 313, 318
"Heavenly Horses," 148, 207, 216–218, 222–225
heavy cavalry. *See also* cavalry
  of Chinese, 235
  of Franks, 236–237, 239, 241–245, 247–249, 254
  of Normans, 237, 254–259, *258*
  stirrups and, 235–237, 243–245, 249, 254, 255

Hecataeus, 157
Hellanicus of Lesbos, 157
Hemmings, Andrew, 27
Henday, Anthony, 340
Henry and Edsel (Bak), 393
Henry VIII, 453
Herodotus, 78–79, 115–116, 134, 148, 157, 158, 163, 165, 169–170, 180–181, 182, 183
Heyd, Volker, 100–101
Hidatsa, xii, 341, 362
Hipparion, 28–29, 51
hippotherapy, 456–457
Historia Francorum (Gregory), 241
The Histories (Herodotus), 115–116
Histories of Alexander the Great (Quintus Curtius), 195
History of the Peloponnesian War (Thucydides), 168
A History of Warfare (Keegan), 127, 428
A History of World Agriculture (Mazoyer and Roudart), 64, 399
The History of the World-Conqueror (Juvaini), 285
Hitler, Adolf, 418, *419*, 427, 433
Hittites, 133, 138–141, 142, 144
Homer, 174
Homestead Acts, 347, 347n, 362
Honeychurch, William, 111, 166–167, 330
Hoofbeats and Society (Lawrence), 455
The Horde (Favereau), 287
Horn, Christian, 67
The Horse (Williams), 24
The Horse, the Wheel, and Language (Anthony), 87, 91
horse fairs, 272
Horse: How the Horse Has Shaped Civilization (Chamberlin), 326
The Horse in the Ancient World (Hyland), 198
The Horse in Human History (Kelekna), 203
The Horse in Motion (film), 325
The Horse in West African History (Law), 135, 309
horsemeat, 49–57, 448–449
Horse Nations (Mitchell), 58, 312, 335
horse racing, 189, 452–453

INDEX

horses
  in agriculture (*See* agricultural practices)
  Centaurian Pact, 2–3, 48, 81–82, 438–439, 459–461
  in cities (*See* urban horses)
  dawn of the horse and early interactions (*See* anatomical adaptations; domestication; evolution of equines; Indo-European domination)
  equine-assisted therapies, 456–457
  Eurasian markets and trade (*See* Mongols; Steppe/Silk Roads)
  evolution and early human interactions (*See* Agricultural Revolutions; anatomical adaptations; domestication; evolution of equines)
  feral (wild) horses, xx, 320, 344–346, 356, 439–450, *447*, 457–458
  forging of empires and warhorses (*See* Alexander the Great; Assyrian Empire; cavalry; chariot warfare; Chinese Empire and civilization; Islamic Empire; medieval Europe; Mongols; Roman Empire; Scythians; World War I; World War II)
  modern replacements for, 389–399, *393*, *395*, *397–398*
  pinnacle of, 376–377, *376*, 388
  return to the Americas (*See* Columbian Exchange; Great Plains)
  sport horses, 451–455
  taxonomy and terminology, xvii–xx

Horses, Oxen and Technological Innovation (Langdon), 262

Horses at Work (Greene), 348, 376

Horses in the Russian Campaign (US Army Historical Division), 426

horse tack and equipment
  harnessing systems and collars, 262–264
  horseshoes, 205, 255, 265–266
  plows, 261–263, 266–268
  saddles, 170–171, 205
  stirrups, 235–237, 243–245, 249, 254, 255
  whippletrees, 264–265

The Horse-World of London (Gordon), 381

Horwitz, Tony, 305

Houyhnhnm, 76

Hoyland, Robert, 252

Hulbert, Richard, 26

Huns and Attila, 225, 229–230, 231–233, 241

Huo Qubing, 219–220, 220n

Huxley, Aldous, 117, 390

Huxley, Thomas Henry, 20–21, 22–25

Hydaspes (battle), *xiv*, 195–196

Hyland, Ann, 130, 147, 198, 256, 265

Hyracotherium, 21–22, 24–27, 30, 40, 41–43

Iarocci, Andrew, 403

Iliad (Homer), 174

Imperial China, 224

Impressment of Horses and Horse-Drawn Vehicles in Time of National Emergency Act, 408

Incan Empire, 311–312, 313–314, 317, 318–319

India, 104–106, 191–192, 195–198, 429

Indian boarding/residential schools, 365, 365n

Indian Ring, 15–16

Indians: A Brief History of a Civilization (Arora), 105

The Indo-European Controversy (Pereltsvaig and Lewis), 102, 103

Indo-European domination
  of China, 106–109
  dairy and, 100–101
  drugs and, 115–117
  exodus from the steppe, 99–101
  of fashion (pants), 117–118
  of Greece, 112
  of India, 104–106
  of language, 90–99, *94*, 109–110, 115, 118–119
  legacy of, 119–120
  of Middle East, 113–114, *113*
  of "Old" Europe, 89–90, 101–104, 105
  steppe (kurgan) hypothesis, 99, 108–109
  of trade along the Steppe Roads, 109–112, 114–117

influenza (equine), 384–387
influenza (human), 409
internal combustion engine, 390–391, 399
International Harvester Company, 396–397
Interstate Highway System, 395
An Introduction to the Principles of Morals and Legislation (Bentham), 373
Iroquois (Haudenosaunee), *xii*, 341
Isabella (queen), 300, 320, 321
Islamic Empire
  academia and inventions, 251–254
  Arabian horses, 130, 239, 246, 249–251, 253–254
  background, 238–239
  cavalry of, 245–247
  Chinese Empire and, 239–240
  Iberian Penninsula and, 240–242, 245–249
  modern horse genetics, 250–251
  polo and, 454
Issus, Battle of, *xiv*, 189–190, 192–193, *192*
Isthmus of Darien, 28
Italy, 295, 429–430, *429*. *See also* Roman Empire

Janis, Christine, 26
Japan, 235, 292, 293, 416, 429–430, *429*, 454
Jaxartes (battle), *xiv*, 194–195
Jeremiah (prophet), 150–151, 169
Jerome (saint), 231
Jim (horse), 414
Joey (horse), 407, 410–411
John of Patmos, 127
John of Plano Carpini, 281, 284, 286–287, 288
Johnston, Velma Bronn "Wild Horse Annie," 443–444, 448
Jones, William, 93–95, 96, 117
Joseph, Brian, 119
Journey Around the Earth (Hecataeus), 157
Juvaini, Ata-Malik, 285

Kadesh, Battle of, *xiv*, 140–141, *141*
Kainai, 340, 358, 361
Kalispel, *xii*, 340
Karunanithy, David, 188, 190
Keegan, John, 127, 233, 291, 301, 428

Keitel, Wilhelm, 433
Kelekna, Pita, 28, 74, 203, 234, 253–254
Ken-L-Ration dog food, 440–442
Kennedy, Hugh, 240, 248
Kentucky Wagon Manufacturing Company, 380
Keyser, Christine, 209
Khazanov, Anatoly, 100
Khmer civilization, 293
Kikkuli, 106, 130, 139, 149
Kilby, Emily, 451
Kiowa, *xii*, 335, 336, 338, 359, 361, 362, 364
knights, 233–234, 242–244, *244*, 249, 254, 259, 261, 298
Koller Papyrus, 140
Korea, 235, 282
Koselleck, Reinhart, 425
koumiss, 78–79
Kristiansen, Kristian, 67, 102, 103
Kublai Khan, 292–293, 300
kurgan (steppe) hypothesis, 99, 108–109
Kurzweil, Ray, 388

Lakota, *xii*, 82, 323, 326, *328*, 336, 341 342, 354, 360, 362, 363, 364
Lakota America (Hämäläinen), 353
Langdon, John, 262, 272
Langdon, Robert, 297
Lanning, Michael, 194
Larpenteur, Charles, 361
La Salle, René-Robert Cavelier de, 336
Las Casas, Bartolomé de, 309, 310, 311
The Last Stand (Philbrick), 341
Law, Robin, 135, 309
Lawrence, Elizabeth Atwood, 455
Leakey, Mary, 28
Leclerc, Georges-Louis, 2
Levin, Jonathan, 387, 396
Levine, Marsha, 83
Levi Strauss & Co., 380–381
Lewis, David Levering, 241
Lewis, Martin, 102, 103, 164
Lewis, Meriwether, 339–340
Li, Chunxiang, 108
Li Bai, 207–208
Liedtke, Walter, 3
The Life of Alexander (Plutarch), 174–175
The Life of Reason (Santayana), 123
Li Guangli, 223–224

Liman von Sanders, Otto, 124, 127
Linnaeus, Carolus, xvii
Lipizzaners, 435–436
Little Book Cliffs Wild Horse Range, 446–447, *447*
Litts, Frank, 442
London, 370–373, 377–378, 381–383, 391–392, 412, *413*, 452–453, 455
Long Depression, 385–386, 386n
Ludendorff, Erich, 410
Lyell, Charles, 96
Lysias, 165
Lysistrata (Aristophanes), 168

Macedonian cavalry. *See* Alexander the Great
The Macedonian War Machine, 359–281 BC (Karunanithy), 188
MacFadden, Bruce, 18
Maheo, 358
Mair, Victor, 118, 213, 230
Mallory, J. P., 97, 100, 108, 213, 230
Man, John, 210, 218
Mandan, xii, 335, 362
Manifest Destiny, 329–331, *329*, 346–349, 351, 364
Man-Kind Dog and Cat Food, 441
manure crises, 368–372, *371*, 379, *380*, 384, 393
Maori, 354, 357
Mapuche (formerly called Araucanians), 321–322, 322n, 323
marijuana, 115–116, 163, 228
Marquesas Islands (Romania), 440
Marsh, Othniel Charles, 11–18, *17*, 22–27, 366
Martel, Charles "the Hammer," 50–51, 236–237, 239, 241–243, 245, 246–249, 254
Martin, Geoffrey, 272–273
Martin, George R. R., 118
Massagetae, 179–181
Massie, Christopher, 404
Maurice (emperor), 236, 241
Maurya Empire, *xv*, 197–199
Mayor, Adrienne, 115, 117, 164, 166
Ma Yuan, 207
Mazappa (Byron), *419*
Mazepa, Ivan, *419*

Mazoyer, Marcel, 64, 399
McCall, Jeremiah, 178, 205
McNeill, William H., 147
McShane, Clay, 374, 377, 392
Mechanized Juggernaut or Military Anachronism? (DiNardo), 420
Medes (and Media), 148, 149, 154–155, 170
medieval Europe
  Agricultural Revolution, 262–263, 266–270
  economic specialization, 270–271
  fairs and festivals, 271–272
  famines and plagues, 274–275, 294, 295–296
  infrastructure projects, 272–273
  manure, 273–274
  post–Black Death, 296–297
  urban clusters and, 269–270
Medieval Warfare in Manuscripts (Porter), 266
Megiddo Plains, battles of, *xiv*, 123–127, *128*, 137, 402
Mendoza, Pedro de, 320
Meriam, Lewis, 365
Meriam Report, 365
Mesopotamia. *See* Assyrian Empire
Messines, Battle of, 403
Mexico, 311–313, 314–317, 318, 333–334, 338, 350, 359
Mexico City, xii, 316, 334
Meyer, Heinz, 421
military. *See* cavalry; chariot warfare
milk, from horses, 78–80, 100
milk wagon horses, *375*, 395–396, 414
Milne, Alan Alexander "A. A.," 411–412
Miohippus, 27
Mishkin, Bernard, 333
Mitanni, 106, 130, 144
Mitchell, Peter, 29, 58, 78, 312, 335, 337, 340, 342, 357, 359, 361–362
Mobei, Battle of, *xv*, 220, 220n
Model T horseless carriage, 128, 391, 396
Modu Shanyu, 215, 215n, 258
Mongo (horse), 445–446
Mongolian wild horse (Przewalski's horses), xviii, xx, 449–450, *451*

Mongols, xv, 279–301
　as "barbarians," 288, 288n
　under Chinggis Khan, 279–284, 289–290, 294
　codified law, 286, 287
　Crusades and, 291–292
　empire expansion, 289–293, 294
　fall of, 293–294
　horsemanship and, 286
　infrastructure projects, 287–289
　under Kublai Khan, 292–293, 300
　legacy of, 293–294, 297–301
　military tactics of, 282, 284–286, 289–290
　under Ögedei, 290–291
　plagues and, 275, 294–296
　trade laws, 287
　tribal affiliations, 284, 284n, 289
Montezuma II (emperor), 314–315
Moore, John, 406–407
Moore, John Trotwood, 28, 28n
More, Thomas, 305–306
Morfi, Juan Agustín, 344
Mormon theology, 13
Morris, Eric, 372–373
Moseley, George Van Horn, 416
Muhammad (prophet), 238, 249–250
Murray, Williamson, 422
musarkisus, 149–150
The Mustangs (Dobie), 346
Muwatalli II, 140
Muybridge, Eadweard, 324–326, *325*

Nahum, 155
Nanda Empire, 196–198
Napoléon, 194
Narváez, Pánfilo de, 311, 315, 333
The Nation magazine, 386–387
NATO pack-transport units, 455–456
The Nature of Horses (Budiansky), 19
Navajo, xii, 334
Nefertiti, 139
Neo-Assyrian Empire, 142, 143–156, 169–170, 179
New Mexico, 334–335
New York City, 367, 368–373, *371*, *375*, 377–379, 381, 383–385, 391, 394
New York City Police Department (NYPD), 455

New Zealand, 354, 357, 429, 440
Nez Perce, xii, 323, 336, 339–340, 364
Nicolle, David, 245, 248
Nietzsche, Friedrich, 86, 86n
Nimrud, 144, *146*, 151
Nineveh, xiv, 144, 150, 151–156, *153*, 155n, 170
Nisaean Plain, xiv, 148–149, 163, 179, 190
Normans, 237, 254–260, *258*

Oakley, Annie, 365
Ögedei, 290–291
Oglala Lakota, 15–16, *328*
Ojibwa (Saulteaux), xii, 340–341
Olander, Thomas, 97
Old Dan the Retired Fire Horse (horse), 414
"Old" Europe, 88, 89–90, 98, 101–104
Olsen, Sandra, 53, 71, 85
Olympic Games, 189, 454–455
Omaha, xii, 336, 340, 341
Oman, Charles, 254
On Airs, Waters, and Place (unknown), 165
100 Decisive Battles (Davis), 248–249
On Horsemanship (Xenophon), 149, 177
On the Cavalry Commander (Xenophon), 177
On the Nature of Animals and Historical Miscellany (Aelian), 205
On the Origin of Species (Darwin), 17–18
On War (Clausewitz), 191
Oregon Trail, 346
O'Reilly, CuChullaine, 384
Orlando, Ludovic, 83–84, 250–251
Orwell, George, 398
Osage, xii, 340
Ostrogoths, 234
O'Sullivan, John, 329
The Other Side of the Frontier (Reynolds), 356
Otto, Nicolaus, 390
The Outline of History (Wells), 154
Overy, Richard, 427, 430, 434, 437
Owen, Richard, 19–22, 24
oxen, 114, 261–263, 266–267, 297

pack-transport units, 455–456
paleontology, 11–17
Pampas, 320–321, 323

INDEX

Panamanian land bridge, 28
Panic of 1873, 385–386
pants, 117–118
Paris, Matthew, 285, 290
Parker, Quanah, 364
Parthians, 222
"Parthian shot," 171, *172*
Parwan, Battle of, 282
patriarchal society, 98
Patton, George, 434, 436
Paulus, Friedrich, 432, 433
Pawnee, *xii*, 335, 336, 338, 340, 341, 342, 362
Peabody, George, 12, 12n, 13
Peabody Museum of Natural History, 12, 16–17
Pearl Harbor, 416
Peloponnesian War, 178, 185–186
Pereltsvaig, Asya, 102, 103, 164
The Perilous Frontier, 225
The Periplus of the Erythraean Sea (trade manual), 227–228
Perissodactyla, xviii–xix, *xix*
Persian Empire
  academia and inventions, 181, 252, 282, 285
  –Akkadian relations, 132
  Alexander the Great and, 174, 177–178, 186–188, 190, 192–193, *192*
  –Amazon relations, 165
  –Assyrian relations, 154–156
  on barbarism, 167
  cavalry of (horsemanship), 179, 181
  on civilization and barbarism, 167
  under Cyrus the Great, 179–181
  under Darius, 181–182
  drug use in, 117
  equine sports, 453–454
  exports and trade, 162, 226, 228–229
  Greco-Persian Wars, 178–180, 183–184, 184n
  horse herd quality, 148
  imperialism and, 179–181, 183–186
  infrastructure projects, 181–183, 204
  Parthians of, 222
  –Scythian relations, 169, 172–173
Peru, 313, 318, 320. *See also* Incan Empire
Peter the Great, 158–160, *159*

Petrarch, 295
Philadelphia, 372, 377–378, *382*, 391
Philbrick, Nathaniel, 341
Philip and Alexander (Goldsworthy), 175
Philip II, 175–176, 178, 185–191, 258, 320
Philipps, David, 16, 30, 304, 336, 444
Piber (Austrian) Breeding Herd, 436
Piggott, Stuart, 114
Piikani, 340, 353–354, 361
Pizarro, Francisco, 311–312, 317, 318–319
plagues, 101–102, 275, 294–297
Plains Indians of North America, 323
Plato, 176, 185
Pliny the Elder, 156, 452
Plutarch, 174–175, 176, 177, 187, 198
Poland, 421–422
police horses, 455–456
Polley, David, 405
polo, 453–454
Polo, Marco, 286, 299–300, *299*
Ponce de León, Juan, 333
Pontic-Caspian Steppe, *xiv*, 77–86. *See also* Indo-European domination; Mongols; Scythians; Steppe/Silk Roads; Xiongnu
Pony Express, 351, 352–353
Porter, Pamela, 266
Porus, 192, 195–196
Post Oak Jim, 360
prehistoric art, 53–55, 53n, *53*, *55*
Premarin and Prempro, 414
Prestwich, Michael, 245
prison horse programs, 457
Proto-Indo-European (PIE) language group, 92–99, *94*, 109–110, 119
Przewalski's horses, xviii, xx, 449–450, *451*
Przhevalsky, Nikolay, xviii, 450
PTSD (post-traumatic stress disorder), 456
public transportation, 379, 381, 391
Pueblo, 310, 333, 335
Purépecha, 310–311, 355

Qian, Sima, 215
Qin dynasty, 212–213
Quapaw, *xii*, 338
Quintus Curtius Rufus, 195

racing horses, 189, 452–453
Ramses II, 138, 140–141
Raney, Walter "Rex," 435
Rank and Warfare Among the Plains Indians (Mishkin), 333
Rau, Gustav, 417
Raulff, Ulrich, 8, 321, 399, 402
Records of the Grand Historian (Shiji) (Sima Qian), 209
Red Cloud, 15, *17*, 360, 365
Redford, Donald, 135
Red Hawk, *328*
Reed, Charles "Hank," 436
Reich, David, 104, 105
Remarque, Erich Maria, 404, 405
Reynolds, Henry, 356
Rice, Condoleezza, 399
Richard "Humanity Dick" Martin, 373
Richard III (king), 7
Richards, Martin, 104
Richardson, William, 21
Riders of the Apocalypse (Dorondo), 417
Rig Veda, 49
Rink, Bjarke, 2
Rochester (NY), 372, 385
Rock, Edward, Jr., 435
rodeos, 454
Rogers, Will, 453
Roman Empire
　academia and cultural impact of, 204–205
　Alexander's influence on, 203–205
　Anglo-Saxon culture and, 259–260
　on barbarism, 231–232
　cavalry and horsemanship of, 205, 206, 230–231
　Chinese Empire and, 206, 226–230, 234
　fall of, 206, 229–234
　horse-specific laws of, 263
　Huns and, 241
　infrastructure projects, 203–204
　military losses, 205–206
　racing events, 452–453
Rommel, Erwin "Desert Fox," 430
Roosevelt, Franklin D., 428
Rose, Kenneth, 34
Rosenwein, Barbara, 300–301
Rothschild, Walter, 74
Roudart, Laurence, 64, 399
Rousseau, Jean-Jacques, 308
Rowson, Alex, 176, 201
The Royal Commentaries of the Inca, 318
Royal Library of Ashurbanipal, 153–154, 155
Royal Mail horses, 381
Royal Society for the Prevention of Cruelty to Animals, 373
Rudenko, Sergei, 162
Ruffo, Giordano, 243, 270
Russell, William, 352
Russia and Soviet Union
　feral horses in, 450
　Mongols and, 289–290, 294
　WWI and, 402, 405–406, 410
　WWII and, 427–429, *429*, 431, 437
Rustichello da Pisa, 300

Sable Island, 457–458
saddle innovations, 170–171, 205
Sallie Gardner (horse), 325–326, *325*
Samarkand, 240, 240n
San Francisco, 383
Santayana, George, 123
Sapir, Edward, 90
Sargon II, 145, 145n, 146, 147–148, 150–151
Sargon of Akkad, 131–132, 145n
Sassoon, Siegfried, 405
Saukamappee, 339, 353–354
Schleicher, August, 95–96
Schleicher's Fable, 96
Schlesser brothers, 441
Schultz, James, 361
Scott, James C., 72
The Scythian Empire (Beckwith), 157
Scythians, 156–173
　Alexander the Great and, 192, 194–195
　Amazon warriors, 164–166, *172*
　–Assyrian relations, 154–156
　background, 156–162, *159–160*, *162*
　as "barbarians," 166–170
　–Cimmerian relations, 156
　drug use, 115–116, 117
　innovations, 170–171
　military campaigns and tactics, 205–206
　–Persian relations, 179–181

INDEX

515

The Scythians (Cunliffe), 156
Sea People, 141–142
Secondary Products Revolution, 60, 75
The Secret History of the Mongols, 282, 286
Seen From Afar, 361
Sennacherib, 143–144, 145, 151–152
Sergeant Reckless (horse), 412
Sewell, Anna, 373, 401
Shalmaneser III, 145
Shaughnessy, Edward, 129
Sherratt, Andrew, 60, 116
Shihuangdi, 212–214, *214*
Shoshone, *xii*, 323, 336, 337, 338–339, 340, 354, 364
Siam Qian, 222
Sidnell, Philip, 177, 188, 259
Siksika, 340, 354
Silk Roads. *See* Steppe/Silk Roads
The Silk Roads (Frankopan), 107
Sima Qian, 209, 212, 213, 218, 223
Simpson, George Gaylord, 24–25, 457
Sioux, *xii*, 15–16, 364. *See also* Lakota
Sits-in-the-Night, 354
Sitting Bull, 364, 365, *366*
skijoring, 454, 454n
Skinner, Robert, 408–409
Slicher van Bath, Bernard, 269
Smith, Sydney, 406
Soares da Silva, Marina, 104
Socrates, 185
Somme, Battle of the, 404
"Song of Sorrow" (Xijun), 222
Song of the Battle of Hastings (*Carmen de Hastingae Proelio*), 256, 257–258
Spanish Jews, 321
Speer, Albert, 417
"spillover" diseases, 75–76, 75n
spiritual machines. *See* urban horses
sports, 172, 451–455
stagecoach services, 351–352, 377
Stalin, Joseph, 428
stallions, xx, 77, 84–85
Stamford Bridge, Battle of, 256
Standage, Tom, 63–64, 375, 384
Stanford, Leland, 324–326, 346
steam engines, 389–390, 391
steppe (kurgan) hypothesis, 99, 108–109
steppe herds. *See* Pontic-Caspian Steppe

Steppe/Silk Roads. *See also* Indo-European domination; Pontic-Caspian Steppe
Alexander and, 206
background, *xiv–xv*, 87, 87n
Black Death and, 295
Chinese Empire and, 216, 219, 224–228, 234–235
Islamic Empire and, 240, 240n
Mongols and, 279–280, 295
opening of, 109
Persia and, 182–183
Polo and, 299–300, *299*
Roman Empire and, 228–234
Strategikon (Maurice), 236, 241
Studebaker Company, 380
Suppiluliuma I, 139–140
Swart, Sandra, 343
Swift, Jonathan, 76

Tacitus, 231
Talas River, Battle of, *xv*, 240
Taliban, 456
tame animals, 63, 69–72
tapirs, xix
The Tarim Mummies (Mallory and Mair), 213
Tarn, William, 189
Taylor, William, 86
Taylor Grazing Act (1934), 442–443
Technology: A World History (Headrick), 64
Tela, 143
Temujin, 282, 283. *See also* Chinggis Khan
Tennyson, Alfred (lord), 7
Tenochtitlán, *xii*, 312, 315–316, *317*
tetanus, 412–414, 413n
therapy with horses, 456–457
Thomas, Georg, 417
Thompson, Francis, 370
Thompson, M. W., 162
thoroughbreds, 163, 251, 417, 435, 436, 453
Thucydides, 168, 185, 186
Thutmose I, 136
Thutmose II, 136
Thutmose III, 125–127, 128, 136, 137–138
Tianma ("Heavenly Horse"), 148, 207, 216–218, 222–225

INDEX

Tiglath-pileser III, 145, 149
Tjaneni, 126–127
Todorova, Henrieta, 102
Tomyris, 179–181
Tours, Battle of, *xiv*, 236, 239–241, 246–249
towns. *See* urban horses
tractors, 396–398, *397–398*
Trader Post Scandal, 15–16
The Travels of Marco Polo (Rustichello), 300
Trevithick, Richard, 389–390
Trojan War, 142
trophic cascade, 58
Troy, 142
The True History of the Conquest of New Spain (Castillo), 314
Tsuut'ina, 354
Tukhachevsky, Mikhail "the Red Napoléon," 424
Tukulti-Ninurta, 148
Tukulti-Ninurta II, 145
Turkoman horses, 163, 251
12 Strong (film), 456

Umayyad Empire, 239, 240, 245, 249, 251–252
United Kingdom. *See* England and British Empire
United States. *See also* Great Plains; World War I; World War II
  current horse population, 438
  feral equines in, 439–448, 457
  military pack–transport units, 456
  racing events, 453
  rodeo and, 454
Ur, 113, *113*
Urartu, 145–147, 148, 156
urban horses, 368–399
  disposal of dead horses, 372–373, 374–375, *375*
  Great Epizootic (1872), 384–388
  manure crises, 273–274, 368–372, *371*, *379*, *380*, 384, 393
  in medieval Europe, 269–274
  mounted police, 455
  pinnacle of, 388
  transition to "horseless carriages," 389–396, *393*, *395*, 399

transportation and commerce dependent on, 375–384, 391–396, *395*
  welfare of, 372–374
Urban II (pope), 261–262
US Federal Aid Road Act (1916), 395
USSR. *See* Russia and Soviet Union
US Wild Horse Inmate Program (WHIP), 457
Ute, *xii*, 335, 336, 337, 340, 362, 364
Utopia (More), 305–306

vaccines, 412–413n, 412–414, *415*
Van Creveld, Martin, 403, 422, 424
Vandals, 233, 234
The Vanishing American (film), 326
Vázquez de Espinosa, Antonio, 320
V battle formation (wedge), 171–172, 193
Vega, Garcilaso de la, 318
Vegetius, 205
vehicle makers, 379–380
Venerable Bede, 248
Victorio, 363–364
Visigoths, 230–231, 233, 240
A Voyage Long and Strange (Horwitz), 305

The Walking Larder (Clutton-Brock), 69
Walla Walla, *xii*, 339
War Horse, 407–408, 410–411
War Horse (DiMarco), 130
Warhorse (Sidnell), 177, 259
warhorses. *See* cavalry; chariot warfare
Warmuth, Vera, 85
War of the Heavenly Horses, 223–225
Warring States, 212
Waterloo, Battle of, 47–48
Watt, James, 389
Weatherford, Jack, 284, 293, 299
Webb, David, 27
Webster, David, 436–437
wedge battle formation, 171–172, 193
Wells, H. G., 61, 154
Wells Fargo, 351, 353
Werth, Alexander, 432
Westreich, Sam, 37
Wheels for the World (Brinkley), 396
Where Have All the Horses Gone? (Levin), 387
White, Lynn, 245
Why the Allies Won (Overy), 427

Wichita, *xii–xiii*, 336, 338, 340
Wild Free-Roaming Horses and Burros Act (1971), 444, 447
wild herds. *See* feral horses
"Wild Horse Annie Act," 443–444
Wild Horse Country (Philipps), 16, 304, 444
Wild West show, 326, *327*, 365–366, 368, 454
Wilkin, Shevan, 100
Willerslev, Eske, 103
William of Normandy, 237
William of Poitiers, 257, 258
William of Rubruck, 79, 289, 298–299
Williams, Wendy, 24, 38, 458
William the Conqueror, 254–260, *258*
William the Conqueror (Douglas), 256
Winnie-the-Pooh, 411–412, *412*
With Buffalo Bill on the U.P. Trail (film), 366
Witsen, Nicolaes, 159
World War I, 400–415
  background, 386, 399, 401–402
  biological warfare, 409–410
  British Expeditionary Force (BEF), 400–401
  calvary operations during, 402–403
  end of, 410
  equine casualties and medical care, 401–402, 405–407, *407*, 411
  equines for noncavalry, 403
  equine supplies, 407–408
  mascots, 411–412, *412*
  medical research on horses during, 412–415, *415*
  Megiddo Plains, 123–127, *128*, 402
  tributes to, 410–411, *413*
World War II, 416–437
  American and Allied mechanization levels, 416, 421–422, 427–429, *429*, 431
  end of, 433–437, *433*

German and Axis mechanization levels, 416–417, *419*, 421–422, 429–433, *429*, *432*
German dependence on horses, 421, 422–427, *426*, 431–433, *432*, 437
German horse eugenics program, 417–418, 435–436
German propaganda, 418–421
Wu (emperor), 207–208, 217–223, *220*
Wuling (king), 210–211
Wusun, 221–222, 221–222n, 225, 231

Xenophon, 149, 155, 174, 175, 176–177, 188, 191, 198, 259
Xijun (princess), 222
Xinjiang, *xiv*, 107–109, 226, 229, 234–235
Xiongnu (Hu or Huns), *xv*
  background, 158, 209–210, 215–216
  as "barbarians," 209
  Chinese responses to, 87–88, 173, 206, 211, 213–216, 219–221, 225
  descendants of, 225, 229–230, 231
  downfall of, 220–221, 225
  Roman cavalry veterans absorbed into, 206
  Sino-Xiongnu War, 219–220

Yakama, *xii*, 339
Yamnaya, 80, 83, 97, 105
Yellowtail, Robert, 456
Young, Brigham, 13
The Young Alexander (Rowson), 176, 201
Yuezhi, 221, 221–222n, 231

zebras, xviii–xix, 29, 73–75, 387
Zhang Qian, 208, 218–219, *220*, 221–222, 224
zoonotic "spillover" diseases, 67, 67n, 75–76, 75n
Zuni, 333

About the Author

Dr. Timothy C. Winegard is a *New York Times* bestselling author of five books including *The Mosquito: A Human History of Our Deadliest Predator*. His works have been published globally in more than fifteen languages. He holds a PhD from the University of Oxford, served as an officer in the Canadian and British Armies, and has appeared on numerous documentaries, television programs, and podcasts. Winegard is an associate professor of history (and head coach of the hockey team) at Colorado Mesa University.